Environmental Isotopes in Hydrogeology

Ian D. Clark
and
Peter Fritz

www.science.uottawa.ca/~eih

Boca Raton　　　　　　　New York

Acquiring Editor: Joel Stein
Project Editor: Albert W. Starkweather
Cover design: Dawn Boyd

Library of Congress Cataloging-in-Publication Data

Clark, Ian D. (Ian Douglas), 1954-
Environmental Isotopes in Hydrogeology / Ian D. Clark and Peter Fritz
   p. cm.
   Includes bibliographical references and index.
   ISBN 1-56670-249-6 (alk, paper)
      1. Radioactive tracers in hydrogeology.   I. Fritz, P. (Peter), 1937-   II. Title
GB1001.72.R34C57
551.49′028—dc21   97-21889
                                                                                                          CIP

This book contains information obtained from authentic and highly regarded sources. Reprinted material is quoted with permission, and sources are indicated. A wide variety of references are listed. Reasonable efforts have been made to publish reliable data and information, but the author and the publisher cannot assume responsibility for the validity of all materials or for the consequences of their use.

Neither this book nor any part may be reproduced or transmitted in any form or by any means, electronic or mechanical, including photocopying, microfilming, and recording, or by any information storage and retrieval system, without prior permission in writing from the publisher.

The consent of CRC Press does not extend to copying for general distribution, for promotion, for creating new works, or for resale. Specific permission must be obtained in writing from CRC Press for such copying.

Direct all inquiries to CRC Press LLC, 2000 Corporate Blvd., N.W., Boca Raton, FL 33431.

Trademark Notice: Product or corporate names may be trademarks or registered trademarks, and are used only for identification and explanation, without intent to infringe.

© 1997 by CRC Press LLC
Lewis Publishers is an imprint of CRC Press

No claim to original U.S. Government works
International Standard Book Number 1-56670-249-6
Library of Congress Card Number 97-21889
Printed in the United States of America 1 2 3 4 5 6 7 8 9 0
Printed on acid-free paper

DEDICATIONS

For Jordan, Ilya and TLH

        IDC

To my family, friends and students who helped me learn what we describe in this book

        PF

THE AUTHORS

Ian D. Clark
Ottawa-Carleton Geoscience Centre
University of Ottawa, Ottawa, Canada
idclark@uottawa.ca   <www.science.uottawa.ca/~geology>

Ian Clark holds a B.Sc. Co-op degree in Earth Sciences and an M.Sc. in Hydrogeology from the University of Waterloo, and received his doctorate degree from the Université de Paris-Sud, Orsay. Dr. Clark's interest in isotope hydrogeology began with his M.Sc. research on the hydrogeology of a geothermal prospect in western Canada. There, the potential for tracing the origin of the thermal waters with environmental isotopes began a long and fruitful collaboration with Peter Fritz, then a professor at Waterloo. Following five years as a hydrogeological consultant, their paths rejoined in Oman, where they undertook an isotope investigation of groundwaters throughout the Sultanate. This program formed the basis of his doctoral research, leading him to Orsay to work with Jean Charles Fontes on a paleo-hydrogeological reconstruction in Oman. His interest in the isotope hydrogeology of arid regions continues with Hani Khoury and Elias Salameh at the University of Jordan. In 1988, Ian joined the Department of Geology at the University of Ottawa to teach and supervise graduate research in isotope hydrogeology. Through collaboration with Bernard Lauriol at the University of Ottawa, he developed a keen interest in permafrost hydrogeology and paleoclimatology in the Canadian Arctic. Other current research activities include studies of hydrogeological analogues for nuclear waste repositories, contaminant hydrogeochemistry and CO_2 in ice cores. Dr. Clark remains in the Department of Geology as Associate Professor and Chair.

Peter Fritz
UFZ Centre for Environmental Research
Leipzig-Halle
gf@gf.ufz.de

Peter Fritz completed his undergraduate, Haupt-Diplom (M.Sc.) and doctorate degrees in geology at the University of Stuttgart. Much of his doctoral research was undertaken in 1962 at the University of Pisa where Professor E. Tongiorgi and Dr. R. Gonfiantini had established one of the world's first environmental isotope laboratories focussing on isotope hydrology. Following a NATO Research Fellowship with Jean Charles Fontes at the University of Paris (Sorbonne), Dr. Fritz spent five years at the University of Alberta in Edmonton as a Research Associate in the Department of Geology. In 1971 he was invited to join the hydrogeological research group being established at the University of Waterloo, and there built up an internationally recognized laboratory and graduate program in environmental isotope research. Dr. Fritz left Waterloo in 1987 to become director of the GSF Institute for Hydrology in Neuherberg, and then moved to Leipzig-Halle to establish the UFZ Centre for Environmental Research. He remains at the UFZ today as Scientific Director. Contributions to research in environmental isotopes include studies on isotope hydrology, work on brines and gases in crystalline rocks, paleoclimatology, methodological developments for radiocarbon dating of groundwaters with dissolved organic carbon, as well as other studies on the sulphur cycle in terrestrial systems. Former students of Dr. Fritz now lead research and teaching careers in over 15 universities worldwide. He is a Fellow of the Royal Society of Canada.

PREFACE

Hydrogeology is one of the most recent and rapidly expanding fields of the geosciences. So too is the complementary field of isotope hydrology. Environmental isotope laboratories are now established in most major universities and research centres. Environmental isotopes have become an integral component of hydrogeological research and applications. Yet, isotope hydrogeology has long been taught without a dedicated textbook that draws together the various aspects of environmental isotopes applied to hydrogeological problems. With this book and the supporting website <www.science.uottawa.ca/~eih> we have tried to provide students and researchers with a synthesis of past and current work, some of the theory behind isotope reactions, the basic equations and fractionation factors routinely used in isotope studies, approaches to solving problems and references to specific areas of research. Case studies are used throughout to emphasize and reinforce the material presented.

Environmental Isotopes in Hydrogeology begins with an introduction of the environmental isotopes, some history and early developments, and the basics of isotope fractionation. For ease of reference, we have compiled the important temperature equations for isotope fractionation and put them in Table 1, found inside the front cover. The book then explores the partitioning of ^{18}O and ^{2}H through the hydrological cycle, and how they serve as tracers for precipitation and groundwater. We then move on to biogeochemical cycles, and particularly the carbon cycle, where ^{13}C and other isotopes are used to identify and quantify reaction pathways in both natural and contaminated landscapes. The methods of dating modern and paleogroundwaters are presented in considerable detail, with case studies and solved problems that show the conjunctive use of various methods. The final chapter is a reference document for field sampling, including summaries of the various methods. Focus is maintained on the principal methods of isotope hydrogeology, although attention is also given to the less routine methods, with references to lead the reader to the relevant literature.

This book serves at an advanced undergraduate and graduate level. Some experience in aqueous geochemistry is helpful, as only the essential background and basic calculations are reviewed. The website for this book presents solutions to the problem sets for students to work through, as well as portions of the text, and links to other important websites. Hard copies of problems with solutions are available through the publisher. The website also provides some useful spreadsheets that can be downloaded and adapted to personal data sets.

Students should be aware of the textbooks by G. Faure *Principles of Isotope Geology* and J. Hoefs *Stable Isotope Geochemistry*, which cover the broader subject of isotopes in geology. Students must also appreciate the fundamental role played by the Hydrology Section of International Atomic Energy Agency (IAEA) as a forum for research. training and publications in isotope hydrology.

Ian Clark and Peter Fritz
Ottawa and Leipzig
June 1997

ACKNOWLEDGMENTS

This book has been constructed from the research and applications of many workers, from Urey, Dansgaard, Craig and Gonfiantini to the latest generation of isotope hydrogeologists. Their efforts and developments are the foundation of the subject. In particular, we wish to acknowledge the contributions of Jean Charles Fontes, who died tragically in 1995 while on a mission for the U.N. in Mali, and John Andrews, who passed away later that year at his home in Reading, U.K.. Their tremendous contributions are evident in the pages of this book. To have known them and learned from them has been immeasurably rewarding.

The data from Oman used in this book are the product of 18 months of groundwater sampling in the wadis and deserts of this country. For this opportunity we thank his Majesty Sultan Qaboos bin Said and the people of Oman.

Textbooks such as this are written for students. We are grateful to those who have contributed to it through their thesis research. We thank our many students over the years, who struggled with earlier versions of these chapters, brought errors and omissions to our attention, worked through the problem sets, and whose enthusiasm for the subject rallied our effort to finally publish it.

CONTENTS

CHAPTER 1: THE ENVIRONMENTAL ISOTOPES 1

Environmental Isotopes in Hydrogeology... 2
    Elements, nuclides, and isotopes 2
    Nucleosynthesis and the birth of the solar system 3
    Early days in isotope research 4
    Why "environmental" isotopes? 5
    Isotopes, ratios, deltas (δ) and permils (‰) 6
Stable Isotopes: Standards and Measurement... 7
    Oxygen-18 and deuterium in waters... 8
    Carbonate, organic carbon and hydrocarbon 9
    Sulphate and sulphide... 11
    Nitrate and reduced nitrogen.. 11
    Chloride 12
    Bromide, lithium and boron 12
    Strontium 13
Isotope ratio mass spectrometry 13
    Gas source mass spectrometry 13
    Solid source mass spectrometry... 15
    δ–Value corrections and conversions 15
Radioisotopes 16
    Tritium 16
    Carbon-14... 18
    Chlorine-36 and iodine-129 19
    Argon-39 20
    Krypton 20
    Uranium series isotopes 20
Isotope Fractionation 21
    Physicochemical fractionation.. 21
    Diffusive fractionation 24
    Isotopic equilibrium 25
    The example of ^{18}O fractionation between water and vapour 26
    Temperature effect on fractionation 27
    Kinetic (nonequilibrium) fractionation... 29
Isotope Fractionation (α), Enrichment (ε) and Separation (Δ).......................... 31
    The example of water-vapour reaction 31
Problems 33

CHAPTER 2: TRACING THE HYDROLOGICAL CYCLE 35

Craig's Meteoric Relationship in Global Fresh Waters 36
Partitioning of Isotopes Through the Hydrological Cycle 37
    Isotopic composition of ocean waters 37

        The atmosphere and vapour mass formation.. 39
        Isotopic equilibrium in water-vapour exchange 39
        Humidity and kinetic (nonequilibrium) evaporation 41
        Deuterium excess "d" in meteoric waters... 43
        Atmospheric mixing and global atmospheric water vapour............................. 46
    Condensation, Precipitation and the Meteoric Water Line.. 46
        Rainout and Rayleigh distillation 47
        Slope of the meteoric water line 49
        Local meteoric water lines.. 51
    A Closer Look at Rayleigh Distillation 55
    Effects of Extreme Evaporation 57
        Evaporation in Lakes 57
        Evaporation of brines 59
    Problems 60

CHAPTER 3: PRECIPITATION 63

    The T–δ^{18}O Correlation in Precipitation 64
        δ^{18}O on the global scale 64
        Latitude effect... 66
        Continental effects 67
    Local effects on T–δ^{18}O 70
        Altitude effect 70
        Seasonal effects.. 71
        Condensation of coastal fog... 73
        Kinetic effects of secondary evaporation.. 74
    Ice Cores and Paleotemperature 75
    Problems 77

CHAPTER 4: GROUNDWATER 79

    Recharge in Temperate Climates 80
        Attenuation of seasonal variations... 80
        Comparing shallow groundwaters with precipitation...................................... 83
        Recharge by snowmelt 85
    Recharge in Arid Regions... 86
        Evaporative enrichment in alluvial groundwaters... 87
        Recharge by direct infiltration... 88
        Soil profiles and recharge rates 89
        Estimating recharge with ^{36}Cl and chloride 92
        Water loss by evaporation vs. transpiration 94
    Recharge from River-Connected Aquifers.. 96
        Time series monitoring in a river-connected aquifer...................................... 96
        The Swiss tritium tracer "experiment" 96
        Water balance with ^{14}C 98
        Recharge from the Nile River.. 98
        Recharge by desert dams.. 99
    Hydrograph Separation in Catchment Studies.. 99
        Example of the Big Otter Creek Basin, Ontario.. 102

        An example from Australia.. 104
    Groundwater Mixing... 104
        Binary and ternary groundwater mixing... 105
        Mixing of groundwaters in regional flow systems 105
        Groundwater mixing in karst systems 107
    Problems 108

CHAPTER 5: TRACING THE CARBON CYCLE...111

    Evolution of Carbon in Groundwaters... 112
    Carbonate Geochemistry... 112
        Activity, concentration and mineral solubility relationships 112
        Atmospheric and soil CO_2.. 115
        Dissolution of soil CO_2 and carbonate speciation 115
        pH buffering and mineral weathering... 117
    Carbon-13 in the Carbonate System.. 119
        Vegetation and soil CO_2... 119
        ^{13}C fractionation in CO_2 – DIC reactions 120
        Evolution of $\delta^{13}C_{DIC}$ during carbonate dissolution 122
        Incongruent dissolution of dolomite... 123
    Dissolved Organic Carbon.. 124
        DOC and redox evolution 126
    Methane in Groundwaters 127
        Biogenic methane 127
        Thermocatalytic methane... 131
        Abiogenic and mantle methane.. 131
        ^{14}C and sources of carbon... 132
    Isotopic composition of carbonates 132
        $\delta^{18}O$ in secondary calcite and paleotemperatures............................. 133
    Problems 134

CHAPTER 6: GROUNDWATER QUALITY 137

    Sulphate, Sulphide and the Sulphur Cycle 138
        Marine sulphate.. 139
        Oxidation of sulphide and terrestrial sulphate 142
        Atmospheric sulphate 144
        Sulphate reduction.. 144
        Sulphate-water ^{18}O exchange 147
    Nitrogen cycling in rural watersheds... 148
        The geochemistry of nitrate.. 149
        Isotopic composition of nitrate... 150
        Nitrate contamination in shallow groundwaters 151
    The "Fuhrberger Feld" Study... 152
        Denitrification and ^{15}N 153
        Sulphate reduction at depth.. 154
    Source of chloride salinity 155
        Ionic ratio indicators... 155
        Chlorine isotopes — $\delta^{37}Cl$ 155

Landfill Leachates 157
Degradation of Chloro-organics and hydrocarbon 159
Sensitivity of Groundwater to Contamination 160
    Temporal monitoring with stable isotopes 161
    Aquitards — impermeable or leaky barriers? 161
    Diffusion across aquitards 163
Summary of Isotopes in Contaminant Hydrogeology 165
    Contamination in agricultural watersheds 166
    Sanitary landfills 167
    Fuel and solvent contaminated sites 167
    Siting hazardous waste facilities 168
Problems 168

CHAPTER 7: IDENTIFYING AND DATING MODERN GROUNDWATERS 171

The "Age" of Groundwater 172
    "Modern" groundwater 172
    The tools for dating groundwater 172
Stable Isotopes 173
Tritium in Precipitation 174
    Cosmogenic tritium 174
    Thermonuclear (bomb) tritium 175
    Nuclear reactor tritium 178
    Geogenic production of ^3H 179
Dating Groundwaters with Tritium 179
    Velocity of the 1963 "bomb peak" 180
    Radioactive decay 181
    Input function for ^3H in groundwater 183
    Time series analysis 184
    Qualitative interpretation of ^3H data 184
    Tritium in alluvial groundwaters — an example from Oman 185
    Deep groundwaters - mixing in fractured rock 186
Groundwater Dating with ^3H - ^3He 186
    Helium–tritium systematics 187
    Applications of the ^3H- ^3He method 188
Chlorofluorocarbons (CFCs) 188
Thermonuclear ^{36}Cl 189
Detecting Modern Groundwaters with ^{85}Kr 191
Submodern groundwater 192
    Argon-39 192
    Silica-32 194
Problems 195

CHAPTER 8: AGE DATING OLD GROUNDWATERS 197

Stable Isotopes and Paleogroundwaters 198
Groundwater Dating with Radiocarbon 200
    Decay of ^{14}C as a measure of time 201
    Production of ^{14}C in the atmosphere 202

Natural variations in atmospheric ^{14}C	203
Anthropogenic impacts on atmospheric ^{14}C	204
The ^{14}C pathway to groundwater in the recharge environment	205
Correction for Carbonate Dissolution	206
Statistical correction (STAT model)	207
Alkalinity correction (ALK model)	208
Chemical mass-balance correction (CMB model)	209
$\delta^{13}C$ mixing ($\delta^{13}C$ model)	210
The effect of dolomite dissolution	212
Matrix exchange (Fontes-Garnier model)	212
Which model do I use?	213
Case study of the Triassic sandstone aquifer, U.K.	215
Some Additional Complications to ^{14}C Dating	217
Matrix diffusion of ^{14}C	217
Sulphate reduction	218
Incorporation of geogenic CO_2	220
Methanogenesis	220
Dilution factors for multiple processes	222
Revisiting the groundwaters in southern Oman	222
Modelling ^{14}C ages with NETPATH	224
^{14}C Dating with Dissolved Organic Carbon (DOC)	225
The initial ^{14}C activity in fulvic acid ($a_o{}^{14}C_{FA}$)	225
Advantages and disadvantages of DOC	226
Case studies for ^{14}C dating with DOC and DIC	227
The Milk River aquifer	227
The Gorleben study, Germany	229
Chlorine-36 and Very Old Groundwater	231
Units of expression for ^{36}Cl data	231
Cosmogenic production of ^{36}Cl	232
Subsurface production of ^{36}Cl	234
Example of the Great Artesian Basin, Australia	235
Summary of ^{36}Cl in groundwater dating	237
The Uranium Decay Series	238
$^{234}U/^{238}U$ disequilibrium	238
Dating with ^{226}Ra and ^{222}Rn	240
^4He and old groundwater	241
Problems	243

CHAPTER 9: WATER - ROCK INTERACTION ... 245

Mechanisms of Isotope Exchange	246
High Temperature Systems	247
Magmatic water and primary silicates	247
The ^{18}O shift in geothermal waters	250
Andesitic volcanism and geothermal waters	252
Subsurface steam separation	253
Geothermometry	253
Low Temperature Water-Rock Interaction	255
Hydration of primary silicate minerals	255
The example of shield brines	256

   Low-temperature exchange in sedimentary formations 258
   Hyperfiltration of isotopes 260
Strontium Isotopes in Water and Rock 260
Isotope Exchange in Gas - Water Reactions 262
   Deuterium shift — exchange with H_2S 262
   ^{18}O exchange between H_2O and CO_2 263
High pH Groundwaters — The Effect of Cement Reactions 264
Problems 265

CHAPTER 10: FIELD METHODS FOR SAMPLING 267

Groundwater 271
   Sample sites 271
   Getting water from the well 272
   Deuterium and oxygen-18 273
   Tritium 273
   Carbon-13 in DIC 274
   Radiocarbon in DIC 275
   Carbon-13 and ^{14}C in DOC 279
   Sulphur-34 and ^{18}O in aqueous sulphur compounds 279
   Nitrate and organic nitrogen 280
   Chloride 281
   Uranium series nuclides 281
Water in the Unsaturated Zone 282
Precipitation 282
   Rain samples for ^{18}O, ^2H and ^3H 282
   Snow and ice ^{18}O, ^2H and tritium 283
Gases 283
   Soil CO_2 283
   Gas in groundwater 284
Geochemistry 285
   Field measurements 285
   Major anions (Cl^-, SO_4^-, NO_3^-, F^-, Br^-) 289
   Major, minor and trace metals 289
   Dissolved organic carbon (DOC) 290

REFERENCES 291

SUBJECT INDEX 312

Chapter 1
The Environmental Isotopes

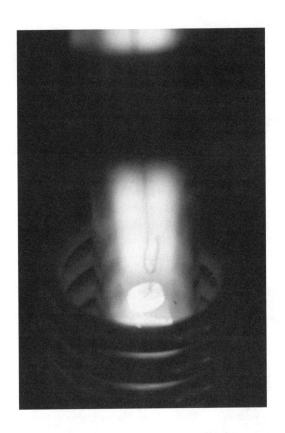

Groundwater moving through the geosphere appears to be a simple enough process, yet a groundwater realm of cryptic underground rivers and channels has remained in our culture since early historic times, sustained by water diviners and rural myths. Surface waters seem easier to understand, their intricacies more apparent because we can see them flow and follow them to their source. But to quench your thirst, they must first be flocculated, sedimented, filtered, limed, chlorinated and often chilled, provided industrial effluents have not already damaged the supply. For groundwaters, the geosphere provides these treatments naturally. Groundwater represents more than 50 times the freshwater resource that surface waters do, yet in North America groundwater is used for less than half of freshwater needs; in Central Europe, groundwater is the dominant source for drinking water.

The dawn of hydrogeology as a science began with Darcy's early experimenting with the plumbing for the fountains of Dijon. Today, the over-exploitation and contamination of this resource has moved groundwater research to the forefront of the geosciences. Nonetheless, like the diviners of historic times, hydrogeologists still wrestle with the questions of groundwater provenance, its renewability and the subsurface processes affecting its quality. These questions become increasingly relevant as we continue to test the limits of groundwater resource sustainability.

Environmental Isotopes in Hydrogeology

Environmental isotopes now routinely contribute to such investigations, complementing geochemistry and physical hydrogeology. The stable isotopic composition of water, for instance, is modified by meteoric processes, and so the recharge waters in a particular environment will have a characteristic isotopic signature. This signature then serves as a natural tracer for the provenance of groundwater. On the other hand, radioisotopes decay, providing us with a measure of circulation time, and thus groundwater renewability. Environmental isotopes provide, however, much more than indications of groundwater provenance and age. Looking at isotopes in water, solutes and solids tells us about groundwater quality, geochemical evolution, recharge processes, rock-water interaction, the origin of salinity and contaminant processes. Let's start with the basics.

Elements, nuclides, and isotopes

The nuclear structure of a nuclide (an isotope-specific atom) is classically defined by its number of protons (Z) which defines the element, and the number of neutrons (N) which defines the isotope of that element. For a given nuclide, the sum of protons and neutrons gives the atomic weight (A), expressed by the notation $^A_Z Nu_N$. For example, most oxygen has 8 protons and 8 neutrons, giving a nuclide with 16 atomic mass units ($^{16}_8 O_8$) while about 0.2% of oxygen has 10 neutrons ($^{18}_8 O_{10}$). In reality, the mass of a nuclide is slightly less than the combined mass of its neutrons and protons. The "missing" mass is expressed as the nuclear binding energy (according to Einstein's mass-energy relationship $E = mc^2$), which represents the amount of energy required to break the nucleus into its constituent nucleons. Conventional notation for a nuclide uses only the elemental symbol and atomic weight (e.g. ^{18}O or ^{34}S).

Whereas the number of neutrons in the nucleus can vary, the range is limited by the degree of instability created by having too many or too few neutrons. Unstable isotopes or radioactive nuclides have a certain probability of decay. Stable isotopes, on the other hand, do not spontaneously disintegrate by any known mode of decay. To date, some 270 stable nuclides and over 1700 radionuclides have been identified. For the light elements (Z up to 20) the greatest stability occurs with a Z:N ratio close to 1, and increases towards 1.5 for heavy elements. In a chart arranged according to Z and N (Fig. 1-1), the stable isotopes of the elements form a stable valley from hydrogen to uranium. Departures from this stable valley produce radionuclides of decreasing stability (shorter half-lifes). Oxygen, for example, has eleven isotopes (^{12}O to ^{22}O) although only the median isotopes, ^{16}O, ^{17}O and ^{18}O, are stable (Fig 1-2). The others are radioactive with half-lives varying from 122 seconds to less than a femtosecond (10^{-15} s).

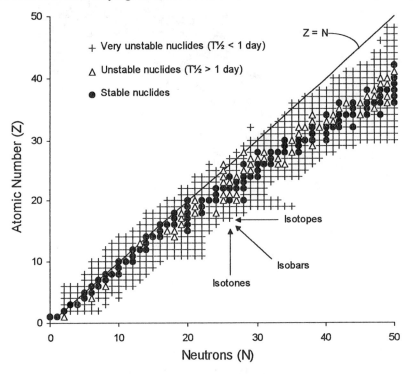

Fig. 1-1 Plot of Z vs. N for nuclides up to tin (Z=50) showing the "stable" valley of the nuclides. The Z : N ratio is 1 for the light nuclides and increases towards 1.5 for the heavier nuclides. Increases or decreases in N for given element produces increasingly unstable isotopes (decreasing T½).

The distribution of stable isotopes reflects the structure of the nucleus. Like electron orbits, the most stable nuclei have filled neutron and/or proton shells. Those nuclides with a "magic number" (2, 8, 20, 28, 50, 82 and 126) of neutrons and/or protons are the most common (e.g. $^{4}_{2}He_{2}$ = 99.99986% of all helium; $^{16}_{8}O_{8}$ = 99.76% of all oxygen; or $^{40}_{20}Ca_{20}$ = 96.9% of all calcium) whereas others have lower abundances (e.g. $^{10}_{5}B_{5}$ = 19.9% of all boron). As the nuclear binding energy occurs between nucleon pairs (protons or neutrons), stable nuclides with even numbers for N and Z, dominate. Thus, 161 of the known stable nuclides have an even N and Z while only 4 have an odd values for N and Z. There are 105 with either an odd N or Z.

Nucleosynthesis and the birth of the solar system

The formation of matter by nucleosynthesis during the birth of our solar system some 5 billion years ago produced most of the stable and radioactive isotopes that exist naturally on earth today. Some, such as technetium (atomic number 43) which has no stable isotopes, have disappeared (it is now synthetically produced within the nuclear fuel cycle). Others with half-lives comparable to the age of the universe (e.g. ^{238}U or ^{232}Th) are still present and can be used to determine the age of our earth, meteorites, and the solar system. These have also generated stable decay products such as ^{207}Pb and ^{208}Pb from ^{235}U and ^{232}Th.

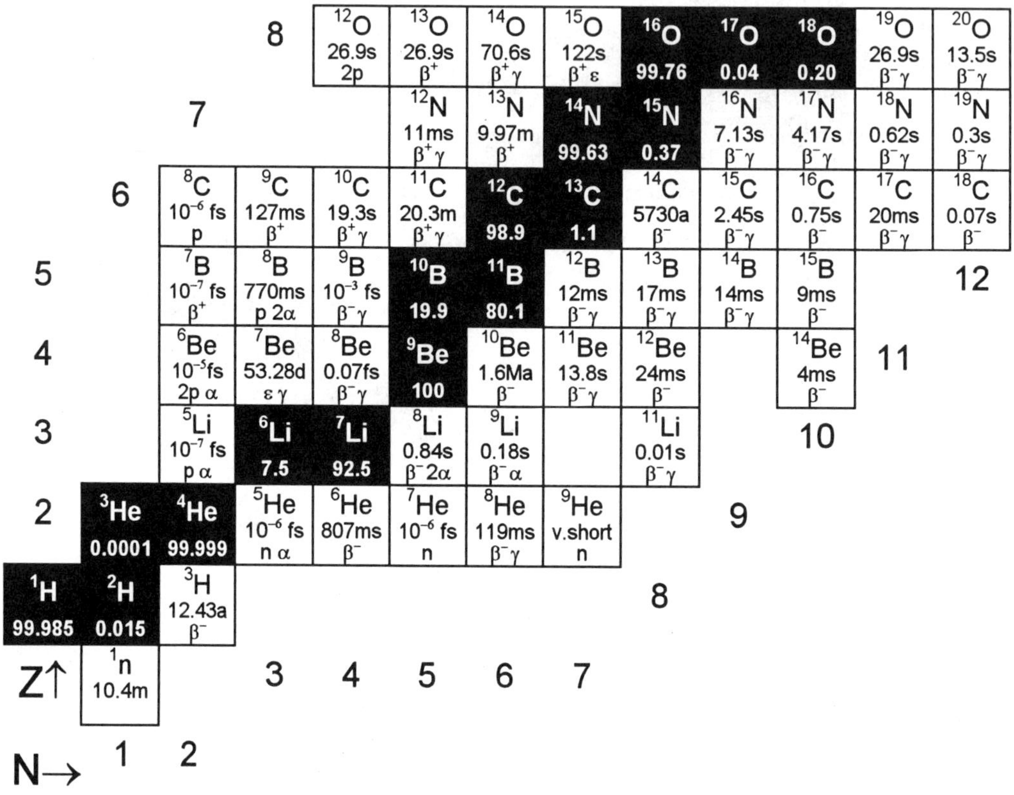

Fig. 1-2 Chart of the light element isotopes showing percent abundances of the stable isotopes (shaded black) and half-lives of radioisotopes (s = second, m = minute, d = day, a = year) with their principal and secondary decay modes, where α = alpha emission (2p and 2n), ε = electron capture, β⁻ = electron (beta) emission, β⁺ = positron, γ = gamma emission, n = neutron emission, p = proton emission (after General Electric Ltd., 1989).

Nucleosynthesis is, however, not only a process of the past: some important environmental isotopes such as tritium, carbon-14, or chlorine-36 are continuously produced by natural neutron fluxes in the upper atmosphere and within rock masses, or within the nuclear fuel cycle and by nuclear weapons testing.

Early days in isotope research

The earliest notion that different *isotopes* (Greek *isos* = equal, *topos* = place) of a given element existed dates to the turn of the century. As Becquerel's discovery of uranium radioactivity, and the Curies' separation of radium, led to an understanding of the decay series of uranium and thorium, questions arose as to why thorium had various atomic weights, and why lead from uranium decay had an atomic weight different from that of common lead.

Further, as chemists discovered the periodic properties of the elements, the fractional masses of many elements posed an enigma. Frederick Soddy first suggested that different kinds of atoms could occupy a place in the periodic table. This was confirmed in the early 20th century by

Francis Aston who developed the positive particle beam experiments of J.J. Thompson into a mass spectrograph. By focusing a stream of positively charged gas ions through a magnetic field onto a photographic plate, he produced separate points of exposure. The mass spectrograph showed that elements with fractional atomic weights, such as chloride at atomic weight 35.45, were comprised of isotopes with different, near-integer masses. In 1932, the discovery of the neutron showed that isotopes of an element differed in the number of this uncharged particle, whose role is to counterbalance the repulsive forces between protons in the atomic nucleus. Since that time, over 2000 isotopes of the 92 naturally occurring elements have been identified.

In 1936, Alfred Nier produced the first precise measurements of isotope abundance ratios. In his mass spectrometer, samples were mounted in the source as solid salts on a thermal ionizing filament, or bled in as a gas and ionized in an electron beam. The positively charged ions were accelerated in an electric field before exiting the source, travelling along a flight tube through a wedge-shaped magnetic field. He measured the amperage of the separated isotopic ion currents (the spectrum of masses) with a single faraday collector. His design remains the basis of stable isotope mass spectrometry today, although modern instruments now have multiple collectors to simultaneously measure several isotope ratios.

With this gift from the physicists for routine measurement of isotope ratios, Earth scientists began to explore the natural variations of isotopes, particularly for carbon, sulphur, oxygen and hydrogen bearing-materials. Thus began a new era in geoscience research with the hydrological cycle and marine paleoclimatic research being the first topics to be investigated (Urey et al., 1951; Epstein and Mayeda, 1953).

Why "environmental" isotopes?

Although all elements present in hydrogeological systems have a number of isotopes, only a few are of practical importance to us. The environmental isotopes are the naturally occurring isotopes of elements found in abundance in our environment: H, C, N, O and S. These are principal elements of hydrological, geological and biological systems. The stable isotopes of these element serve as tracers of water, carbon, nutrient and solute cycling. They are also light elements. As a consequence, the relative mass differences between their isotopes are large, imparting measurable fractionations during physical and chemical reactions. For example, ^{2}H has 100% more mass than its sister isotope ^{1}H, whereas the two stable isotopes of bromine (^{81}Br and ^{79}Br) have a mass difference of only 2.5%. Radioactive environmental isotopes are also important in hydrogeology. From their decay we have a measure of time and so environmental radionuclides such as ^{14}C and ^{3}H can be used to estimate the age or circulation of groundwater.

The family of environmental isotopes is growing as new methods allow the routine analysis of additional isotopes. Accelerator mass spectrometry (AMS) analysis has brought ^{36}Cl into mainstream isotope hydrogeology. Refinements in solid source mass spectrometry and inductively coupled plasma mass spectrometry (ICP-MS) allows high precision measurement of the isotopes of trace elements such as U, Th, Li and B. The major stable environmental isotopes used in hydrogeology are presented in Table 1-1.

Environmental isotopes are now used to trace not only groundwater provenance, but also recharge processes, subsurface processes, geochemical reactions and reaction rates. Their importance in studies of biogeochemical cycles and soil-water-atmosphere processes is increasingly being recognized, and new applications in contaminant hydrogeology are being made.

Table 1-1 The stable environmental isotopes

Isotope	Ratio	% natural abundance	Reference (abundance ratio)	Commonly measured phases
^{2}H	^{2}H/^{1}H	0.015	VSMOW ($1.5575 \cdot 10^{-4}$)	H_2O, CH_2O, CH_4, H_2, OH^- minerals
^{3}He	^{3}He/^{4}He	0.000138	Atmospheric He ($1.3 \cdot 10^{-6}$)	He in water or gas, crustal fluids, basalt
^{6}Li	^{6}Li/^{7}Li	7.5	L-SVEC ($8.32 \cdot 10^{-2}$)	Saline waters, rocks
^{11}B	^{11}B/^{10}B	80.1	NBS 951 (4.04362)	Saline waters, clays, borate, rocks
^{13}C	^{13}C/^{12}C	1.11	VPDB ($1.1237 \cdot 10^{-2}$)	CO_2, carbonate, DIC, CH_4, organics
^{15}N	^{15}N/^{14}N	0.366	AIR N_2 ($3.677 \cdot 10^{-3}$)	N_2, NH_4^+, NO_3^-, N-organics
^{18}O	^{18}O/^{16}O	0.204	VSMOW ($2.0052 \cdot 10^{-3}$) VPDB ($2.0672 \cdot 10^{-3}$)	H_2O, CH_2O, CO_2, sulphates, NO_3^-, carbonates, silicates, OH^- minerals
^{34}S	^{34}S/^{32}S	4.21	CDT ($4.5005 \cdot 10^{-2}$)	Sulphates, sulphides, H_2S, S-organics
^{37}Cl	^{37}Cl/^{35}Cl	24.23	SMOC (0.324)	Saline waters, rocks, evaporites, solvents
^{81}Br	^{81}Br/^{79}Br	49.31	SMOB	Developmental for saline waters
^{87}Sr	^{87}Sr/^{86}Sr	^{87}Sr = 7.0 ^{86}Sr = 9.86	Absolute ratio measured	Water, carbonates, sulphates, feldspar

Isotopes, ratios, deltas (δ) and permils (‰)

The variations in numbers of neutrons in an element provides for the different masses (atomic weights) of the element and the molecules of which they may be a part. For example, heavy water, $^2H_2^{16}O$, has a mass of 20 compared to normal water, $^1H_2^{16}O$, which has a mass of 18. Molecules with differences in mass have different reaction rates. This leads to the isotope partitioning or *fractionation* described by Urey (1947).

Stable environmental isotopes are measured as the ratio of the two most abundant isotopes of a given element. For oxygen it is the ratio of ^{18}O, with a terrestrial abundance of 0.204%, to common ^{16}O which represents 99.796 of terrestrial oxygen. Thus the $^{18}O/^{16}O$ ratio is about 0.00204. Fractionation processes will of course modify this ratio slightly for any given compound containing oxygen, but these variations are seen only at the fifth or sixth decimal place.

Measuring an absolute isotope ratio or abundance is not easily done and requires some rather sophisticated mass spectrometric equipment. Further, measuring this ratio on a routine basis would lead to tremendous problems in comparing data sets from different laboratories. However, we are mainly interested in comparing the variations in stable isotope concentrations rather than actual abundance, and so a simpler approach is used. Rather than measuring a true ratio, an apparent ratio can easily be measured by gas source mass spectrometry. The apparent ratio differs from the true ratio due to operational variations (machine error, or m) and will not be constant between machines or laboratories or even different days for the same machine. However, by measuring a known reference on the same machine at the same time, we can compare our sample to the reference. Isotopic concentrations are then expressed as the difference between the measured ratios of the sample and reference over the measured ratio of the reference. Mathematically, the error (m) between the apparent and true ratios is cancelled. This is expressed using the delta (δ) notation:

$$\delta^{18}O_{sample} = \frac{m(^{18}O/^{16}O)_{sample} - m(^{18}O/^{16}O)_{reference}}{m(^{18}O/^{16}O)_{reference}}$$

As fractionation processes do not impart huge variations in isotope concentrations, δ-values are expressed as the parts per thousand or permil (‰) difference from the reference. This equation then becomes the more familiar:

$$\delta^{18}O_{sample} = \left(\frac{(^{18}O/^{16}O)_{sample}}{(^{18}O/^{16}O)_{reference}} - 1 \right) \cdot 1000 \text{ ‰ VSMOW}$$

VSMOW is the name of the reference used, in this case Vienna Standard Mean Ocean Water. A δ–‰ value that is positive, say +10‰, signifies that the sample has 10 permil or 1% more ^{18}O than the reference, or is enriched by 10‰. Similarly, a sample that is depleted from the reference by this amount would be expressed as $\delta^{18}O_{sample} = -10$‰ VSMOW.

Stable Isotopes: Standards and Measurement

A basis of environmental isotope geochemistry is the global comparison of data sets, which demands standardization of measurements between laboratories. Over the past few decades, appropriate materials have been established as internationally recognized isotope references or standards. These standards are limited in quantity and so cannot be used by laboratories on a routine basis. The data from their mass spectrometers then leave the laboratory referenced to these internationally recognized materials. The appropriate reference standards for the stable environmental isotopes are given in Table 1-1, along with their abundance ratio.

Two organizations collaborate on the calibration, cataloguing and distribution of these materials; the United Nations International Atomic Energy Agency (IAEA, or "the Agency" as it has come to be referred to), and the National Institute of Standards and Technology (NIST), formerly the National Bureau of Standards (NBS). Material is available through either of these sources:

National Institute of Standards and Technology
Standard Reference Materials Program
Room 204, Building 202
Gaithersburg, Maryland 20899-0001
USA
Phone: 301-975-6776
Fax: 301-948-3730
E-Mail: srminfo@enh.nist.gov
Website: <www.nist.gov>

International Atomic Energy Agency
Section of Isotope Hydrology
Wagramerstrasse 5, P.O. Box 100
A-1400 Vienna
Austria
Phone: 43-1-206021735
Fax: 43-1-20607
E-Mail: iaea@iaea1.iaea.or.at
Website: <www.iaea.or.at>

The protocol for reporting isotope data has seen some confusion over the past few decades due to the introduction and calibration of new reference materials. The Commission on Atomic Weights and Isotopic Abundances of the International Union of Pure and Applied Chemistry, at their 38th General Assembly in 1995, made recommendations for reporting ^2H, ^{13}C and ^{18}O data that have since been adopted by the isotope community (Coplen, 1996). The International Atomic Energy Agency has summarized calibration details for most reference materials in the IAEA technical document 825 (IAEA, 1995).

The methods for converting samples from their natural materials (water, calcite, sulphate, etc.) to a gas or solid for isotope ratio mass spectrometry (IRMS) are continually being developed and improved. The more routine methods are described here and in Chapter 10. Laboratories also have their own "home-grown" adaptations of published methods, with their own specific requirements for sample size and preservation. Most isotope laboratories can be found through the website <**http://beluga.uvm.edu/geowww/isogeochem.htm**>. This site hosts a forum for discussion and information on isotope geochemistry. A directory maintained at Syracuse University <**www.geochemistry.syr.edu/cheatham/InstrPages.html**> provides links to isotope and geochemistry laboratories world wide.

Oxygen-18 and deuterium in waters

Craig (1961a) introduced Standard Mean Ocean Water (SMOW) as the standard for measurements of ^{18}O and ^{2}H in water. In fact, Craig's SMOW never existed. It was a hypothetical water calibrated to the isotopic content of NBS-1, a water sample taken from the Potomac River and catalogued by the former National Bureau of Standards. SMOW provided an appropriate reference for meteoric waters, as the oceans are the basis of the meteorological cycle. Craig (1961a) defined SMOW according to:

$$\left(\frac{^{18}O}{^{16}O}\right)_{SMOW} = 1.008 \left(\frac{^{18}O}{^{16}O}\right)_{NBS-1} = (1993.4 \pm 2.5) \cdot 10^{-6}$$

$$\left(\frac{^{2}H}{^{1}H}\right)_{SMOW} = 1.050 \left(\frac{^{2}H}{^{1}H}\right)_{NBS-1} = (158 \pm 2) \cdot 10^{-6}$$

Subsequently, the IAEA prepared a standard water from distilled seawater that was modified to have an isotopic composition close to SMOW. This reference is identified as VSMOW (Vienna Standard Mean Ocean Water).

Measurements on VSMOW show that:

$$\left(\frac{^{18}O}{^{16}O}\right)_{VSMOW} = (2005.2 \pm 0.45) \cdot 10^{-6} \quad \text{(Baertschi, 1976)}$$

and $\left(\frac{^{2}H}{^{1}H}\right)_{VSMOW} = (155.76 \pm 0.05) \cdot 10^{-6}$ (Hageman et al., 1970)

VSMOW has been the internationally accepted reference for ^{18}O and ^{2}H in waters for almost three decades. From these absolute abundance measurements it would seem that VSMOW is more than 5‰ enriched in ^{18}O over SMOW, and some 14‰ depleted in ^{2}H. However, comparison of SMOW-referenced data sets with VSMOW data sets shows that this is not the case. Rather, this difference reflects the uncertainty of absolute abundance measurements.

When VSMOW replaced Craig's SMOW standard, VSMOW came to be expressed simply as SMOW. This practice is now no longer acceptable and VSMOW is the correct reference to use. Some groups prefer V-SMOW to VSMOW. The non-hyphenated acronym used by NIST and many journals, has been adopted in this book.

For waters that are highly depleted from ocean water, a second water standard was distributed by the IAEA. This is Standard Light Antarctic Precipitation, or SLAP. Its value with respect to VSMOW was established on the basis of an inter-laboratory comparison organized by the IAEA (Gonfiantini, 1978):

$$\delta^{18}O_{SLAP} = -55.50‰ \text{ VSMOW}$$

and $\quad \delta^2H_{SLAP} = -428.0‰$ VSMOW

Mass spectrometers do not like water, which is a "sticky" gas and loves to condense throughout the high vacuum inlet and ion source. Consequently, ^{18}O in water is measured by equilibrating the water first with CO_2 and then analyzing the CO_2. The sample should have pH < 4.5 to ensure the fast exchange of oxygen between water and carbon dioxide, through the reaction:

$$CO_2 + H_2O \leftrightarrow H_2CO_3$$

The $\delta^{18}O$ value of the water sample is derived from that of the CO_2, with which there is an equilibrium. The fractionation between CO_2 and H_2O has been examined by many authors. The average fractionation factor (α, discussed below) of these values is 1.0412 (Friedman and O'Neil, 1977) which is now in standard use. Thus, CO_2 in equilibrium with water is about 41.2‰ enriched in ^{18}O. Analytical precision on $\delta^{18}O$ values is usually better than ±0.2‰.

Deuterium in water is measured by reducing the water to elemental hydrogen, H_2, using zinc (Coleman et al., 1982; Florkowski, 1985), although uranium ovens are still used. A new method has been developed involving H_2–H_2O exchange with a platinum powder catalyst. As the all the water is reduced and all hydrogen is converted to hydrogen gas, no isotope fractionation occurs and mass spectrometric measurement of the $^2H/^1H$ ratio requires no correction for the preparation technique used. For deuterium, the analytical error is usually ±1.0‰.

Carbonate, organic carbon and hydrocarbon

The range of oxidation states of carbon makes it a fundamental element of the biosphere and hydrosphere. Carbon-13 traces carbon sources and reactions for a multitude of inter-reacting organic and inorganic species. The paleotemperature scale developed in the early 1950s using the $^{18}O/^{16}O$ ratio in marine carbonates adopted PDB as the international reference material (Urey et al., 1951). PDB was the internal calcite structure (rostrum) from a fossil *Belemnitella americana* from the Cretaceous Pee Dee Formation in South Carolina. In 1957, Craig formally introduced PDB as the standard for both ^{13}C and ^{18}O in carbonate minerals. It has subsequently been adopted as the ^{13}C standard for all carbon compounds, including CO_2, dissolved inorganic carbon species (DIC), dissolved organic carbon (DOC), cellulose and other fixed-C solids (CH_2O), organic liquids, methane and other hydrocarbons. However, VSMOW is the standard for measurements of ^2H or ^{18}O in organic molecules (e.g. CH_4, CH_2O etc.).

Before the limited PDB supply was exhausted, Friedman et al. (1982) used it to calibrate a crushed slab of white marble of unknown origin, designated as NBS-19:

$$\delta^{18}O_{NBS-19} = -2.20‰ \text{ PDB}$$

$$\delta^{13}C_{NBS-19} = +1.95‰ \text{ PDB}$$

The IAEA in Vienna has subsequently defined the hypothetical VPDB (considered as identical to PDB) as the reference against which all $\delta^{13}C$ measurements and carbonate-$\delta^{18}O$ are reported.

The measurement of isotopes in carbonate minerals is done on CO_2 gas that is normally produced by acidification, a method developed by McCrea (1950). Carbon dioxide is produced from carbonate minerals by reaction with 100% phosphoric acid (H_3PO_4) at 25°C (Urey et al., 1951). The conversion of calcite to CO_2 follows the reaction:

$$CaCO_3 + 2H^+ \rightarrow CO_2 + H_2O + Ca^{2+}$$

This conversion is quantitative for C (all the carbon is transferred), and thus as long as all of the CO_2 is recovered, there can be no fractionation for ^{13}C. The $\delta^{13}C$ measured for the gas is then the same as the calcite. However, as one of the three oxygen atoms is lost from the $CaCO_3$, ^{18}O will fractionate and the $\delta^{18}O$ of the gas will differ considerably from the calcite. Fortunately, this fractionation is constant at a given temperature. As long as the acid does not exchange oxygen with the CO_2, the isotopic contents of a given carbonate can be determined. For this reason, water-free orthophosphoric acid (H_3PO_4) is used. The carbonate-CO_2 fractionation differs for each carbonate mineral and the appropriate factor must be used. Many have been measured by Sharma and Clayton (1965). These and others are given in Table 1-2. For samples with intimate mixtures of different carbonate minerals such as calcite and dolomite, Al-Aasm et al. (1990) present a method for selective extraction and measurement.

Table 1-2 The $10^3 \ln\alpha^{18}O_{CO_2\text{-Carb}}$ fractionation factors for some carbonate minerals during conversion to CO_2 with phosphoric acid (after Friedman and O'Neil, 1977)

Carbonate	T°C	$10^3 \ln\alpha^1$
$CaCO_3$ - Calcite	25	10.20
$CaCO_3$ - Aragonite	25	10.29
$CaMg(CO_3)_2$ - Dolomite	25	11.03
$CaMg(CO_3)_2$ - Dolomite	25	11.71[2]
$SrCO_3$ - Strontianite	25	10.43
$BaCO_3$ - Witherite	25	10.91
$FeCO_3$ - Siderite	25	10.1175[3]
$FeCO_3$ - Siderite	50	10.1075[3]
$MgCO_3$ - Magnesite	50	11.53

1. This expression and the fractionation factor, α, are defined below.
2. Rosenbaum and Sheppard, 1986.
3. Carothers et al., 1988.

Both VPDB and VSMOW are recognized international standards for ^{18}O. While waters are exclusively referenced to VSMOW, carbonates can refer to either. VPDB was originally introduced for paleoclimatic studies, where the ^{18}O content of carbonate was used as a paleotemperature scale. However, the use of carbonate isotopes has gone far beyond this field, and in water-carbonate studies it is common to express $\delta^{18}O$ data for carbonate against the VSMOW scale. Conversion is also necessary when deriving information about the $\delta^{18}O$ content of the water in which a carbonate has formed. The conversion chart in Fig. 1-3 or the following equations can be used (Coplen et al., 1983).

$$\delta^{18}O_{VSMOW} = 1.03091 \cdot \delta^{18}O_{VPDB} + 30.91$$

$$\delta^{18}O_{VPDB} = 0.97002 \cdot \delta^{18}O_{VSMOW} - 29.98$$

The conversion chart is interesting as it shows that PDB must have precipitated from a seawater with $\delta^{18}O$ very close to VSMOW at 25°C. This is evident because the $\delta^{18}O$ value of CO_2 produced by reaction of VPDB calcite with 100% H_3PO_4 at 25°C is +0.22‰ relative to CO_2 equilibrated with VSMOW water. The analytical precision on both ^{13}C and ^{18}O measurements is usually better than ±0.15‰.

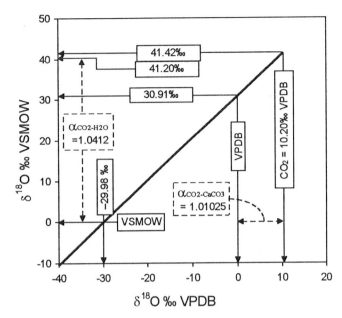

Fig. 1-3 Conversion chart for ^{18}O between VSMOW and VPDB, with fractionation factors for 25°C. The bold line equates values on the VPDB scale to values on the VSMOW scale according to the two reciprocal equations:
$\delta^{18}O_{VSMOW} = 1.03091 \cdot \delta^{18}O_{VPDB} + 30.91$ ‰, and $\delta^{18}O_{VPDB} = 0.97002 \cdot \delta^{18}O_{VSMOW} - 29.98$ ‰.

Sulphate and sulphide

Sulphur species can be sampled from groundwater as sulphate (SO_4^{2-}) or sulphide (H_2S or HS^-) for measuring ^{34}S contents and, for sulphate, ^{18}O. Both sulphate and sulphide mineral phases can also be analysed. Sampling sulphate and sulphide minerals requires no field preparation. Samples can be collected from outcrops, spring vents, as well as fracture surfaces and aquifer rocks in drill cores and cuttings.

Analyses are done either on SO_2 or SF_6. Corrections for ^{18}O in the SO_2 are necessary as this contributes to the mass-66 peak which is used to measure $^{34}SO_2$. Such is not the case with SF_6, as F has only one stable isotope, and so contributes no additional equal-mass ions.

The international standard against which $\delta^{34}S$ values are referenced is the troilite (FeS) phase of the Cañon Diablo meteorite (CDT), which has a $^{34}S/^{32}S$ abundance ratio of 0.0450. Oxygen-18 in sulphate is referenced to VSMOW. Like ^{18}O and ^{13}C, measurements are reported as δ ‰ difference from the standard with an analytical precision of better than or about ±0.3‰.

Nitrate and reduced nitrogen

Nitrogen-15 has been used for some time in hydrogeological studies, particularly for insights to sources and fate of nitrate and other N-contaminants in groundwaters. More recently, ^{18}O in NO_3^- has been developed as a complementary tool in such studies (e.g. Böttcher et al., 1990). The preparation procedures for ^{15}N (Chapter 10) are rather complex and involve reduction to NH_4^+ followed by oxidation to N_2 gas.

The internationally adopted reference for ^{15}N analyses is atmospheric nitrogen (AIR), which comes from this well mixed reservoir with a very reproducible mass ratio of $3.677 \cdot 10^{-3}$. Other reference materials are available through the IAEA and NIST (IAEA, 1995). As it is readily available, laboratories can also use it as their working standard. Analytical precision reflects the complicated preparation procedures and is usually not better than about ± 0.5 ‰. Oxygen in nitrate is converted to CO_2 for isotope measurement and is referenced to VSMOW.

Chloride

Chlorine exists in nature with two common stable isotopes: ^{35}Cl and ^{37}Cl. No other stable isotopes exist and, among the radioactive chlorine isotopes, only ^{36}Cl accumulates to measurable quantities because it has a relatively long half-life (301,000 yr). In geochemical studies, chloride has always been considered to be a conservative element. It is very soluble, and even in the biosphere it usually remains in its 1– valence state. Combining its nonreactive behaviour with its relatively small mass difference between the two stable isotopes, we can see why chlorine isotopes are not highly fractionated in nature. However, studies are now showing measurable fractionations that can be used to distinguish sources and mixing of groundwater. Also, the ^{37}Cl content of chloro-contaminants (solvents, etc.) in groundwater can fingerprint the source and may develop into a powerful tool in contaminant hydrology.

Mass spectrometric analyses are done with purified methyl chloride gas produced from Cl^- which is stripped from solution. The accepted reference for chlorine isotopes is Standard Mean Ocean Chloride (SMOC). As there was no significant variation from the data reported by Kaufmann et al. (1984) for a world survey of seawater samples, all local seawater standards are referenced to SMOC. The methyl chloride used as reference standard in the mass spectrometer is calibrated against SMOC regularly. The total analytical precision must be better than ± 0.1 ‰ in order to record the very small natural variations.

Bromide, lithium and boron

Interest in ^{37}Cl measurements for salinity studies has created interest in another halide, Br^-. Although its greater mass and conservative geochemical behaviour suggest that fractionation will be minimal, measurements of $^{81}Br/^{79}Br$ ratios may yet provide insights into the origin of salinity and evaporation processes. At this time, methods are still under development for measurement at the precision needed to observe the minor fractionation that this geochemically conservative element is expected to have.

Similarly, ^{6}Li and ^{11}B have become interesting isotopes for tracing the origin of salinity in groundwaters. You and Chan (1996) developed a method analyzing Li_3PO_4 by solid source mass spectrometry, as opposed to the established $Li_2BO_2^+$ method, for determination of δ^6Li. Measurements are referenced to the NBS L-SVEC Li_2CO_3 standard.

Measurement of isotope ratios can also be done by inductively coupled plasma mass spectrometry (ICP-MS). ICP-MS allows precise measurement of elemental abundances by mass, using an ionized gas stream passing from a high temperature argon plasma into a quadropole mass spectrometer. High resolution now allows discrimination between isotopes of certain elements. The abundance ratio of ^{11}B to ^{10}B is now measured this way (Gregoire, 1987).

Strontium

Strontium isotopes in rocks have provided insights into tectonic processes over geological time and can even be used to date sedimentary rocks (Viezer, 1989). They are also relevant to hydrogeological investigations since rocks of different ages can have significantly differing $^{87}Sr/^{86}Sr$ ratios. Since a significant portion of ^{87}Sr originates from the decay of ^{87}Rb, rocks rich in Rb (usually closely associated with K-rich rocks) tend to have higher $^{87}Sr/^{86}Sr$ ratios. Water that has significantly interacted with the rock matrix may "adopt" the rock signature in its dissolved Sr^{2+}, which reflects its flow path. Analyses are done by solid source mass spectrometry and results are expressed as $^{87}Sr/^{86}Sr$ ratios. The National Bureau of Standards distributes a reference strontium carbonate with $^{87}Sr/^{86}Sr = 0.701$.

Isotope ratio mass spectrometry

Measuring mass differences of molecular compounds can be done by a variety of mass spectrometer designs. The basis of isotope ratio mass spectrometry (IRMS) is to bend a beam of charged molecules in a magnetic field into a spectrum of masses (Fig. 1-4). The beam of charged molecules is usually generated by thermal ionization of a solid sample (solid source), or by ionizing a gaseous sample (gas source). The source design depends upon the isotopes of interest. Solid sources are best for high molecular weights such as strontium, uranium and lead, but also serves for lithium, which is analysed as lithium tetraborate. Most light elements are converted to gas.

Gas source mass spectrometry

In 1947, Alfred Nier developed the first dual-inlet, double-collector gas-source mass spectrometer. The double collector allowed the simultaneous measurement of two isotopes and the dual inlet allowed ratio measurement on both a sample and a standard by alternating between inlets. Gas source mass spectrometry has since become the measurement technique of preference for isotope ratios of most of the light elements (e.g. H, C, N, O and S) because of its relative simplicity and because the use of international standards allows comparison of data bases from different laboratories. A host of preparation methods have been developed and improved to convert different sample compounds to an appropriate gas including CO_2, SO_2, H_2 and N_2.

A heated tungsten-coated iridium (thoria) filament inside the source block cavity ionizes a laminar stream of gas entering the ultra-high vacuum source (Fig. 1-4). The gas molecules are stripped of one electron, producing positive ions (e.g. CO_2^+) which are then accelerated through a voltage gradient and focused into the flight tube upon exiting the source. The ionization efficiency varies between 0.01 and 0.1% for different instruments. The ion beam bends as it passes through the field of a magnet installed over the flight tube. Here, the beam separates into a spectrum of masses according to the isotopes present. Each mass beam continues to the ion detectors where preset faraday cup collectors measure each ion current. By collecting two or three ion beams simultaneously, the ion currents can be expressed as mass ratios. For example, CO_2 would contribute three principal peaks at mass 44 ($^{12}C^{16}O_2$), mass 45 ($^{13}C^{16}O_2$ or $^{12}C^{17}O^{16}O$) and mass 46 ($^{12}C^{16}O^{18}O$). A dual-inlet system allows the mass spectrometer to alternately measure ratios in the sample and a working or laboratory standard. Thus, the extreme fractionation imparted during ionization in the source is resolved. The early mass spectrometers suffered from drifting electronics, which precluded accurate abundance measurements. These instabilities have since been overcome with solid-state and fibre optic signal transfer systems.

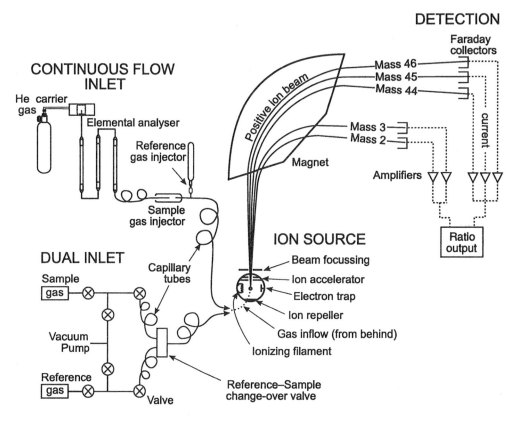

Fig. 1-4 Schematic of a gas source isotope ratio mass spectrometer (IRMS), showing both continuous flow and dual inlets. The continuous flow inlet here is shown with a sample combustion and gas chromatograph configuration. Capillary tubes ensure laminar, non-fractionating gas flow. Example shows mass range of CO_2 gas, and includes the short radius flight tube for H_2 found on many designs. Other mass ranges (for SO_2 and N_2) are attained by either additional fixed-position faraday collectors or by adjusting the beam. For manufacturers details, see <http://beluga.uvm.edu/geowww/isogeochem.htm>.

Quadropole mass spectrometers employ a method of isotope separation and measurement that offers economies in size and cost, but lack the precision of a Nier-type instrument. However, the portability of quadropole mass spectrometry allowed the Viking mission to perform isotope analyses of Martian soil and atmosphere.

The direction of gas source mass spectrometry is now towards continuous flow systems where the sample and standard gases are carried into the mass spectrometer source in a stream of helium gas (Fig. 1-4). In this way, pulse injections of sample gas can be analysed, reducing volume constraints and sample size. Measurement in the nano-mole (10^{-9} moles of gas) size range is now possible, opening a host of research possibilities. These systems also greatly increase sample through-put when configured with automated sample preparation systems. In particular, an elemental analyzer on the front end of a continuous flow mass spectrometer allows the analysis of solid, liquid, and mixed matrix samples which can be combusted and the gases separated on a chromatographic column. Helium carries the sample through the column and into the mass spectrometer. Continuous flow mass spectrometers configured with a laser ablation sampling system provide researchers with a tool for micro-analysis of sulphides and carbonates, with a spatial resolution in the order of 10 to 50 µm.

Solid source mass spectrometry

Solid source mass spectrometry is an important method of analysis for elements that are not readily converted to a gas. While the principle of bending an ionized beam into a mass spectrum with a magnetic field is the same, the source is quite different. Samples are converted to their elemental state and coated onto a (usually iron) filament which is then loaded into the source cavity. The filament is heated, ionizing the coating and emitting a charged ion beam. Thermal ionization mass spectrometry or TIMS is routinely used for isotopic analysis of strontium, uranium, thorium and even bromine in hydrogeology. Applications of TIMS in isotope geology are much broader, including, for example, the isotopes of osmium, neodymium, and lead.

Finally, accelerator mass spectrometry (AMS) must be mentioned as a method that offers high-precision measurements of isotope ratios that are exceedingly low. AMS has been developed largely for radioisotopes, including radiocarbon ($^{14}C/^{12}C \approx 10^{-11}$), ^{36}Cl ($^{36}Cl/^{35}Cl \approx 10^{-15}$) and ^{129}I, for which counting methods require both large sample size and long counting time. The AMS method is described later in this chapter.

δ–Value corrections and conversions

The raw δ–measurements produced by a typical triple-collector, gas-source mass spectrometer represent the mean for a set of six to ten sample/reference mass ratio comparisons, ($\overline{R}$), calculated from:

$$\delta \overline{R} = \left(\frac{\overline{R}_{sample}}{\overline{R}_{reference}} - 1 \right) \cdot 1000 \text{ ‰ Reference}$$

Here, the $\delta \overline{R}$ value could be for mass ratio 45/44 ($\delta^{45}\overline{R}$), or 46/44 ($\delta^{46}\overline{R}$) in the case of CO_2. To produce a useful value, it must be corrected for isobaric interferences — contributions from equal-mass ions that add to the mass peak of interest. The value must then be converted from the laboratory reference or working standard (WS) to an international reference, like VPDB. Gonfiantini (1981) presents the detailed calculations for these corrections and conversions.

For CO_2, the corrections are complicated because of the interest in both ^{18}O and ^{13}C. The mass ratio for ^{13}C (45/44) is dominated by $^{13}C^{16}O_2$, although small amounts of $^{12}C^{17}O^{16}O$ will also contribute. The amount of $^{12}C^{17}O^{16}O$ can be determined from the measurement of $\delta^{18}O$ because the fractionation of ^{17}O and ^{18}O is proportional. Unfortunately, the calculation is not so simple. As $\delta^{18}O$ is measured from the 46/44 mass ratio peak, a correction has to be made for $^{12}C^{17}O^{17}O$ and $^{13}C^{17}O^{16}O$. To correct raw δ–values measured from the mass-45 (^{13}C) and mass-46 peaks (^{18}O) for trace isotopes, Craig (1957) developed two now universally accepted relationships:

$$\delta^{13}C = 1.0676 \, \delta^{45}\overline{R} - 0.0338 \, \delta^{46}\overline{R} \text{ ‰ VPDB}$$
$$\delta^{18}O = 1.0010 \, \delta^{46}\overline{R} - 0.0021 \, \delta^{45}\overline{R} \text{ ‰ VPDB}$$

These corrections are applied in conjunction with the conversion from the lab reference to the international reference. One starts, for example, with the raw sample δ–‰ values from the mass spectrometer measured against the lab reference gas or working standard (i.e. $\delta^{45}\overline{R}_{sample-WS}$, and $\delta^{46}\overline{R}_{sample-WS}$), and the δ–value for the lab working standard referenced against the international standard ($\delta^{13}C_{WS-VPDB}$ and $\delta^{18}O_{WS-VPDB}$). The corrected and converted sample values ($\delta^{13}C_{sample-VPDB}$ and $\delta^{18}O_{sample-VPDB}$)) are calculated according to the following steps:

1. Determine the "uncorrected" δ-values for the lab working standard against VPDB ($\delta^{45}\overline{R}_{WS-VPDB}$ and $\delta^{46}\overline{R}_{WS-VPDB}$) according to the inverse of Craig's equations:

$$\delta^{45}\overline{R}_{WS-VPDB} = (\delta^{13}C_{WS-VPDB} + 0.0338 \cdot \delta^{18}O_{WS-VPDB}) / 1.0676$$

$$\delta^{46}\overline{R}_{WS-VPDB} = (\delta^{18}O_{WS-VPDB} + 0.0021 \cdot \delta^{13}C_{WS-VPDB}) / 1.0010$$

2. The $\delta\overline{R}_{sample}$ values are then expressed as if they had been measured using VPDB as the reference gas (which is only theoretically possible, since VPDB is not a gas). For this we use the expression for converting δ-values referenced to a working standard to an international standard:

$$\delta^{45}\overline{R}_{sample-VPD} = \frac{\delta^{45}\overline{R}_{sample-WS} \cdot \delta^{45}\overline{R}_{WS-VPDB}}{1000} + \delta^{45}\overline{R}_{WS-VPDB} + \delta^{45}\overline{R}_{sample-WS}$$

$$\delta^{46}\overline{R}_{sample-VPDB} = \frac{\delta^{46}\overline{R}_{sample-WS} \cdot \delta^{46}\overline{R}_{WS-VPDB}}{1000} + \delta^{46}\overline{R}_{WS-VPDB} + \delta^{46}\overline{R}_{sample-WS}$$

3. The raw $\delta\overline{R}_{sample}$ values, now converted to the international standard, are corrected for the equal-mass ion effects using the equations of Craig (1957) given above.

Other gases also require correction. During the measurement of δ^2H, only the 3/2 mass ratio is measured. However, the H_3^+ ion is also created in the source. The production of this ion, however, is proportional to the production of H_2^+ and the pressure of H_2. Most mass spectrometers now automatically compensate electronically for this error. Contributions from ^{18}O in SO_2 interfere with the mass 66 peak when measuring $\delta^{34}S$. Correction is simpler than for CO_2 and requires that the ^{18}O content of the lab working SO_2 gas be known.

Radioisotopes

The radioisotopes routinely employed in hydrogeology include tritium and carbon-14. Measurement of chlorine-36 is becoming increasingly available to non-specialists, while the use of other radioisotopes like argon-39 and the krypton (^{85}Kr and ^{81}Kr) is restricted to a limited number of research laboratories due to complications in sampling, analysis and interpretation. For the noble gases, huge volumes of water must be vacuum extracted in the field for analysis in low-level counters. Details on these radionuclides are given in Table 1-3. Sampling methods are covered in detail in Chapter 10.

Tritium

Tritium, ^3H, is a short-lived isotope of hydrogen with a half-life of 12.43 years. It attracted considerable interest during the era of thermonuclear bomb testing. Dr. R. Brown with Atomic Energy of Canada Limited (AECL) was the first to begin monitoring ^3H fallout from atmospheric weapons tests. His data for Ottawa precipitation begins in 1952 and documents the dramatic increases in atmospheric ^{3}H produced during the ensuing two decades of hydrogen bomb testing.

Table 1-3 The environmental radioisotopes

Isotope	Half-life (years)	Decay mode	Principal Sources	Commonly measured phases
^{3}H	12.43	β^-	Cosmogenic, weapons testing	H_2O, CH_2O
^{14}C	5730	β^-	Cosmogenic, weapons testing, nuclear reactors	DIC, DOC, CO_2 $CaCO_3$, CH_2O
^{36}Cl	301,000	β^-	Cosmogenic and subsurface	Cl^-, surface Cl-salts
^{39}Ar	269	β^-	Cosmogenic and subsurface	Ar
^{85}Kr	10.72	β^-	Nuclear fuel processing	Kr
^{81}Kr	210,000	ec	Cosmogenic and subsurface	Kr
^{129}I	$1.6 \cdot 10^7$ yr	β^-	Cosmogenic, subsurface, nuclear reactors	I^- and I in organics
^{222}Rn	3.8 days	α	Daughter of ^{226}Ra in ^{238}U decay series	Rn gas
^{226}Ra	1600	α	Daughter of ^{230}Th in ^{238}U decay series	Ra^{2+}, carbonate, clays
^{230}Th	75,400	α	Daughter of ^{234}U in ^{238}U decay series	Carbonate, organics
^{234}U	246,000	α	Daughter of ^{234}Pa in ^{238}U decay series	UO_2^{2+}, carbonate, organics
^{238}U	$4.47 \cdot 10^9$	α	Primordial	UO_2^{2+}, carbonate, organics

β^- - beta emission.
α - alpha emission.
ec - electron capture.

Small but measurable amounts of tritium are also produced naturally in the stratosphere by cosmic radiation on ^{14}N. Both natural and anthropogenic tritium enter the hydrological cycle via precipitation. Its presence in groundwater provides evidence for active recharge. As it is part of the water molecule, it is the only direct water dating method available.

Tritium has also gained importance in the medical field to tag compounds in biological reactions, although the concentrations used here exceed environmental concentrations by several orders of magnitude. Environmental concerns limit the use of artificial ^{3}H as a tracer in hydrological studies.

Tritium concentrations are expressed as absolute concentrations, using tritium units (TU) and so no reference standard is required. One TU corresponds to one ^{3}H atom per 10^{18} atoms of hydrogen. For 1 litre of water, its radioactivity is equivalent to 0.12 Bq (1 Becquerel = 1 disintegration/second), or 3.2 pCi/l (1 pCi is 10^{-12} Curies and a Curie is the radioactivity of 1 gram of ^{226}Ra; 1 Curie = $3.7 \cdot 10^{10}$ Bq). Groundwaters today seldom have more than 50 TU and are typically in the <1 to 10 TU range.

Tritium is measured by counting β^- decay events in a liquid scintillation counter (LSC). A 10 mL sample aliquot is mixed with the scintillation compound that releases a photon when struck by a β^- particle. Photomultiplier tubes in the counter convert the photons to electrical pulses that are counted over a several-hour period. Results are calculated by comparing the count to those of calibrated standards and blanks. Increased precision is gained through concentration by electrolytic enrichment of ^{3}H in the water before counting, or by conversion to propane (C_3H_8) for gas proportional counting. Direct liquid-scintillation counting carries a precision of ±7 TU, whereas with enrichment and LSC this is improved to better than ±0.8 TU. With propane synthesis, a precision of ±0.1 TU can be obtained.

Carbon-14

Willard Frank Libby earned the Nobel Prize in 1960 for his earlier work developing radiocarbon (^{14}C) as a tool for archeological dating. Its long half-life of 5730 years makes it useful for late Quaternary chronology and is now also extensively applied in hydrogeology to date groundwater. Natural production in the upper atmosphere is balanced by decay and burial to maintain a steady-state atmospheric $^{14}CO_2$ activity of about 13.56 disintegrations per minute (dpm) per gram of C, or about 1 ^{14}C atom per 10^{12} stable C atoms. However, the high neutron fluxes associated with the explosion of nuclear devices also produced large quantities of radiocarbon so that by 1964 the atmospheric concentrations in the northern hemisphere had almost doubled. This radiocarbon has now been almost "washed out" but can be found both in plant materials and the oceans.

Dating of groundwater with radiocarbon cannot be done on the water molecule itself but must rely on the dissolved inorganic and organic carbon in the water (DIC and DOC). Both forms enter groundwater from atmospheric $^{14}CO_2$ via the soil zone. Old groundwater can be dated if sufficient time has passed for measurable decay of the initial ^{14}C activity (Chapter 8).

Unlike tritium, ^{14}C activities are referenced to an international standard known as "modern carbon" (mC). The activity of "modern carbon" is defined as 95% of the ^{14}C activity in 1950 of the NBS oxalic acid standard. This is close to the activity of wood grown in 1890 in a fossil-CO_2-free environment and equals 13.56 dpm/g carbon. Thus, measured ^{14}C activities are expressed as a percent of modern carbon (pmC). Like stable isotopes, ^{14}C fractionates during organic and inorganic phase transformations and reactions. To maintain universality for dating purposes, ^{14}C activities must be normalized to a common $\delta^{13}C$ value of –25‰. Oxalic acid has a $\delta^{13}C$ value of –19.3‰ (Craig, 1961). Because the fractionation for ^{14}C is slightly more than twice (2.3× greater, Saliege and Fontes, 1984) that of ^{13}C, the enrichment amounts to:

$$2.3\ (\delta^{13}C_{sample} + 25)\ ‰$$

While this may not be significant for organic samples for which $\delta^{13}C$ is close to –25‰, it does affect other materials. For example, let's take a carbonate with:

$\delta^{13}C$ value of :   1.5 ‰
and ^{14}C activity of :   **65 pmC**

Normalization to the standard value of –25‰ for ^{13}C will then correct the ^{14}C value by:

$$2.3\ (–25 – 1.5) = –61.0\ ‰\ \text{or} –6.1\%$$

or by a factor of 0.939. To normalize the analysis, it must be multiplied by this factor, thus correcting for the elevated initial ^{14}C gained through isotopic fractionation. The corrected activity is then:

$$65 \times 0.939 = \mathbf{61.0\ pmC}.$$

Carbon-14 was first measured by gas-proportional counters containing the sample as CO_2 gas within a cathode cylinder. Decay events are counted by the potential loss when β^- particles strike an inner anode wire. External events (background radiation) registered by the counting cylinder are also registered in similar tubes arranged in a surrounding bundle. Thus, simultaneously recorded events are considered external and subtracted to increase precision. Gas-proportional counting is also used for other gaseous radioisotopes (e.g. ^{39}Ar and ^{85}Kr).

For ^{14}C, proportional counting has been largely replaced by less expensive liquid scintillation counting, with the sample converted to benzene (C_6H_6). Organic and inorganic material is converted to CO_2 which is reacted under vacuum with hot lithium metal to produce lithium carbide (Li_2C_2). This is then reacted with water to give acetylene (C_2H_2) which is then converted to benzene by reaction on a heated vanadium or chromium oxide catalyst. The benzene is mixed with a scintillation compound and placed with blanks and standards in a counter for several hours. The limiting background activities can be as low as ±0.5 pmC. For both methods of analysis, between 1 and 3 g of carbon is necessary, depending on the age. Dilution with ^{14}C-free CO_2 allows sample sizes of less than 1 g to be analysed (with proportional losses of precision). For groundwaters, this involves stripping the DIC from upwards of 25 L of sample by raising the pH above 11 with NaOH and adding an excess of $BaCl_2 \cdot 2H_2O$ to precipitate $BaCO_3$ (and whatever sulphate is present).

The use of compounds with a high CO_2 absorption such as Carbasorb® have been tried in conjunction with liquid scintillation counting (Qureshi et al., 1989). The advantage of avoiding the time-consuming step of synthesizing benzene is countered by a decrease in analytical precision (±3 to 5 pmC), but is reasonable for some groundwaters where corrections for geochemical reaction in the aquifer give even greater uncertainties of age.

Scintillation counting for ^{14}C is now eclipsed by accelerator mass spectrometry (AMS), for which sample size is dramatically reduced and yet can attain even greater ages with its high precision. The group of elements now routinely measured by AMS include ^{10}Be, ^{14}C, ^{26}Al, ^{36}Cl, ^{41}Ca and ^{129}I. Less than 5 mg of carbon (< 10 cc of CO_2) can be analysed with precision better than 0.5% of the sample activity for samples younger than a few thousand years, and less than 5% of the sample activity up to 40,000 years B.P. Reasonable dates up to 60,000 years have been measured on solid samples. Groundwater samples of less than 1 L can now be analysed, which simplifies field procedures enormously.

Two preparation methods exist: one in which acetylene is converted to graphite, and the second, in which CO_2 is converted directly to graphite in the presence of hydrogen. The graphite target is then mounted in the source of the accelerator, where a cesium beam is fired at it to release negatively charged C atoms through a room-sized, high-voltage (2 to >10 MV) particle accelerator. The high energy of the ions allows separation from interfering molecular ions. This separation permits counting ^{14}C on an individual ion basis for comparison with the ^{12}C beam.

Chlorine-36 and iodine-129

Like ^{14}C and ^{3}H, both ^{36}Cl and ^{129}I are produced by cosmic radiation in the upper atmosphere, and as a consequence of weapons testing in the atmosphere during the 1960s and 70s. These radionuclides are incorporated into the hydrological cycle by dry fallout or rainfall. Their long half-lifes (Table 1-3) make them suitable for dating old groundwaters and tracing sources of salinity.

The low production rate of ^{36}Cl and its rapid assimilation by the marine chloride pool results in an exceedingly low abundance in the hydrosphere. While ^{36}Cl was measured in rain as early as 1960 by β^- decay counting of Cl extracted from an 8000-L sample, it was the recent expansion of AMS facilities which allowed practical applications in hydrogeological studies. However, the relatively greater atomic mass of chlorine requires higher energies for determination by AMS, and so it is analysed in only a few centres worldwide.

Chlorine-36 is measured as a fraction of total Cl with a detection limit generally better than 10^{-15}. Because of its low concentration in waters, ^{36}Cl is generally expressed in terms of atoms per litre, with values on the order of 10^7 to 10^9. Samples containing a few tens of mg Cl can be analysed, although high salinity (high Cl⁻) and sulphate (^{36}S interference) samples pose problems with the detection limit.

Earlier methods for the analysis of ^{129}I included β^- decay counting, neutron activation to ^{130}I ($T_{½}$ = 12.36 h) and negative-ion mass spectrometry (Delmore, 1982). AMS is now widely applied due to the high precision ($^{129}I/I < 10^{-13}$) and reduced sample size (2 to 10 mg I) (Elmore, 1980).

Argon-39

Argon isotope applications in hydrology are based on the production of ^{39}Ar in the atmosphere. Its half-life of 269 years potentially fills the gap between tritium and radiocarbon. It is produced in the upper atmosphere by cosmic radiation on ^{40}Ar and follows the hydrological cycle like other noble gases. However, its low atmospheric activity of ^{39}Ar (0.112 dpm, Loosli and Oeschger, 1979) requires that at least 2 L STP of Ar be extracted from several m^3 of water for gas-proportional counting. No reference samples exist and only very few laboratories possess the sophisticated equipment required for extraction and counting.

Krypton

Krypton has two isotopes which are of potential interest for hydrogeological investigations: ^{81}Kr and ^{85}Kr with half-lives of 210,000 years and 10.46 years, respectively. Thus, ^{81}Kr is of interest for the recognition of very old groundwater, whereas ^{85}Kr has a half-life similar to that of tritium. Unfortunately, neither are easily sampled (>>1 m^3 of water is usually required) or measured.

The determination of ^{81}Kr relies on atom-counting under extreme experimental conditions. "Fortunately", nuclear activities have led to a continuous rise in ^{85}Kr concentrations, signifying that both sample volume and counting efforts could be reduced. Current atmospheric ^{85}Kr levels now exceed 60 dpm/cm^3 STP Kr (Loosli et al., 1991). As indicated, ^{81}Kr is measured by advanced mass spectrometry, whereas the activity of the β^- emitter ^{85}Kr is measured in low-background, high-pressure gas proportional counters.

Uranium series isotopes

The dissimilarities in geochemistry of U and Th and the complementary half-lifes of their radioisotopes provide one of the most reliable dating methods for Quaternary materials. The method is based on measurements of the $^{234}U/^{238}U$ and $^{230}Th/^{234}U$ activity ratios in a sample, which will be equal to 1 if secular equilibrium has been reached. The degree of disequilibrium provides a measure of time. Until recently, the $^{234}U/^{238}U$ and $^{230}Th/^{234}U$ activity ratios were determined by α-spectrometry. Thermal ionization mass spectrometry (TIMS) now provides higher precision analyses with mg-size samples (Edwards et al., 1987; Bard et al., 1993).

Isotope Fractionation

Although it was shown early this century that stable isotopes of oxygen and hydrogen existed, it was not until much later that their abundance and variation in the hydrosphere were investigated. Friedman, in 1953, first noted that in precipitation a change in concentration of $H_2{}^{18}O$ was accompanied by a change in ^2HHO. In 1961, Craig published his landmark finding that these two isotopes are partitioned by meteorological processes in a rather predictable fashion. Subsequent work has shown how isotopes are partitioned through other systems, such as ^{13}C in the carbon cycle.

How are environmental isotopes partitioned? Thermodynamic fractionation is a fundamental process. Here we will explore the details of isotope fractionation, using ^{18}O and ^2H in the hydrological cycle as an example. Fractionation of ^{13}C, ^{34}S, ^{15}N and other isotopes will be discussed in other chapters.

Isotope fractionation occurs in any thermodynamic reaction due to differences in the rates of reaction for different molecular species. The result is a disproportionate concentration of one isotope over the other on one side of the reaction. It is expressed by the fractionation factor α which is the ratio of the isotope ratios for the reactant and product:

$$\alpha = \frac{R_{reactant}}{R_{product}}$$

e.g. $\quad \alpha^{18}O_{water-vapour} = \dfrac{(^{18}O/^{16}O)_{water}}{(^{18}O/^{16}O)_{vapour}}$

There is more than one way to fractionate isotopes in a thermodynamic reaction. We can have physicochemical reactions under equilibrium conditions or nonequilibrium (kinetic) conditions, and we can have fractionation by molecular diffusion.

Physicochemical fractionation

The basis for isotope fractionation was first formally developed by Harold Urey who presented his landmark paper on *The Thermodynamic Properties of Isotopic Substances* in 1946 before the Chemical Society (Urey, 1947). In this paper, Urey showed that isotope fractionation can be represented as an exchange of isotopes (e.g. ^{16}O and ^{18}O) between any two molecular species or phases that are participating in a reaction.

Reactions can be simple physical changes of state:

$$H_2O_{water} \leftrightarrow H_2O_{vapour}$$

$$\alpha^{18}O_{water-vapour} = \frac{(^{18}O/^{16}O)_{water}}{(^{18}O/^{16}O)_{vapour}}$$

chemical transformations:

$$CO_2 + H_2O \leftrightarrow H_2CO_3$$

$$\alpha^{13}C_{CO_2-H_2CO_3} = \frac{(^{13}C/^{12}C)_{CO_2}}{(^{13}C/^{12}C)_{H_2CO_3}}$$

and can (and in most cases, do) involve intermediary reaction steps:

$$CO_{2(g)} \leftrightarrow CO_{2(aq)} + H_2O \leftrightarrow H_2CO_3 \leftrightarrow H^+ + HCO_3^-$$

$$\alpha^{18}O_{CO_2-HCO_3} = \frac{(^{18}O/^{16}O)_{CO_2}}{(^{18}O/^{16}O)_{HCO_3}}$$

Isotope exchange reactions generally involve a series of reaction steps through which the isotope passes in establishing the exchange between two species. The exchange reaction can then be expressed simply as exchange between these two species. Isotope exchange reactions can be aqueous, e.g.:

$$^2HS^- + {}^1H_2O \leftrightarrow {}^1HS^- + {}^2H^1HO$$

mineral-solution:

$$CaC^{16}O_3 + H_2^{18}O \leftrightarrow CaC^{18}O^{16}O_2 + H_2^{16}O$$

gas-solution:

$$^1H^2H + {}^1H_2O \leftrightarrow {}^1H_2 + {}^1H^2HO$$

The basis of physicochemical fractionation is the difference in the strength of bonds formed by the light vs. the heavier isotopes of a given element. Differences in bond strength for isotopes of the same element provide for differences in their reaction rates. This difference in bond energy is illustrated by the diagram in Fig. 1-5. The zero-point energy is essentially the minimum potential energy of a molecular bond in a vibrating atom. It takes energy to push them closer together, or to separate them from this "comfort zone" or ideal interatomic distance. The energy needed to break them apart (a requirement if they are to react) differs for isotopically different molecules. The heavy isotope will have a stronger bond and require greater energy to dissociate than a light isotope. Consequently, lighter nuclei react more quickly.

In a reaction that proceeds both forward and backward at essentially the same rate (the condition for chemical equilibrium), bonds are continually breaking and reforming. Statistically, we can expect the stronger bonds to survive longer, and so it is usually the heavy isotopic species that are partitioned into the more condensed phase, i.e. into the solid phase in mineral-solution reactions, or into the aqueous phase in vapour-liquid reactions.

The thermodynamic development of Urey relates the difference in dissociation energy of a molecule to its *partition functions* (Q), which are based on three molecular movements or energy frequencies: vibrational, translational and rotational. Q can be defined for each of these three movements of a molecule:

$$Q = \sigma^{-1} m^{3/2} \sum e^{-E/kT}$$

where: σ = a symmetry value
m = mass

E = the energy state, and the summation is taken over all energy states from the zero-point to the energy of the dissociated molecule (J·mole^{-1})
k = Boltzmann constant (gas constant per molecule)
 = n · 1.380658 · 10^{-23} JK^{-1}
T = thermodynamic temperature K

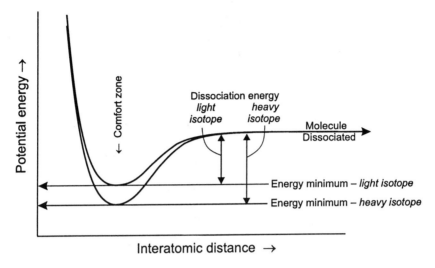

Fig. 1-5 The potential energy - interatomic distance relationship for heavy and light isotopes of a molecule. The dissociation energy differs for the two isotopes, and affects reaction rates. Isotope fractionation during physicochemical reactions arises from this difference.

Partition functions are defined for each isotope in a given exchange reaction at a given temperature. Their calculation is greatly simplified when only the ratio of partition functions for two isotopically different (but chemically identical) species are considered. Because we express isotope fractionation as a ratio, this is suited to our work.

$$\left(\frac{Q^*}{Q}\right) = \left(\frac{Q^*}{Q}\right)_{trans} \cdot \left(\frac{Q^*}{Q}\right)_{rot} \cdot \left(\frac{Q^*}{Q}\right)_{vib}$$

As translational and rotational frequencies are nearly identical for isotopic species, only vibrational frequency differences effectively partition isotopes. Using the generic isotope (*) exchange reaction:

$$X + Y^* \rightarrow X^* + Y$$

we can define the thermodynamic constant for the reaction as:

$$K = \frac{[X^*] \cdot [Y]}{[X] \cdot [Y^*]} = \frac{[X^*]}{[X]} \cdot \frac{[Y]}{[Y^*]} = \frac{\left(\frac{[X^*]}{[X]}\right)}{\left(\frac{[Y^*]}{[Y]}\right)}$$

The square brackets denote thermodynamic concentration (activity). The concentration of the isotope on either side of the exchange reaction is determined by its partition function Q and so the partition function ratio is the thermodynamic constant K. For exchange reactions involving only one isotopic species (most reactions of interest), the exponents become 1 and the partition function ratios can be related to the more familiar isotope ratios R:

$$K = \frac{\left(\frac{[X^*]}{[X]}\right)}{\left(\frac{[Y^*]}{[Y]}\right)} = \frac{(Q_X^* / Q_X)}{(Q_Y^* / Q_Y)} = \frac{R_X}{R_Y} = \alpha_{X-Y}$$

Note that if there was no distinction between the reaction of the normal and the isotopic species in the reaction, the thermodynamic constant would be equal to 1. Accordingly, there would be no fractionation.

Diffusive fractionation

A third reaction type where fractionation of isotopes occurs is the diffusion of atoms or molecules across a concentration gradient. This can be diffusion within another medium (e.g. CO_2 diffusing through a static air column, or Cl^- diffusing through clay) or diffusion of a gas into a vacuum. Fractionation arises from the differences in the diffusive velocities between isotopes.

During diffusion into a vacuum, the steady-state fractionation is the ratio of the velocities of the two isotopes, which can be calculated according to:

$$v = \sqrt{\frac{kT}{2\pi m}}$$

where: v = molecular velocity (cm · s^{-1})
k = Boltzmann constant (gas constant per molecule)
   = n · 1.380658 · 10^{-23} JK^{-1}
m = molecular mass (e.g. 7.3665 · 10^{-26} kg for $^{12}C^{16}O_2$; *CRC Handbook of Chemistry and Physics*)
T = absolute temperature K

The fractionation factor is given by Graham's Law, which equates the ratio of diffusive velocities to the mass ratio:

$$\alpha_{m^*-m} = \frac{v^*}{v} = \frac{\sqrt{\frac{kT}{2\pi m^*}}}{\sqrt{\frac{kT}{2\pi m}}} = \sqrt{\frac{m}{m^*}}$$

In the more common case where a gas or solute is diffusing through another medium, the mass of the medium (in this case, air) must be taken into account, using the equation:

$$\alpha = \left[\frac{m^*(m + 28.8)}{m(m^* + 28.8)} \right]^{1/2}$$

where 29 is average mass of air (79% N_2 and 21% O_2, i.e. 0.79 × 28 + 0.21 × 32 = 28.8).

By the nature of diffusive fractionation, it is possible to establish a steady-state diffusion regime, but not an equilibrium (equal forward and backward reaction rates), as this would be a completely mixed system with no fractionation. Thus, diffusion is a kinetic fractionation, which is discussed in more detail below.

Isotopic equilibrium

Isotope fractionation is generally considered for equilibrium conditions for which reliable fractionation factors can be calculated or experimentally measured. However, the condition of isotopic equilibrium requires that:

- Chemical equilibrium exists, such that forward and backward reaction rates are the same (i.e. no net forward or backward reaction).

- The chemical reactions have proceeded in both forward and backward directions enough times to mix the isotopes between reactant and product reservoirs.

- Both reactant and product reservoirs must be well mixed. If not, then isotopic equilibrium will exist only for the immediately produced reactants and products in the vicinity of the reaction site (e.g. air-water interface for vapour-water equilibrium).

For example, chemical equilibrium between CO_2 and water is expressed as:

$$CO_2 + H_2O \leftrightarrow H_2CO_3$$

Isotopic equilibrium is requires the exchange of oxygen atoms during hydration and dehydration of CO_2, at which time C–O bonds are forming and breaking. This has to occur several times to ensure reaction of all three oxygen atoms in the carbonic acid. The exchange of ^{18}O between CO_2 and water then looks something like:

$$C^{16}O_2 + H_2^{18}O \leftrightarrow H_2C^{18}O^{16}O_2 \leftrightarrow C^{18}O^{16}O + H_2^{16}O$$

The equilibrium fractionation of ^{18}O in this reaction is:

$$\alpha^{18}O_{CO_2(g)-H_2O} = \frac{[^{18}O/^{16}O]_{CO_2}}{[^{18}O/^{16}O]_{H_2O}} \approx 1.04$$

which means that CO_2 is about 40‰ more enriched in ^{18}O than water. A global-scale example is atmospheric CO_2 which is derived from various sources, yet has $\delta^{18}O \approx 40 \pm 2$‰ VSMOW. As the ^{18}O content of seawater is essentially 0 ± 1‰ VSMOW and we can conclude that exchange with the oceans is the principal control on $\delta^{18}O_{CO_2(atm)}$.

To attain isotopic equilibrium between reservoirs of reactants and products, such as atmospheric CO_2 and ocean water, one reservoir has to react completely (in this case, the much smaller atmospheric CO_2 reservoir). Complete reaction is aided by mixing within the reservoirs.

In most systems, complete physicochemical equilibrium is seldom maintained, and a net transfer of reactants to products occurs. Yet, if the reaction is close to physicochemical equilibrium, it can still have an isotopic equilibrium. For example, we can have isotopic equilibrium during precipitation of calcite from $Ca-HCO_3$ water. Yet, to have calcite precipitation, we must have chemical disequilibrium, i.e. supersaturation with respect to calcite. Similarly, rain and snow are considered to form in isotopic equilibrium with the vapour mass, although to have rain, the in-cloud temperature must have fallen below the dew point and the system is out of thermodynamic equilibrium. Nonetheless, isotopic equilibrium can be achieved providing the net forward reaction does not greatly exceed the rate of back reaction, through which isotopic equilibrium is established.

The example of ^{18}O fractionation between water and vapour

Fractionation between water and vapour is fundamental to the hydrological cycle and plays an important role in partitioning ^{18}O and ^{2}H between the various reservoirs (oceans, vapour, rain, runoff, groundwater, snow and ice). Let's start at the beginning of the hydrological cycle with the evaporation of water to form a vapour mass.

Evaporation is the transfer or flux (J) of liquid water to water vapour:

$$H_2O_{(l)} \xrightarrow{J_e} H_2O_{(v)}$$

while condensation is simply the reverse reaction:

$$H_2O_{(v)} \xrightarrow{J_c} H_2O_{(l)}$$

If we were to start with a tub of water at constant temperature in a closed system with a dry atmosphere, both of these reactions will proceed, although the evaporative flux, J_e, would be high, and the flux of vapour back to the water, J_c, would initially be infinitely low. To evaporate, a water molecule must break its hydrogen bond with the liquid water. The energy to do this represents the water's vapour pressure. Naturally, at higher temperatures, it is easier for water molecules to break this bond, and so vapour pressure increases with temperature. On the other hand, the condensation flux (J_c) relates to the concentration of water molecules in the gas phase above the water. As the air approaches 100% humidity, J_c increases towards J_e. At equilibrium, the two fluxes are equal:

$$H_2O_{(l)} \underset{J_c}{\overset{J_e}{\rightleftarrows}} H_2O_{(v)}$$

and there is no net evaporation or condensation.

Now let's look at how different isotopes of water behave during evaporation. The rate of evaporation (J_e) at a given temperature is a function of the vapour pressure of H_2O, which is a function of the hydrogen bond strength between the polar water molecules. Since a $^{18}O-H$ bond between molecules is stronger than a $^{16}O-H$ bond, the $H_2^{18}O$ has a lower vapour pressure than

$H_2^{16}O$. In fact, as shown by Szapiro and Steckel (1967), the vapour pressures for $H_2^{18}O$ and $H_2^{16}O$ differ by almost 1% at 20°C. The result is that the reaction:

$$H_2^{16}O_{(l)} \xrightarrow{J_e} H_2^{16}O_{(v)}$$

will proceed almost 1% faster than the reaction:

$$H_2^{18}O_{(l)} \xrightarrow{J_e} H_2^{18}O_{(v)}$$

and so we can say that $^{16}J_e > {}^{18}J_e$. The greater vapour pressure or flux of $H_2^{16}O$ leads to an enrichment of ^{16}O in the vapour phase. Conversely, heftier $H_2^{18}O$ accumulates in the liquid phase.

This continuous accumulation of light isotopes in the vapour phase is eventually balanced by an increase in the condensation flux, $^{16}J_c$. The flux of ^{16}O and ^{18}O from vapour to water ($^{16}J_c$ and $^{18}J_c$) is a function of their concentration, and not the strength of their hydrogen bonds. As ^{16}O becomes preferentially accumulated in the vapour, $^{16}J_c$ increases over $^{18}J_c$. At the point where the two fluxes for the two isotopes are balanced ($^{16}J_c = {}^{16}J_e$, and $^{18}J_c = {}^{18}J_e$) the system has reached isotopic equilibrium, and condensation fluxes for each isotope balance their respective evaporative fluxes. At this point of thermodynamic equilibrium, the greater vapour pressure of ^{16}O causes an overall enrichment in the vapour, and a depletion for ^{18}O. All this discussion applies equally for ^2H, although the magnitude of the isotope effects are different.

Temperature effect on fractionation

Isotope fractionation is strongly dependent on the temperature of the reaction. This is not surprising considering that isotope fractionation is a thermodynamic reaction. Recall from above that the fractionation factor α for two reacting compounds is derived from the isotope partitioning functions Q:

$$\alpha_{X-Y} = \frac{R_X}{R_Y} = \frac{(Q_X^* / Q_X)}{(Q_Y^* / Q_Y)}$$

Recall also that Q is an expression of bond strength or dissociation energy of the oscillating diatomic molecule. This dissociation energy varies with temperature for a given isotopic species:

$$Q = \sigma m^{3/2} \sum e^{-E/kT}$$

At low temperature, the energy term becomes less significant, so Q is defined by mass differences and Q-ratios are quite different from 1. At very high temperatures, kT becomes large with respect to bond strength, and the Q-ratios become close to 1. Accordingly, the fractionation factor α is close to 1 at high temperature. With decreasing temperature, α departs from unity, and isotope partitioning between the reactants and products occurs. Cross-overs from this general trend can occur between the two temperature extremes, due to anharmonic terms in the partition functions, but this is seldom of concern in the temperature range of most hydrological systems. Bottinga (1969) and Criss (1991) derive and clarify the temperature dependences of partition function ratios at low and high temperature.

The fractionation of ^{18}O between water and vapour, for example, is of the order of 11.5‰ (expressed as $10^3 \ln\alpha^{18}O_{w-v}$) at freezing, and decreases to about 5‰ at 100°C (Fig. 1-6). Deuterium decreases from about 106‰ at 0°C to only 27‰ at 100°C. Values for these fractionation factors are given in Table 1-4.

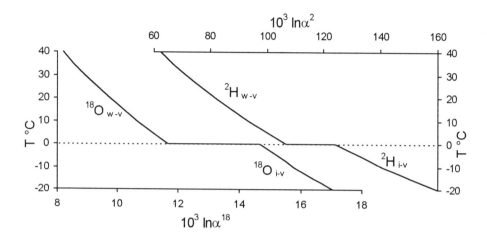

Fig. 1-6 Fractionation of ^{18}O and ^2H for water-vapour (from equations of Majoube, 1971) and ice-vapour (Majoube, 1971 and O'Neil, 1968) for the temperature range of –30 to +50°C.

Table 1-4 Values for fractionation relationships for ^{18}O and ^2H in water-vapour-ice reactions

Water-vapour[1]			Water-ice		
T°C	$10^3\ln\alpha^{18}O_{w-v}$	$10^3\ln\alpha^2H_{w-v}$	T°C	$10^3\ln\alpha^{18}O_{w-v}$	$10^3\ln\alpha^2H_{w-v}$
–10	12.8	122	0[2]	3.1	19.3
0	11.6	106	0[3]	2.8	20.6
5	11.1	100			
10	10.6	93			
15	10.2	87			
20	9.7	82	Ice-vapour		
25	9.3	76	T°C	$10^3\ln\alpha^{18}O_{w-v}$	$10^3\ln\alpha^2H_{w-v}$
30	8.9	71	0[4]	14.7	126
40	8.2	62	0[5]	14.4	127
50	7.5	55			
75	6.1	39			
100	5.0	27			

1 Majoube (1971): $10^3\ln\alpha^{18}O_{w-v} = 1.137(10^6/T^2) - 0.4156(10^3/T) - 2.0667$
$10^3\ln\alpha^2H_{w-v} = 24.844(10^6/T^2) - 76.248(10^3/T) + 52.612$
2 O'Neil (1968)
3 Suzuoki and Kumura (1973)
4 Combination of Majoube (1971) and O'Neil (1968)
5 Combination of Majoube (1971) and Suzuoki and Kumura (1973)

Equilibrium fractionation factors can be determined at different temperatures through experimentation, and can be calculated from the partitioning functions. Estimates of fractionation in our example of the water-vapour-ice system have been well constrained through many investigations over the past several decades. A comparison of results gives a good idea of how well defined temperature equations really are. Horita and Wesolowski (Fig. 1-7) show that experimental work supports the relationship determined by Majoube in 1971.

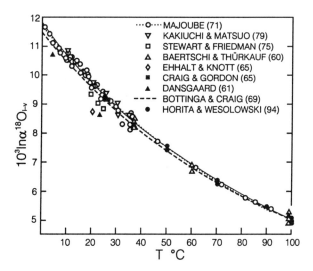

Fig. 1-7 Comparison of fractionation factors for ^{18}O in water-vapour exchange from various workers. Regression of these data (solid line) gives $10^3 \ln\alpha^{18}O_{l-v} = -7.685 + 6.7123\,(10^3/T) - 1.6664\,(10^6/T^2) + 0.35041\,(10^9/T^3)$. Reprinted from Horita and Wesoloski (1994) with permission from Elsevier Science Ltd, The Boulevard, Langford Lane, Kidlington OX5 1GB, UK.

An inspection of the equations for fractionation factors (Table 1-4) shows that the fractionation term is expressed as the natural logarithm of α, multiplied by 10^3. Why $10^3 \ln\alpha$? This is because the correlation between α and absolute temperature is represented by an equation of the form:

$$\ln\alpha_{X-Y} = aT^{-2} + bT^{-1} + c$$

which produces a linear graph when plotting $\ln\alpha$ against the inverse of absolute temperature. As α is generally close to 1 for most isotope reactions, $\ln\alpha$ is very close to 0 and so multiplying by 10^3 fits with the ‰ convention for δ-values.

The temperature-fractionation equations of hydrogeological interest can be constructed from the equation constants given in Table 1 inside the front cover of this book, as well as in subsequent chapters. They include $T-10^3 \ln\alpha$ relationships compiled by Friedman and O'Neil (1977) plus recent refinements and additions. For illustration, some of the important isotope fractionation factors are plotted with temperature in Fig. 1-8.

Kinetic (nonequilibrium) fractionation

A sudden change in temperature or the addition or removal of a reactant can move a system far from thermodynamic equilibrium, and the forward reaction rate accelerates. By consequence, reverse reaction through which isotope exchange occurs diminishes. These are the conditions of kinetic fractionation. The effects will vary depending on the reaction pathway, and can either diminish mass-discrimination during fractionation or enhance it.

In the case of calcite precipitation under equilibrium conditions, calcite is enriched by about 1‰ over the bicarbonate in solution. However, as the rate of calcite precipitation is accelerated, the difference decreases towards 0 (Turner, 1982). Similarly, if water freezes very rapidly, the 3‰ equilibrium enrichment of $\delta^{18}O$ in the ice with respect to the water (Table 1-4) is reduced towards 0.

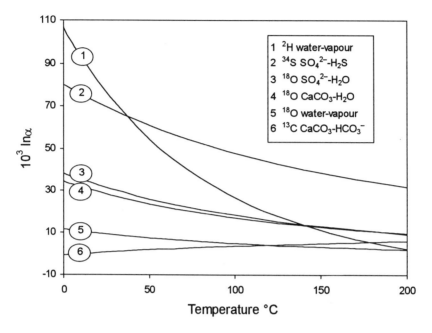

Fig. 1-8 Fractionation factors for some useful isotope exchange reactions plotted as a function of temperature.

Both are examples of accelerated reaction which is accompanied by kinetic fractionation that reduces mass discrimination, i.e.:

$$\text{Reactant} \xrightarrow{\alpha \to 1} \text{Product}$$

When a reaction involves two or more products, two reaction pathways are available, either the light isotope reacts or the heavy isotope reacts. The pathway that dominates depends on the dissociation energies. For example, as bicarbonate water freezes, one out of two HCO_3^- ions reacts with Ca^{2+}, while the other dehydrates to form CO_2:

$$2HCO_3^- + Ca^{2+} \to CO_2 + CaCO_3 + H_2O$$

During rapid freezing ($HCO_3^- + H^+ \to CO_2 + H_2O$) the weaker ^{12}C–O bonds are preferentially broken, producing a ^{13}C-depleted CO_2 product and an enriched calcite. The lack of back reaction (to the left in this case) precludes isotopic equilibrium between the product CO_2 and $CaCO_3$. Values up to $\delta^{13}C = +17‰$ have been measured for calcite produced under such conditions (Clark and Lauriol, 1992).

Bacterially mediated reactions (essentially irreversible reactions) can also be considered as kinetic reactions, although the role of enzymes and related biochemistry provide complicated reaction pathways. These are usually redox reactions and are associated with very large and, for specific conditions, reproducible isotope fractionation effects. Some important biologically-mediated reactions include photosynthesis ($\delta^{13}C_{CO_2} - \delta^{13}C_{CH_2O} \approx -17‰$ for C_3 plants), sulphate reduction ($\delta^{34}S_{SO_4} - \delta^{34}S_{H_2S} \approx 20$ to $30‰$) and methanogenesis ($\delta^{13}C_{CH_4} - \delta^{13}C_{CO_2} \approx -80‰$).

Isotope Fractionation (α), Enrichment (ε) and Separation (Δ)

Along with the fractionation factor α, some additional parameters and relationships are useful in isotope exchange calculations. The first is the relationship between α and the δ–‰ values for two compounds involved in an isotope fractionation reaction. The isotopic difference or isotope separation, Δ, between the two compounds, whether or not they are co-reacting, is simply:

$$\Delta_{X-Y} = \delta_X - \delta_Y$$

However, the fractionation factor for isotope exchange between these two compounds is derived from the δ–‰ values according to:

$$\alpha_{X-Y} = \frac{1 + \dfrac{\delta_X}{1000}}{1 + \dfrac{\delta_Y}{1000}} = \frac{1000 + \delta_X}{1000 + \delta_Y}$$

The enrichment factor, ε, can be used to express this isotopic difference in ‰ notation, and is defined by the same formulation as the δ–‰ values:

$$\varepsilon_{X-Y} = \left(\frac{R_X}{R_Y} - 1\right) \cdot 10^3 = (\alpha - 1) \cdot 10^3$$

The isotope enrichment factor ε is comparable to the isotope separation value Δ, and both are expressed as ‰ values.

The example of water-vapour reaction

Using the example of water-vapour exchange, ^{18}O fractionation can be determined from measured δ-values for water and water vapour and compared with thermodynamic fractionation factors. For a water at 25°C with $\delta^{18}O = 0.00‰$ VSMOW, and water vapour with $\delta^{18}O = -9.30‰$ VSMOW, the separation (Δ) is:

$$\Delta^{18}O_{\text{water-vapour}} = \delta^{18}O_{\text{water}} - \delta^{18}O_{\text{vapour}} = 9.30‰$$

For this case, the isotopic enrichment (ε):

$$\alpha^{18}O_{\text{water-vapour}} = \frac{(^{18}O/^{16}O)_{\text{water}}}{(^{18}O/^{16}O)_{\text{vapour}}} = \frac{1000 + \delta^{18}O_{\text{water}}}{1000 + \delta^{18}O_{\text{vapour}}} = 1.0093$$

and

$$\varepsilon^{18}O_{\text{water-vapour}} = \left(\frac{(^{18}O/^{16}O)_{\text{water}}}{(^{18}O/^{16}O)_{\text{vapour}}} - 1\right) \cdot 1000 = 9.30‰$$

From Fig. 1-6, fractionation of ^{18}O between water and vapour is:

$$\alpha^{18}O_{\text{water-vapour}} = 1.0093 \; @ \; 25°C \qquad \text{and so:} \quad 10^3 \ln\alpha_{\text{water-vapour}} = 9.26‰$$

32   Chapter 1   The Environmental Isotopes

This all seems like simply playing around with ε, δ, Δ, and $10^3\ln\alpha$ values to express the same thing — isotope fractionation. However, there are differences, and these differences become more critical when fractionation is strong or with δ-values that are large. It is important to remember that there are two fractionation systems — a real system for which we measure δ-values of reactants and products in the real world, and a theoretical system at equilibrium for which we calculate (or experimentally derive) α or $10^3\ln\alpha$ values. The Δ- and ε-values serve as approximations to compare measured δ-values with theoretical fractionation factors for determining reaction temperature or degree of equilibrium in isotopic systems.

For reactions where the fractionation is small, the error in these approximations is acceptably small (usually varying in the fourth decimal place). However, where α is significant, $10^3\ln\alpha$ should be used. There is also an effect of reversing the two species in the fractionation factor. By expressing fractionation as Y over X rather than X over Y, not only the sign, but also the error in the approximations change. The calculations in Table 1-5 show the sensitivity of these approximations. The deviation from equality between these expressions of fractionation is more evident in Fig. 1-9.

Table 1-5 Variation between ε and $10^3\ln\alpha$ as expressions of isotopic fractionation between two species X and Y

δ_X ‰	δ_Y ‰	α_{X-Y}	Δ_{X-Y} ‰	ε_{X-Y} ‰	$10^3\ln\alpha_{X-Y}$ ‰	Discrepancy ‰
0	0	1.00000	0	0.00	0.00	0.00
10	–10	1.02020	20	20.20	20.00	0.05
20	–20	1.04042	40	40.82	40.00	0.82
30	–30	1.06186	60	61.86	60.00	1.86
0	–20	1.02041	20	20.41	20.20	0.41
0	–40	1.04167	40	41.67	40.82	1.67
0	–60	1.06383	60	63.83	61.88	3.83
–20	–40	1.02083	20	20.83	20.62	0.83
–40	–60	1.02128	20	21.28	21.05	1.28
–60	–80	1.02174	20	21.74	21.50	1.74

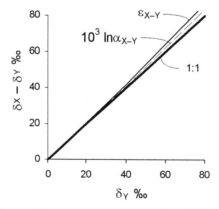

Fig. 1-9 Deviation of calculated isotope differences between compounds X and Y from 1:1. Here, X is given a fixed value of 0‰, and Y has increasingly more negative values. The 1:1 slope is for $\delta_X - \delta_Y$. As the isotopic difference between the two compounds increases, so does the deviation in the calculation of this difference.

When interchangeably using the expressions Δ, ε, and $10^3\ln\alpha$ for isotope fractionation, remember that they are approximations:

$$\varepsilon_{X-Y} \approx 10^3 \ln\alpha_{X-Y} \approx \Delta_{X-Y}$$

These expressions are useful when fractionation is small and values are close to the reference, but can bring errors to your calculations when fractionation is great and isotope separations are large. Take, for example, ^2H exchange between the water and dissolved hydrogen sulphide, which has an enormous fractionation:

$$\alpha^2H_{H_2O-H_2S} = 2.37 \text{ (Galley et al. 1972)}$$

Expressing this fractionation as $10^3\ln\alpha$ gives:

$$10^3\ln\alpha^2H_{H_2O-H_2S} = 863‰$$

whereas using the enrichment expression yields:

$$\varepsilon^2H_{H_2O-H_2S} = (\alpha - 1) \cdot 10^3 = 1370‰$$

yet calculating a Δ-value for $\delta^2H_{H_2S}$ in equilibrium with $\delta^2H_{H_2O} = 0‰$ using $\alpha^2H_{H_2O-H_2S} = 2.37$ and the equation from above:

$$\alpha^2H_{H_2O-H_2S} = \frac{1000 + \delta^2H_{H_2O}}{1000 + \delta^2H_{H_2S}}$$

gives $\delta^2H_{H_2S} = -578‰$

and so, $\Delta^2H_{H_2O-H_2S} = \delta^2H_{H_2O} - \delta^2H_{H_2S} = 0 - (-578) = 578‰$

In this case, using ε^2H or $10^3\ln\alpha$ to represent the difference in δ^2H values for H_2S and H_2O is very misleading. This example is discussed in more detail in Chapter 9.

In the following chapters, we attempt to show how isotope fractionation and partitioning can be used to interpret isotope data from natural settings. The objective is, of course, not to simply account for the isotope distributions that we observe, but to use this tool to elucidate and quantify the processes and reactions in our hydrogeological system.

Problems

1. What is the relative enrichment or depletion of VSMOW, in δ-‰ notation, relative to the average terrestrial abundance of ^{18}O given in Table 1-1? What about VPDB?

2. Determine the $\delta^{18}O$ value for VPDB on the VSMOW scale. Assuming that the original standard PDB is the same as VPDB, was it precipitated in isotopic equilibrium with VSMOW? Explain.

3. The University of Ottawa laboratory working standard for water has $\delta^{18}O_{ws} = +11.3‰$ VSMOW. A sample measured in the laboratory has a raw value of $-13.5‰$ WS. What is its value against VSMOW? Against SLAP?

4. What is the stable daughter nuclide produced by the β^- decay of ^{14}C? What about ^{10}Be?

5. Write geochemical reactions for each of the following processes through which isotope exchange can occur, (noting the intermediary species, if any). Determine the fractionation factors for 25° and 100°C. Describe a condition for each where equilibrium would not be attained?

6. Determine the fractionation of $^2H_{H_2}$, $^3He/^4He$ and $^{13}C_{CO_2}$ for steady–state diffusion in air between a well-mixed reservoir and an open atmosphere with negligible concentrations of these gases. Which gas experiences the greatest fractionation?

7. For the isotopic separation between two species X and Y, we can use one of three expressions: ε_{X-Y}, $\delta_X - \delta_Y$, or $10^3 \ln\alpha_{X-Y}$. At what isotope separation, and what deviation from the reference, does the difference between these expressions start to give an error greater than the analytical precision for (i) ^2H, (ii) ^{13}C, and (iii) ^{18}O?

8. Isotope fractionation effects are expressed as α-values and isotope abundances are given as permil differences from a reference (δ-value). Establish the relationship between α and δ in the form of a simple equation.

9. Hydrogen gas produced from water is strongly depleted in ^2H (which is the basis of enriching ^3H in water by hydrolysis) and has a fractionation factor $\alpha_{H_2O-H_2} = 3.76$ at 25°C. What will be the δ^2H for H_2 produced from water with $\delta^2H_{H_2O} = -75‰$ at 25°C? At 50°C? Is the enrichment factor $\varepsilon^2H_{H_2O-H_2}$ a reasonable representation of $\Delta^2H_{H_2O-H_2}$?

10. Isotopic exchange exists between three components of a geochemical system. The fractionation factors α_{1-2} and α_{2-3} are known. Can we assume that $\alpha_{1-2} \times \alpha_{2-3} = \alpha_{1-3}$? Derive the relationship to show whether it is or is not. What about $\varepsilon_{1-2} + \varepsilon_{2-3} = \varepsilon_{1-3}$?

Chapter 2
Tracing the Hydrological Cycle

While undertaking a regional survey of nitrate in shallow alluvial groundwater in the Fraser Valley near Vancouver, B.C., the project hydrogeologist queried one homesteader about the quality of his well water. "Why, it's crystal clear . . . it's recharged by the glaciers up there on Mount Baker." Following his gaze to Baker, some 60 km south on the far side of the Fraser River, the hydrogeologist speculated on the improbability of the fellow's claim. In such a case, environmental isotopes serve nicely to trace the origin of the groundwaters and reconcile the two opinions.

Stable isotopes in water, ^{18}O and ^2H, are affected by meteorological processes that provide a characteristic fingerprint of their origin. This fingerprint is fundamental to investigating the provenance of groundwater.

Craig's Meteoric Relationship in Global Fresh Waters

The flux of moisture from the oceans and its return via rainout and runoff is, on an annual basis and global scale, close to a dynamic equilibrium. Only major climatic shifts will seriously change the storage of fresh water in glacial, groundwater and surface water reservoirs. Within the existing cycle, however, significant variability does exist and, as only about 10% of the moisture evaporating from the oceans reaches the land surface, one might expect considerable variation in the terrestrial portion of the hydrological cycle. Evaporation from the oceans, rainout, re-evaporation, snow and ice accumulation and melting, and runoff vary under different climatic regimes. Each step of this cycle partitions ^{18}O and ^2H amongst the different freshwater reservoirs.

Considering the complexity of the hydrological cycle, it may be surprising that ^{18}O and ^2H behave at all predictably. In 1961, Harmon Craig published his finding that $\delta^{18}O$ and δ^2H in fresh waters correlate on a global scale. Craig's "global meteoric water line" defines the relationship between ^{18}O and ^2H in worldwide fresh surface waters:

$$\delta^2H = 8\,\delta^{18}O + 10\ ‰\ \text{SMOW} \qquad \text{(Craig, 1961b)}$$

Note that the term "meteoric" is used in isotope hydrology rather than "meteorological", probably for ease of pronunciation, and has nothing to do with fiery objects from space. Also note the use of SMOW as a reference. VSMOW was not introduced until over a decade later. Subsequent monitoring of the stable isotopic composition of precipitation world wide (IAEA Global Network for Isotopes in Precipitation – GNIP) has refined this relationship, shown in Fig. 2-1. The importance of Craig's observation is that the isotopic composition of meteoric waters behaves in a predictable fashion.

Craig's line is only global in application, and is actually an average of many local or regional meteoric water lines which differ from the global line due to varying climatic and geographic parameters. Local lines will differ from the global line in both slope and deuterium intercept. Nonetheless, Craig's paper is perhaps the most often cited in environmental isotope hydrology as his global meteoric water line (GMWL) provides a reference for interpreting the provenance of groundwaters.

A key observation made by Craig was that isotopically depleted waters are associated with cold regions and enriched waters are found in warm regions. This partitioning was soon recognized as a tool for characterizing groundwater recharge environments, and is now the basis of groundwater provenance studies.

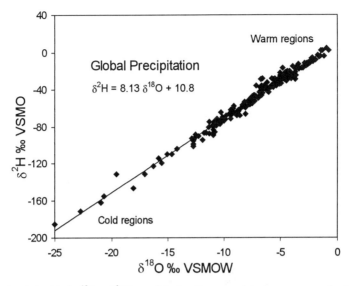

Fig. 2-1 The meteoric relationship for ^{18}O and ^{2}H in precipitation. Data are weighted average annual values for precipitation monitored at stations in the IAEA global network, compiled in Rozanski et al. (1993).

Partitioning of Isotopes Through the Hydrological Cycle

The meteoric relationship of ^{18}O and ^{2}H arises from fractionation during condensation from the vapour mass. However, it is a Rayleigh distillation during rainout that is responsible for the partitioning of ^{18}O and ^{2}H between warm and cold regions. The evolution of the $\delta^{18}O$ and $\delta^{2}H$ composition of meteoric waters begins with evaporation from the oceans.

Isotopic composition of ocean waters

The isotopic composition of modern seawater is close to VSMOW although it has varied considerably over geologic time. Carbonates precipitated from Archean and Proterozoic oceans which were up to 8‰ lighter in ^{18}O than those from modern seawater, indicating that the early earth's oceans were isotopically depleted and warmer than today (Veizer et al., 1989; 1992). The oceans became gradually enriched through Proterozoic and Phanerozoic time by exchange with isotopically enriched crustal rocks (Wadleigh and Veizer, 1992). This evolution of seawater occurs mainly through the alteration of basalts at mid-oceanic ridges and release of crustal fluids along subduction zones, both of which contribute ^{18}O to the oceans (Lawrence, 1989).

Shackleton and Opdyke (1973) show that the growth and decay of ^{18}O-depleted ice sheets during the late Cenozoic has imparted significant variations on seawater. These variations are recorded by the ^{18}O in calcite foraminifera which grow in equilibrium with the prevailing seawater and provide the SPECMAP record of global ice (Fig. 2-2). Corrected for mass-balance and temperature considerations, the difference between the highest $\delta^{18}O$ values during maximum glaciation and the lowest values during interglacial times is 1.5 to 2.0‰. Ocean sediment ($\delta^{18}O$) and water data are available through the U.S. National Oceanic and Atmospheric Administration at <**http://nodc.noaa.gov**>.

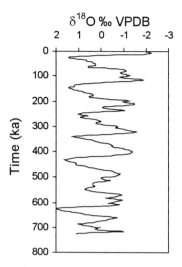

Fig. 2-2 The SPECMAP record of the $\delta^{18}O$ composition of seawater during the Quaternary Period (data from Imbrie et al., 1984). Increases in ocean water ^{18}O represent periods of extensive accumulation of isotopically depleted glacial ice.

Although Craig's standard mean ocean water was selected as the reference for $\delta^{18}O$ values, the true mean $\delta^{18}O$ value of ocean waters is estimated to be 0.5‰ VSMOW with corresponding enrichments in δ^2H. Observed variations correlate with salinity (Fig. 2-3). The salinity of seawater with $\delta^{18}O$ = 0‰ VSMOW is about 34,500 ppm and varies between 33,500 and 37,600 ppm. This correlates with its stable isotope composition of –0.5‰ in Antarctic Bottom Water to 1.3‰ in North Atlantic Surface Waters (Ferronsky and Brezgunov, 1989). These variations in salinity and stable isotopes are attributable to evaporation at the ocean surface. Craig (1966) reported values for the Red Sea up to $\delta^{18}O$ = 1.8‰ with salinity = 40,200 ppm. Along continental margins, ocean water is often diluted by runoff to values less than VSMOW. For example, seawater in the Georgia Strait between Vancouver Island and the mainland has $\delta^{18}O$ values as low as –10.5‰ due to dilution by Fraser River discharge (Dakin et al., 1983).

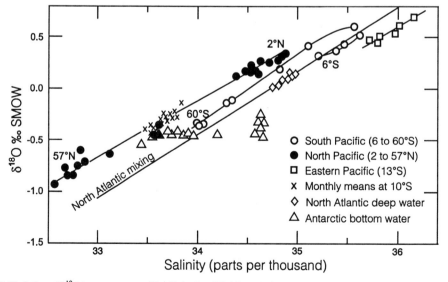

Fig. 2-3 Variation of $\delta^{18}O$ in ocean water with salinity (modified from Craig and Gordon, 1965).

The atmosphere and vapour mass formation

The atmosphere extends over 400 km above the Earth's surface. It comprises the troposphere, which extends from the surface up to 16 km a.s.l. at the equator (~8 km at the poles), the stratosphere (to 55 km), the mesosphere (to 80 km), the thermosphere (to 400 km) and the exosphere transition into outer space. Many nuclear reactions relevant to hydrogeology take place within the upper stratosphere and beyond, including the production of ^{14}C and tritium.

Although it is the thinnest of the five layers, the troposphere contains over 90% of the atmosphere's mass, and all of its weather. Temperature within the troposphere decreases with altitude to about –60°C at the base of the stratosphere. This *lapse rate* averages about 6° per kilometre (10°C/km for dry adiabatic expansion, until the dew point is reached). It reverses in the stratosphere where temperatures at the upper limits (55 km) reach 0°.

Beyond fluctuations of a few percent, the pressure distribution in the atmosphere follows a fairly predictable decrease with altitude. At 5500 m, pressure is 50% of that at sea level. By 18,000 m, it has dropped to 10%. High-velocity air currents within the stratosphere (jet streams) are important in the hemispherical distribution of volcanic dust, aerosols and anthropogenic components (tritium, CFCs). However, very little water vapour circulates to the stratosphere, and so weather is essentially confined to the troposphere (from Greek *tropos* = turning).

The evaporative flux of ocean water to the troposphere is mainly controlled by temperature, which greatly affects the moisture carrying capacity of air (Fig. 2-4). Mean annual sea surface temperatures (SST) vary from near 0°C in the polar regions to greater than 29°C in the eastern equatorial Pacific. The greatest supply (>70%) of vapour to the troposphere is from evaporation over the warm subtropical seas. The isotopic evolution in the hydrological cycle begins here with evaporation and the formation of vapour. Isotope exchange and fractionation during this stage is the subject of the following sections.

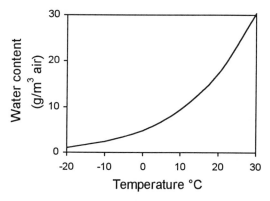

Fig. 2-4 The exponential increase in the moisture carrying capacity of vapour-saturated air from –20 to 30°C. This relationship can be calculated from the equation (with T in °C): Vapour (g/m^3) = 0.0002 T^3 + 0.0111 T^2 + 0.321 T + 4.8.

Isotopic equilibrium in water-vapour exchange

The difference in the vapour pressures of $H_2^{18}O$ and ^2HHO imparts disproportional enrichments in the water phase during evaporation. This difference accounts for ^2H enrichment in water which is roughly 8 times greater than for ^{18}O, under equilibrium conditions. As we saw in Chapter 1, these equilibrium fractionations are temperature-dependent and measurable.

Accordingly, if the evaporation of ocean water was an equilibrium reaction, the isotopic composition of atmospheric water vapour and precipitation could be easily determined. This is not the case, but it is instructive to do the calculations. For oceans at 25°C, the equilibrium isotopic composition of water vapour should be:

$$\delta^{18}O_{vapour} = \delta^{18}O_{seawater} + \varepsilon^{18}O_{v-w} = 0.0 + (-9.3) = -9.3‰$$

$$\delta^{2}H_{vapour} = \delta^{2}H_{seawater} + \varepsilon^{2}H_{v-w} = 0.0 + (-76) = -76‰$$

In cooler (10°C) regions, equilibrium water vapour would have an isotopic composition of $\delta^{18}O$ = –10.6‰ and $\delta^{2}H$ = –93‰. Over high-latitude seas, vapour would have an isotopic composition as low as $\delta^{18}O$ = –11.6‰ and $\delta^{2}H$ = –106‰. We see that on a $\delta^{18}O$-$\delta^{2}H$ diagram (Fig. 2-5), these vapour masses do not fall on the global meteoric waterline (GMWL). They are also different from North Pacific marine vapour and from the value of global mean precipitation of Craig and Gordon.

If we then cool these vapour masses and produce precipitation according to equilibrium fractionation, the precipitation is also displaced from the GMWL, but on a line originating at seawater and following a positive slope greater than 8. These calculated values for rain are more enriched than seawater, and very much more enriched than we would normally observe in precipitation and surface waters.

Craig and Gordon (1965) estimate the mean isotopic composition of worldwide precipitation to be $\delta^{18}O$ = –4‰ and $\delta^{2}H$ = –22‰, quite different from our calculations. Evidently, the primary formation of atmospheric water vapour and precipitation are more complicated processes, involving *kinetic* evaporation and mixing.

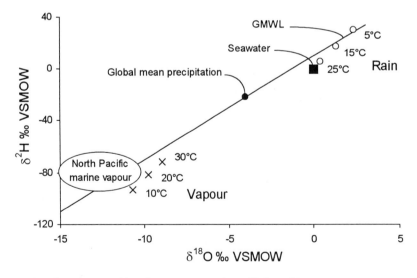

Fig. 2-5 The calculated isotopic composition of water vapour (×) in equilibrium with seawater at temperatures of 10, 20 and 30°C. Circles (o) are for precipitation that would form by cooling the vapour mass by 5°, under equilibrium conditions. Note that in this "equilibrium" model precipitation data plot on a line passing through the origin (seawater), but do not plot on the global meteoric waterline (GMWL), nor on the mean of global precipitation. North Pacific marine vapour (shown) sampled by Craig and Gordon (1965) from latitudes between 0° and 32°N falls on the GMWL, indicating nonequilibrium vapour formation.

Humidity and kinetic (nonequilibrium) evaporation

The rate of evaporation limits vapour-water exchange and so limits the degree of isotopic equilibrium. Increased rates of evaporation impart a kinetic or nonequilibrium isotope effect on the vapour. Kinetic effects are affected by the surface temperature, wind speed (shear at the water surface), salinity, and most importantly, humidity. At lower humidities, water-vapour exchange is minimized, and evaporation becomes an increasingly nonequilibrium process.

Isotope effects during nonequilibrium evaporation from surface water bodies has been examined theoretically and experimentally (e.g. Craig and Gordon, 1965; Gonfiantini, 1965; Merlivat, 1970; Gat, 1971). Gonfiantini (1986) provides an excellent summary of this work. The principal observation is that water becomes progressively enriched in both ^{18}O and ^{2}H during evaporation (Fig. 2-6). However, the humidity, h, affects this progressive enrichment. If there is no exchange with the vapour phase (h remains close to 0%), the enrichment follows a Rayleigh distillation. This is an exponential function that describes the progressive partitioning of (in this case) heavy isotopes into the water reservoir as it diminishes in size:

$$R = R_o f^{(\alpha-1)}$$

R_o is the initial isotope ratio in the water, R is the ratio when only a fraction, f, remains, and α is the equilibrium fractionation factor during evaporation. Rayleigh processes are important in other hydrogeological processes and are discussed in greater detail later in this chapter.

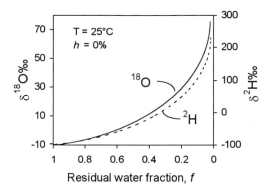

Fig. 2-6 Enrichment of ^{18}O and ^{2}H in distilled water during evaporation of water under controlled conditions. Humidity h is 0% and temperature is 25°C. As the fraction of water remaining approaches 0, a Rayleigh distillation causes an exponential increase in the heavy isotopes. Under conditions of increasing humidity, exchange with the vapour phase reduces the exponential enrichment. Under conditions of high humidity, a steady state value is reached due to complete exchange with the vapour mass.

The model for water-vapour transfer involves a series of steps through the water surface and boundary layer into the open atmosphere (Fig. 2-7). The boundary layer is a microns-thick atmosphere over the liquid water interface with virtually 100% water saturation. This layer is in isotopic equilibrium with the underlying water column.

Between the boundary layer and the mixed atmosphere above is a transition zone through which water vapour is transported in both directions by molecular diffusion. It is within this layer that nonequilibrium enrichment arises. Molecular diffusion is a fractionating process due to the fact that the diffusivity of $^{1}H_2^{16}O$ in air is greater than $^{2}H^{1}H^{16}O$ or $H_2^{18}O$.

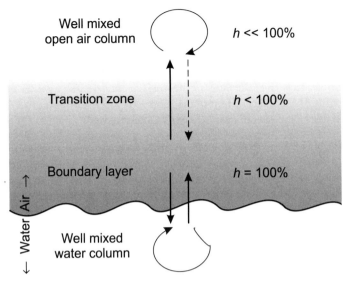

Fig. 2-7 Model for nonequilibrium evaporation over a water body. Arrows indicate relative fluxes of water between the mixed water column and the boundary layer, and between the boundary layer and the well mixed air column. Differences in the rate of diffusion of ^{18}O to ^{16}O and ^{2}H to ^{1}H impart a "kinetic" isotope depletion in the overlying air column.

Under conditions of high humidity (h near 1), diffusion in both directions through the transition zone is equal and no net diffusive fractionation occurs. However, when h is low, net diffusion from the boundary layer to the atmosphere will occur. This depletes the boundary layer (and water surface) in the more diffusive $^{1}H_{2}^{16}O$ species and enriches these reservoirs in the heavy isotopes, $^{2}H^{1}H^{16}O$ and $H_{2}^{18}O$. The maximum kinetic enrichment that this diffusive effect can impart is calculated from our formula for diffusive fractionation of a gas in air (Chapter 1). In this case it is calculated for ^{18}O in H_2O_v from the boundary layer (bl) to the vapour reservoir of the open air water vapour (v):

$$\alpha^{18}O_{bl\text{-}v} = \left(\frac{20}{18} \cdot \left(\frac{18 + m_{air}}{20 + m_{air}}\right)\right)^{1/2}$$

Using the molecular mass of $H_2^{18}O$ (20), $H_2^{16}O$ (18), and air (m_{air} = 28.8 for 79% N_2 and 21% O_2), gives $\Delta\epsilon^{18}O_{bl\text{-}v}$ = 1.0323. Converting this to permil enrichment, we see that for ^{18}O, the water vapour would be depleted from the water column by a kinetic factor ($\Delta\epsilon$) as large as:

$$\Delta\epsilon^{18}O_{bl\text{-}v} = (\alpha^{18}O_{bl\text{-}v} - 1) \cdot 10^3 = 32.3\text{‰}$$

Similarly, ^{2}H in the water vapour would be depleted by a maximum of:

$$\Delta\epsilon^{2}H_{bl\text{-}v} = (\alpha^{2}H_{bl\text{-}v} - 1) \cdot 10^3 = 16.6\text{‰}$$

In reality, these kinetic effects are not so extreme because humidity is seldom close to 0%. Gonfiantini (1986) describes the kinetic effect in terms of humidity with the following relationships:

$$\Delta\epsilon^{18}O_{bl\text{-}v} = 14.2\,(1-h)\text{ ‰}$$

$$\Delta\epsilon^{2}H_{bl\text{-}v} = 12.5\,(1-h)\text{ ‰}$$

The total fractionation between the water column and the open air is then the sum of the fractionation factor for equilibrium water-vapour exchange ($\varepsilon_{l\text{-}v}$) and the kinetic factor ($\Delta\varepsilon_{bl\text{-}v}$). For ^{18}O, this would be:

$$\delta^{18}O_l - \delta^{18}O_v = \varepsilon^{18}O_{l\text{-}v} + \Delta\varepsilon^{18}O_{bl\text{-}v}$$

Remember to keep the subscripts in the right order. These calculations are presented as the enrichment of the water with respect to the vapour. The depletion in the vapour with respect to the water is the reciprocal or *negative* enrichment.

On a $\delta^{18}O - \delta^{2}H$ diagram, the result is a deviation from the meteoric water line along a line with a lower slope, s, which depends largely on the relative humidity, h. Gonfiantini presents the mathematical basis for this relationship, shown in Fig. 2-8. At values of low h, kinetic evaporation is maximized and s will be low. Gat (1971) calculates that for very low relative humidities (in the order of 0.25), s will be close to 4, whereas for h closer to 0.75, s is greater than 5. Not until h reaches 0.9 does the slope approach 8.

During evaporative enrichment of water, the vapour will have a reciprocal depletion, and plot on the same evaporative line, but opposite the initial composition of the water (i.e. on the left side of the meteoric water line) (Fig. 2-9).

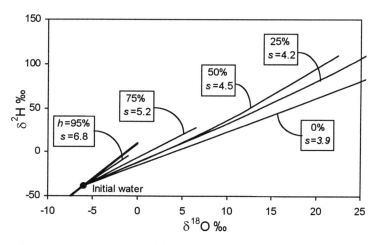

Fig. 2-8 Isotopic enrichment in evaporating water and the effect of humidity. Slopes are approximations of early portion of each curve near the GMWL (heavy line) (from Gonfiantini, 1986).

This discussion of nonequilibrium isotope effects caused by evaporation gives us a basis for studying isotope partitioning in the hydrological cycle. Later in this chapter, the effects of extreme evaporation will be explored, in the context of water balance in lakes, and the concentration of solutes in brines.

Deuterium excess "d" in meteoric waters

Using this understanding of kinetic evaporation, the displacement of the global meteoric water line from seawater now makes sense. Under conditions of strong (kinetic) evaporation, the fractionation for both ^{18}O and ^{2}H is greater than at equilibrium, although the kinetic fractionation of ^{18}O exceeds that of ^2H. In the case of an infinitely large initial reservoir (i.e. the oceans), kinetic isotope effects are observed not in the water but in the vapour (Fig. 2-9).

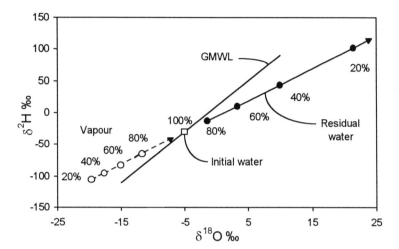

Fig. 2-9 Isotopic evolution of a finite volume of water during evaporation with 50% humidity. Solid points are the residual water at given percentages of the initial reservoir. Open points are the isotopic composition of the accumulating vapour reservoir. The vapour is initially depleted, but evolves toward the isotopic composition of the initial water.

Fig. 2-10 shows the isotopic composition of water vapour evaporated from seawater at various humidities. The calculated isotopic composition of the first rain from this vapour mass is also shown. The rain is enriched over the vapour, and is positively displaced along a line with slope $s = 8$. This slope is due to equilibrium condensation in the cloud, a process that is discussed in the following section.

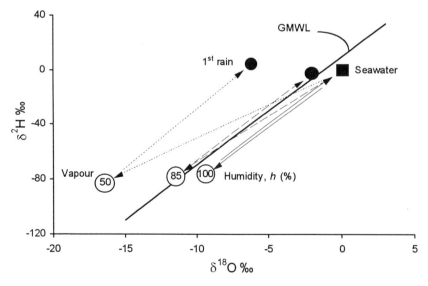

Fig. 2-10 Kinetic isotope effects during evaporation of seawater to form vapour (open circles with h in %) for various humidities at 25°C, together with the 1st rain (filled circles) formed by equilibrium condensation. When humidity is less than 100%, excess deuterium is found in the rain.

Under conditions of 100% humidity, the vapour is in isotopic equilibrium with seawater, and the first rain plots on a line through seawater. When humidity is low ($h = 50\%$) the vapour is strongly depleted and the precipitation plots well above the GMWL. When humidity is about 85%, precipitation plots very close to the global meteoric water line. Accordingly, global atmospheric water vapour forms with an average humidity just slightly greater than 85%, and produces precipitation on a line that is displaced from seawater by +10‰ for δ^2H. This is why Craig's meteoric water line for global precipitation has a deuterium excess of 10‰.

Dansgaard (1964) first proposed the use of the value, d, to characterize the deuterium excess in global precipitation. The value d is defined for a slope of 8, and is calculated for any precipitation sample as:

$$d = \delta^2H - 8\delta^{18}O$$

Fig. 2-11 shows the relationship between humidity and the deuterium excess. On a global basis, d averages about 10‰, but regionally it varies due to variations in humidity, wind speed and sea surface temperature (SST) during primary evaporation. SST is important as it can affect the humidity contrast Δh between the ocean surface and the upper air column. Rozanski et al. (1993) show that d correlates poorly with latitude, although seasonal variations are observed. Table 2-6 provides examples of stations where d deviates significantly from the global average.

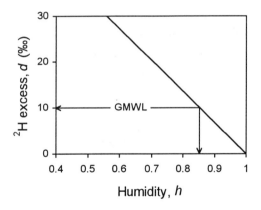

Fig. 2-11 The deuterium excess parameter, d, as a function of humidity, h, during kinetic evaporation from the ocean surface. Only a minor variation in d occurs with a change in temperature (after Merlivat and Jouzel, 1979).

Table 2-6 Examples of extreme values for deuterium excess d, with T and $\delta^{18}O$, from selected localities (data are weighted annual averages, from Rozanski et al., 1993)

Location	*T°C*	*$\delta^{18}O$*	*d–‰*
Bethel, Alaska	−1.9	−12.1	1.2
Argentine Island (65°S)	−4.0	−10.8	1.8
Flagstaff, Arizona	7.5	−8.1	3.5
Edmonton, Alberta	3.2	−17.1	5.3
Halley Bay, Antarctica	−18.2	−20.7	7.5
Canton Island, central Pacific	28.3	−3.7	7.7
Thule, Greenland	−11.6	−22.7	8.6
Perm, Russia	2.4	−13.0	14.0
Perth, Australia	18.6	−3.9	16.2
Kabul, Afghanistan	11.5	−7.2	17.2
Tirat Yael, northern Israel	16.5	−6.7	23.2
Rabba, southern Jordan	—	−4.8	23.6

Atmospheric mixing and global atmospheric water vapour

Craig and Gordon (1965) estimate from model considerations that the mean isotopic composition of global precipitation is $\delta^{18}O = -4‰$ and $\delta^2H = -22‰$, a value which plots on Craig's global meteoric water line. They go on to propose that the composition of atmospheric waters is not a simple process of vapour formation through non-equilibrium evaporation of ocean water. Rather, the formation of water vapour masses over the oceans is a more complicated process of mixing between evaporated water vapour and residual atmospheric water vapour following condensation and rain at sea. Their model (Fig. 2-12) combines kinetic evaporation from the oceans and atmospheric mixing. Mass is conserved (net evaporation = net precipitation) and the mean isotopic composition of evaporated ocean water equals the global mean for precipitation.

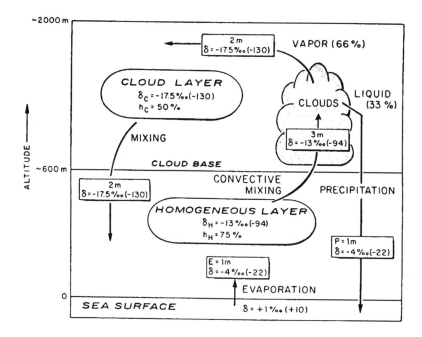

Fig. 2-12 The Craig and Gordon (1965) model for the isotopic composition of atmospheric water vapour over the oceans. The δ-values are given for ^{18}O (δ^2H in parentheses), the relative mass fluxes during evaporation and precipitation are given in metres of water, E = evaporation, P = precipitation, h = humidity, H = Homogeneous layer, C = Cloud layer.

Condensation, Precipitation and the Meteoric Water Line

The preceding section shows that the formation of atmospheric vapour masses is a nonequilibrium process due to the effects of low humidity and mixing of different vapour masses. However, the reverse process — condensation to form clouds and precipitation — takes place in an intimate mixture of vapour and water droplets with 100% humidity. Here, equilibrium fractionation between vapour and water dominates (see Table 1-4), although some in-cloud processes play a role in determining the $\delta^{18}O$ and δ^2H composition of precipitation. The isotopic evolution of precipitation during rainout is largely controlled by temperature.

Early studies on the isotopic composition of precipitation were made by Dansgaard (1953), Epstein and Mayeda (1953) and Friedman (1953). Later, the International Atomic Energy

Agency (IAEA), in collaboration with the World Meteorological Organization (WMO), established the Global Network of Isotopes in Precipitation (GNIP) — a world-wide selection of meteorological stations at which samples are collected to monitor the $\delta^{18}O$–δ^2H composition of precipitation.

The data produced by this network are essential for environmental isotope hydrology. They are available on the world wide web at <**www.iaea.or.at:80/programs/ri/gnip/gnipmain.htm**>. The regression line for the long-term averages of $\delta^{18}O$ and δ^2H measured for precipitation at the 219 stations in the network adds some precision to Craig's line:

$$\delta^2H = 8.17 \ (\pm\ 0.07)\ \delta^{18}O + 11.27\ (\pm\ 0.65)\ ‰\ \text{VSMOW} \qquad \text{(Rozanski et al., 1993)}$$

This line is the true global meteoric water line, as it is based on precipitation and not on surface waters. The global meteoric water line is essentially an average of local or regional meteoric water lines which vary in both slope and intercept according to meteorological conditions. A major result of the IAEA-WMO survey is a rather comprehensive picture of the global distribution of stable isotope contents in precipitation. The global distribution of ^{18}O derived from these data is presented in Chapter 3.

Rainout and Rayleigh distillation

The only way to produce rain is through the cooling of a vapour mass. *If the temperature doesn't drop, it will not rain.* Cooling occurs by adiabatic expansion (no loss of enthalpy) as warm air rises to lower pressures, or by radiative heat loss. When the dew point is passed (the temperature at which humidity is 100%), water vapour condenses to maintain thermodynamic equilibrium, and it will rain (or snow). If the temperature stabilizes or warms, condensation stops or reverses and humidity drops.

As an air mass follows a trajectory from its vapour source area to higher latitudes and over continents, it cools and loses its water vapour along the way as precipitation, a process known as "rainout." Within the cloud, equilibrium fractionation between vapour and the condensing phases preferentially partitions ^{18}O and ^2H into the rain or snow (recall in Fig. 1-6). Along the trajectory of the air mass, the process of rainout *distills* the heavy isotopes from the vapour. The vapour then becomes progressively depleted in ^{18}O and ^2H according to a Rayleigh–type distillation. Isotopically enriched rain is forming and falling from a diminishing vapour mass, and the residual vapour becomes isotopically depleted. Subsequent rains, while enriched with respect to the remaining vapour, will be depleted with respect to earlier rains from the same vapour mass. Rainout is then an evolution towards colder, isotopically–depleted precipitation.

This strong correlation between temperature and $\delta^{18}O$–δ^2H controls the position of precipitation on the meteoric water line. From this correlation we can derive isotope effects due to seasons, altitude, latitude, continentality and paleoclimates — the basis of isotope hydrology. We can model this temperature — isotope evolution during rainout according to the same Rayleigh distillation equation presented for evaporation (page 41):

$$R = R_o f^{(\alpha-1)}$$

In this case, R_o is the vapour's initial isotope ratio ($^{18}O/^{16}O$ or $^2H/^1H$) and R would be the ratio after a given proportion of the vapour had reacted (i.e. rained out). The fraction f is the residual vapour reservoir in the cloud. The fractionation factor α is for equilibrium water-vapour exchange at the prevailing in-cloud temperature (Table 1-4). Equilibrium fractionation during condensation is generally the case in clouds, where $h = 100\%$. Only in the super-cooled air masses at high latitudes will kinetic effects become important (Fisher, 1991). In a Rayleigh function, α is expressed as product-reactant (i.e. α_{l-v}). In the case of ^{18}O, this would be $\alpha^{18}O_{l-v} = 1.0094$ at 25°C, but will increase as the temperature drops. Rayleigh processes in other isotope systems are presented in more detail at the end of this chapter. Here we will use a simplified version of the Rayleigh equation to calculate $\delta^{18}O$ evolution during rainout (Fig. 2-13).

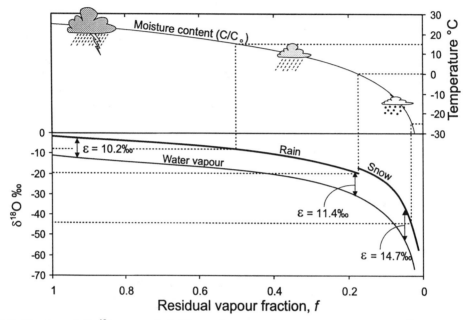

Fig. 2-13 The change in the ^{18}O content of rainfall according to a Rayleigh distillation, starting with $\delta^{18}O_{vapour} = -11\%$, temperature = 25°C, and a final temperature of –30°C. Note that at 0°C, fractionation between snow and water vapour replaces rain-vapour fractionation. The fraction remaining has been calculated from the decrease in moisture carrying capacity of air at lower temperatures, starting at 25°C (from calculations for Fig. 2-4). Dashed lines link $\delta^{18}O$ of precipitation with temperature of condensation.

At any point along a rainout trajectory, the residual fraction of water vapour $\delta^{18}O_{v\,(f)}$ will have an isotopic composition calculated as:

$$\delta^{18}O_{v\,(f)} \approx \delta_o{}^{18}O_v + \varepsilon^{18}O_{l-v} \cdot \ln f$$

If the initial vapour has $\delta_o{}^{18}O_v = -10\%$, then for $f = 0.9$ (10% loss to rain) and $\varepsilon^{18}O_{l-v} = 9.4\%$, its isotopic composition will have evolved to $\delta^{18}O_{v\,(f)} = \delta^{18}O_{v(0.9)} = -11.0\%$. At the point where 50% of the original vapour has rained out, $\delta^{18}O_{v(0.5)} = -16.5\%$, and when only 10% water vapour remains, it has become depleted to $\delta^{18}O_{v(0.1)} = -31.6\%$. This simple calculation has not taken into account the minor increase in $\varepsilon^{18}O_{l-v}$ at the lower temperatures of condensation.

The rain produced by the vapour mass at any given f is:

$$\delta^{18}O_{rain(f)} \approx \delta^{18}O_{v(f)} + \varepsilon^{18}O_{l-v(T)}$$

where ε is a function of temperature T. This models the evolution of $\delta^{18}O_{rain}$ with rainout. To observe the evolution of $\delta^{18}O_{rain}$ with dropping temperature requires that we link the decreasing vapour reservoir during rainout with the decreasing capacity of air to hold moisture as temperature drops (Fig. 2-4).

Fig. 2-13 shows the isotopic evolution of rain and snow which forms from a continuously cooling vapour mass. The upper curve is the decreasing content of vapour in the air as temperature drops. The lower curves are the $\delta^{18}O$ values for the vapour and precipitation during rainout. In this case, the Rayleigh distillation is corrected for the fact that the fractionation factor $\alpha^{18}O_{l-v}$ increases as the temperature drops. Note the offset for precipitation when the condensation temperature drops below freezing, due to difference between rain-vapour and snow-vapour fractionation at 0°C.

The correlation between temperature and $\delta^{18}O$ is evident in this diagram where strong depletions in precipitation occur at advanced stages of rainout (low T and low residual fraction f). A similar evolution takes place for ^2H during rainout. The result is the evolution towards depleted values on a $\delta^{18}O$–δ^2H diagram (Fig. 2-14), essentially along the meteoric water line for global precipitation.

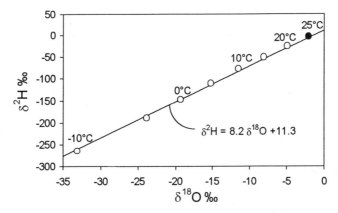

Fig. 2-14 Evolution of ^{18}O and ^2H during rainout. The depletion begins in this example at 25°C with a water vapour evaporated from seawater (VSMOW) at a humidity of 85%. The trend follows very closely the meteoric relationship observed for rainfall monitored by the IAEA (Gat, 1980).

An example of the gradual depletion in ^{18}O and ^2H during rainout was observed during a 12-hour storm event in the Sultanate of Oman in 1986 (Fig. 2-15). Over the duration of the rain, the $\delta^{18}O$ value in the rain dropped from –3‰ to less than –6‰. As both isotopes follow this depletion, these rain samples still plot on a line with slope near 8, trending from enriched values early in the storm to the lighter end of the meteoric water line by the end of the storm.

Slope of the meteoric water line

The intimate mixture of vapour and condensation in clouds tends to preserve isotopic equilibrium and the meteoric relationship between $\delta^{18}O$ and δ^2H. So, as the in-cloud temperature drops and rainout proceeds, the rain will have an isotopic composition controlled by equilibrium fractionation with the vapour and will plot on the MWL with slope close to 8.

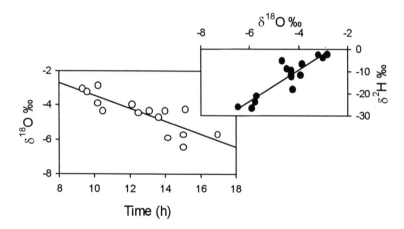

Fig. 2-15 Rainfall from a single storm, February 1, 1986, in Muscat, Oman. A characteristic Rayleigh depletion with time as the storm proceeds is evident in the time-$\delta^{18}O$ diagram. The regression line for the $\delta^{18}O$-δ^2H plot is $\delta^2H = 7.1\ \delta^{18}O + 19.1$, a line very close to the average meteoric water line for Northern Oman, and close to that for arid Mediterranean-type climates.

Why a slope of ~8? This relates to the ratio of the equilibrium fractionation factors for ^2H and ^{18}O:

$$s \approx \frac{10^3 \ln \alpha^2 H_{l\text{-}v}}{10^3 \ln \alpha^{18} O_{l\text{-}v}} = 8.2 \quad \text{at } 25°C$$

A rigorous analysis would show that this equation is also only an approximation, albeit a close one (J. Gat, pers. comm., 1996), and sufficient for our purposes here. The actual slope will vary with the average temperature of condensation (Fig. 2-16).

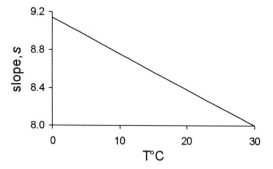

Fig. 2-16 Correlation of temperature and slope s ($s = \alpha^2H_{l\text{-}v}/\alpha^{18}O_{l\text{-}v}$) using the fractionation factors of Majoube (1971). Only at higher temperatures does the equilibrium slope approach 8.

As shown above, the slope of the meteoric relationship between ^{18}O and ^2H in global precipitation is very close to 8. This slope can be affected by evaporation that occurs *after* condensation. If rain is falling through a dry air column above the ground, some will evaporate, imparting a kinetic fractionation on the drop. Friedman et al. (1962) first showed that evaporation during rainfall would shift the water away from the meteoric water line along an evaporation slope less than 8. Dansgaard (1964) described such evaporation of rain as the *amount effect* on the isotopic composition of precipitation, although in-cloud phenomena and mixing also appear to play a role.

As one would expect, this amount effect is best observed in arid regions. For example, the low slope for rainfall data from Bahrain in the Arabian Gulf is clearly affected by secondary evaporation during rainfall. This effect on *s* is greatest for light rains. Once the air column becomes water-saturated, such evaporation is diminished. Excluding the data for rainfall events less than 20 mm orients the slope of the LMWL closer to the equilibrium value of 8:

$\delta^2H = 6.3\, \delta^{18}O + 11.6$ (all rains, Bahrain)

$\delta^2H = 7.8\, \delta^{18}O + 13.6$ (rains >20 mm, Bahrain)

Data from the Sultanate of Oman shows this nonlinearity of the $\delta^{18}O$-δ^2H relationship due to the influence of evaporation during rainfall (Fig. 2-17). Using data for rainfalls greater than 20 mm yields a slope closer to the "equilibrium" value of 8. Note that secondary evaporation also decreases the deuterium excess and intercept.

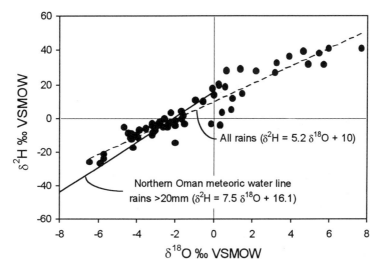

Fig. 2-17 The δ^2H - $\delta^{18}O$ relationship in rainfall sampled from northern Oman. The enriched values demonstrate the amount effect of light rains, due to evaporation of rain drops falling through hot dry air.

Local meteoric water lines

Craig's global meteoric water line is essentially a global average of many local meteoric water lines, each controlled by local climatic factors, including the origin of the vapour mass, secondary evaporation during rainfall and the seasonality of precipitation. These local factors affect both the deuterium excess and the slope (Fig. 2-18). For regional or local investigations, it is important to compare surface water and groundwater data with a local meteoric water line (LMWL). However, it is not always possible to rigorously monitor precipitation over a representative period of time (at least 1 year), and meteoric water lines must be borrowed from the closest available monitoring station.

In regions where strong deuterium excesses are noted, the Eastern Meteoric Water Line (EMWL), developed for precipitation in the Eastern Mediterranean region, is often used:

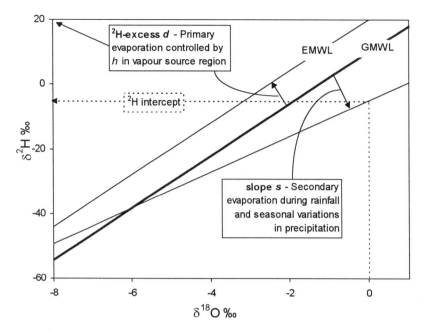

Fig. 2-18 Principal factors affecting the slope s and deuterium excess d of meteoric water lines. Note the distinction between deuterium excess, which is calculated for a slope of 8, and the deuterium intercept, determined with the actual slope and for $\delta^{18}O = 0$.

$$\delta^2H = 8\delta^{18}O + 22 ‰ \quad \text{Eastern Mediterranean Meteoric Water Line (EMWL)}$$
$$\text{(Gat and Carmi, 1970)}$$

Meteoric waters in temperate settings also deviate from the GMWL. A 7-year study of precipitation across Canada (Fritz et al., 1987a) shows the diversity of local meteoric water lines across a spectrum of climatic regimes (Fig. 2-19). Amount weighted data from five stations are presented in Table 2-7, including Ft. Smith in Arctic Canada (Michel, 1977). Both the slope and δ^2H intercept for local meteoric water lines deviate significantly from the global meteoric water line on an annual and seasonal basis.

For the interior stations, the summer MWLs are influenced by secondary evaporation during rainfall, giving slopes less than 8 (7.4 and 7.5). However, in winter, these MWLs are close to the global line as evaporation during snowfall is negligible. The eastern stations are influenced by tropical air streams from the warm Gulf of Mexico and Atlantic Ocean and consequently have higher δ^2H-intercept values than the western stations, which receive their weather from the higher latitude Pacific Westerlies (Fig. 2-19).

The low d values for Victoria reflect high humidity during formation of the Pacific vapour masses. The d values in eastern Canada trend towards the global average of 10, although in summer d and δ^2H-intercept values are lower due secondary evaporation effects. Despite the strong regional and seasonal differences in meteoric water lines, the average of all these data produce a meteoric water line for Canada (excluding precipitation data from Arctic Canada) which is close to GMWL:

$$\delta^2H = 7.75\,\delta^{18}O + 9.83 \quad \text{Canadian meteoric water line}$$

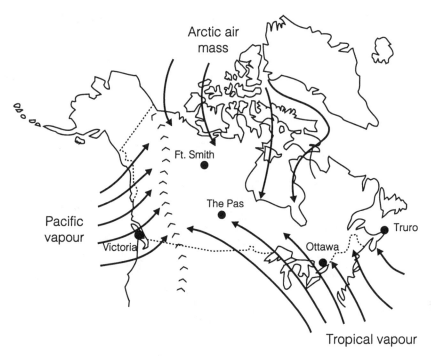

Fig. 2-19 Meteoric regimes in Canada. The effect of these systems on precipitation is seen in the local meteoric water lines for stations across the country. The Arctic air stream brings no moisture, except locally in the north during summer.

Table 2-2 Meteoric water lines for stations from varying climates across Canada (from Fritz et al., 1987a).

Station	Climate	Season	$\delta^{18}O$ vs. δ^2H Slope	Intercept	d
Victoria	Pacific marine	Year	7.8	2.9	3.6
		Summer	8.3	3.9	1.6
		Winter	7.5	−1.6	3.9
Le Pas	Western interior	Year	7.6	0.6	5.6
		Summer	7.4	−3.0	3.5
		Winter	8.0	11.3	10.4
Ottawa	Eastern interior	Year	7.6	6.5	12.2
		Summer	7.5	4.8	11.0
		Winter	7.9	11.0	14.7
Truro	Atlantic marine	Year	7.4	5.6	10.6
		Summer	7.8	8.3	8.2
		Winter	7.4	5.2	12.8
Fort Smith	Arctic continental	Year	7.5	−4.9	0.3

The composite nature of the GMWL is illustrated in Fig. 2-20 which plots meteoric water lines from various areas. These lines are drawn only over the range of the precipitation data they represent.

Some observations:

- Continental stations (Ottawa and Vienna) are close to the GMWL and have a wide range due to strong seasonal variations in temperature.

- The marine stations, including Midway Island (tropical Pacific at 28°N 177°W) and Victoria (west coast of Canada), have narrower ranges of data than the continental stations

54 *Chapter 2 Tracing the Hydrological Cycle*

due to the moderating maritime effect on temperature. The more negatively placed Victoria line reflects its more northerly latitude.

- St. Helena (equitorial Atlantic at 16°S 6°W) and the monsoon region of southern Oman (17°N 54°E) have very limited ranges which give poorly defined slopes. Such precipitation represents the first stage of rainout and has evolved little from the original moisture source.

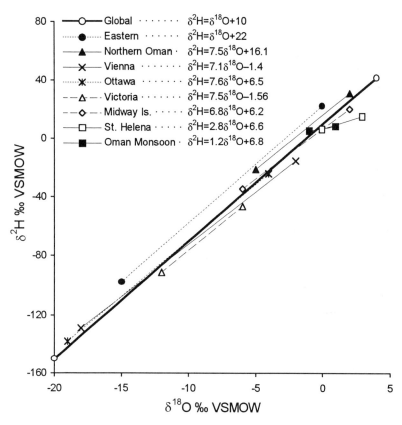

Fig. 2-20 Regional and local meteoric water lines from various weather stations (Global — Craig, 1961a; Eastern — Gat and Carmi, 1970; Victoria, Ottawa — Fritz et al., 1987a; Vienna, Midway Is., St. Helena — Rozanski et al., 1993; Oman — Clark, 1987).

The conclusion from this investigation of meteoric water lines? Any given locality will have a characteristic local meteoric water line (LMWL) which may be quite different from the global line. A local line can reflect the origin of the water vapour and subsequent modifications by secondary processes of re-evaporation and mixing. Any detailed study of groundwater recharge using $\delta^{18}O$ and δ^2H should attempt to define as best as possible the LMWL.

A final point about the statistical comparison of field data to a local meteoric water line is in order. Using a large precipitation data set will improve the confidence limits of the meteoric water line, allowing a better interpretation of groundwater data. Confidence limits can be calculated by standard statistical methods available on most computer spreadsheet programs. The solution to problem 5 at the end of this chapter constructs the 95% confidence limits for the line fit to a precipitation data set as an example.

A Closer Look at Rayleigh Distillation

Rayleigh-type evolution occurs for a host of isotope reactions. The example of a depleting water vapour mass during rainout is only one. Another is the change in $\delta^{18}O$ and δ^2H during the freezing or evaporation of water. The $\delta^{34}S$ of sulphate is enriched by a Rayleigh process during the reduction of SO_4^{2-} to H_2S in aquifers. The evolution of $\delta^{13}C_{DIC}$ during methanogenesis by CO_2 reduction, or $\delta^{15}N$ during denitrification are other examples. In all cases, an exponential enrichment or depletion occurs in the residual reservoir of the reactant as it is converted to the product. For an ideal distillation, the reactant reservoir must be finite and well mixed, and does not re-react with the reaction product.

The general form of a Rayleigh distillation equation states that the isotope ratio (R) in a diminishing reservoir of the reactant is a function of its initial isotopic ratio (R_o), the remaining fraction of that reservoir (f) and the equilibrium fractionation factor for the reaction ($\alpha_{product-reactant}$):

$$R = R_o f^{(\alpha-1)}$$

For the example of ^{18}O in evaporating water, this would be:

$$\left(\frac{^{18}O}{^{16}O}\right) = \left(\frac{^{18}O}{^{16}O}\right)_o f^{(\alpha_{v-l}-1)}$$

This form of the Rayleigh equation is more easily used when expressed in δ–‰ notation. Converting the isotope ratios to δ values gives:

$$\ln\left(\frac{\delta + 1000}{\delta_o + 1000}\right) = (\alpha - 1) \cdot \ln f = \frac{\varepsilon}{1000} \cdot \ln f$$

When the δ-values are small (i.e. close to the reference VSMOW), this equation can be simplified to:

$$\delta - \delta_o \cong 10^3 (\alpha - 1) \cdot \ln f$$

or $\quad \delta - \delta_o \cong \varepsilon \cdot \ln f$

Using a generic isotope with an arbitrary initial value ($\delta_{o\text{-react}}$) and fractionation factor ($\varepsilon_{prod\text{-}react}$), a Rayleigh distillation can be plotted (Fig. 2-21).

The evolution observed in the reactant reservoir during a Rayleigh distillation will also be seen in the product reservoir (Fig. 2-21). However, we can consider this evolution in the reaction product as either an instantaneous product (closed system) or as an accumulating product reservoir (open system). In a closed system, the product is instantaneously removed from the system and does not mix with subsequent reaction products (e.g. ice from the freezing of water). In an open system, the product forms a growing and well-mixed reservoir (e.g. H_2S accumulating in a groundwater during sulphate reduction).

The isotopic composition of the product reservoir under open system conditions is derived from the isotopic composition of the entire system, which is equal to the initial value of the reactant reservoir, $\delta_{o-react}$. At any point during the reaction, the isotopic composition of the two reservoirs obeys the mass-balance equation:

$$\delta_{o-react} = f \cdot \delta_{react} + (1-f) \cdot \delta_{prod-reservoir}$$

Recalling that:

$$\delta_{react} = \delta_{o-react} + \varepsilon \cdot \ln f$$

the mean isotopic composition of the product reservoir ($\delta_{prod-reservoir}$) can be determined:

$$\delta_{prod-reservoir} = \delta_{react} - \frac{\varepsilon_{prod-react} \ln f}{(1-f)}$$

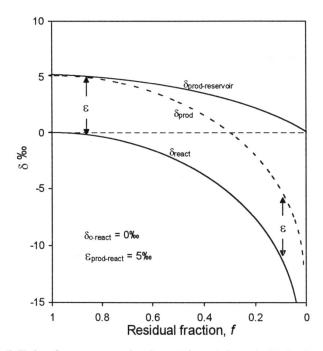

Fig. 2-21 Rayleigh distillation of a reactant reservoir as it reacts to completion under (a) closed system conditions represented by the δ_{prod} curve (i.e. the product is removed from reacting or mixing), and (b) under open system conditions where the product reservoir is well mixed, represented by the $\delta_{prod-reservoir}$ curve.

Note that as $f \rightarrow 1$, $\delta_{prod-reservoir} \rightarrow \delta_{react} + \varepsilon$, but be careful as the equation is undefined for $f = 1$. The isotopic evolution of the product reservoir (Fig. 2-21) eventually equals the original isotopic composition of the reactant, thus preserving the isotopic mass balance of the system.

The freezing of water is a good example of a closed system reaction. As a given reservoir of water begins to freeze, the first ice to form will be enriched in ^{18}O by about 3‰ (see Table 1-4) and the water will be marginally depleted. As freezing continues, the ice formed from the continually depleted water preserves this trend towards increasingly lower δ-values.

Michel (1986) shows nice Rayleigh distillation curves for ice in a frost blister formed under closed system conditions. The frost blister formed by continuous downward freezing in a water-filled cavity overlying permafrost. Samples from a vertical profile through a metre of ground ice (Fig. 2-22) show a systematic trend towards depleted values, reflecting the evolution of the water from which the ice was forming. Data displaying a Rayleigh distillation can then be used to calculate the fractionation factor for the reaction, which in turn can be diagnostic of the process. In the case of Fig. 2-22, the calculated α values are slightly lower than for equilibrium fractionation, indicating minor kinetic (diffusion) effects during freezing. From a data set such as this, the $\delta^{18}O$ of the water precursor can then be determined from the calculated fractionation factor and the $\delta^{18}O$ value of the initial ice ($\delta^{18}O_w = \delta_o^{18}O_i - \epsilon^{18}O_{i-w}$), or from the amount-weighted mean of all the samples ($\sum_{1}^{n}[\delta^{18}O_n \cdot fraction_n]$).

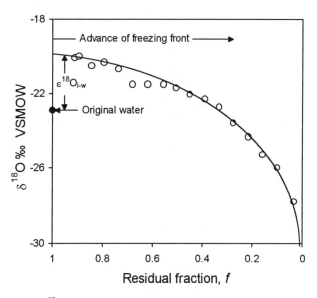

Fig. 2-22 Rayleigh distillation of ^{18}O during freezing of water in the fixed volume reservoir of an ice blister in a permafrost landscape in the Yukon Territory, Canada (data from Michel, 1986).

Effects of Extreme Evaporation

The kinetic effects of evaporation in meteoric processes were introduced earlier. Here we examine more closely the effects of strong evaporation on surface waters. These include the extensive evaporation of fresh waters on one hand and the evaporation of seawater and brines on the other. In the latter case, solute concentration plays an important role.

Evaporation in Lakes

Isotope effects during evaporation provide a tool to study water balances in lakes with long mean residence times. Gonfiantini (1986) presents the calculations for such mass balances (Fig. 2-23) for the case of constant volume. When the isotope content of evaporating water is plotted against the residual fraction, we see that enrichment is asymptotic towards higher δ-values at humidities below about 50%. When humidity is greater than about 50%, the enrichment approaches a steady-state value which is, as will be shown below, strongly dependent on the salinity of the residual water. At very high salinities, the enrichment is reversed.

The humidity is critical in determining the fraction of water lost by evaporation vs. outflow. To determine h, pan evaporation studies are important (Welhan and Fritz, 1977). In such studies, a pan of the local water is set out in the study area and left to evaporate. Samples of the residual water are taken periodically for analysis of $\delta^{18}O$ and δ^2H as it evaporates to completion. It is essential that the pan site is chosen to represent the same wind and humidity conditions that are present over the lake — a requirement that is not easily fulfilled. From the isotope data, h can be calculated. An example of a pan evaporation study is presented later in Chapter 4.

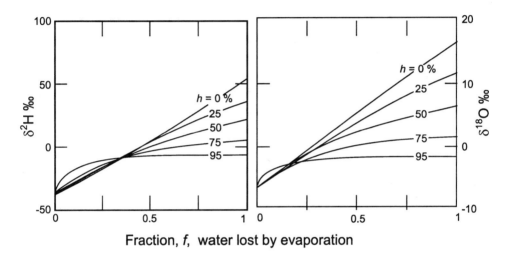

Fig. 2-23 The isotopic composition of lake water for varying fractions of water loss by evaporation, assuming well mixed conditions and constant volume (inflow = outflow + evaporation). For $f = 1$ evaporation = inflow and no outflow exists. Calculations are made for various relative humidities (from Gonfiantini, 1986).

Fontes et al. (1970) show strong evaporative loss for Lake Chad, where inflow from the Chari River ($\delta^{18}O = -3$ to $-5‰$) becomes highly enriched in the lake ($\delta^{18}O > 10‰$) with a clear correlation with salinity. Dinçer (1968) shows successive enrichment in three lakes in arid Anatolia (Turkey) with varying losses from evaporation vs. outflow. The enrichments in ^{18}O vary between 6.4‰ for 40% evaporative loss, 7.9‰ for about 60% loss and 10.4‰ for 100% loss to evaporation.

When evaporation of a freshwater lake basin proceeds to completion (no residual water and no leakage), the isotopic enrichments can be most extreme. Fontes and Gonfiantini (1967a) measured $\delta^{18}O$ and δ^2H in an ephemeral pond during evaporation under the extremely arid conditions in Saharan Algeria. They relate these data to the residual water fraction according to exponents of f (Fig. 2-24). The absence of solutes in this example allows the isotopic enrichments to become very extreme. The empirical exponent was calculated to give a straight line fit to the data and is considered to depend mainly on relative humidity. The isotopic contents of the residual water follow an exponential enrichment as f approaches 0 according to a Rayleigh distillation function.

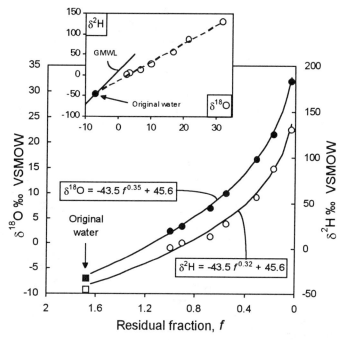

Fig. 2-24 The strong enrichment in both ^{18}O and ^{2}H observed in an ephemeral lake during evaporation under extremely arid conditions in the Sahara desert (Gonfiantini and Fontes, 1967). Data (circles) are regressed with an exponential function and extrapolated to determine the original isotopic composition of the water, shown in square symbols (δ^{18}O = –7‰; δ^2H = –46‰). This gives residual water fractions greater than 1 and accounts for evaporative water loss prior to sampling. Inset shows the low slope of this enrichment trend (s = 4.5) due to evaporation under conditions of low humidity.

Evaporation of brines

Restricted basins show under conditions of extreme evaporation δ^{18}O values which increase asymptotically to a steady-state value controlled by the influx of fresh water, influx of ocean water, and relative humidity. However, as Fig. 2-25 shows, the effects of solutes are important. The Dead Sea brines, which encrust the shores with salt, has a salinity exceeding 23% (7 × seawater), and yet a δ^{18}O value approaching only 4.5‰ VSMOW (Horita and Gat, 1989). The Red Sea brines, which have solute concentrations over 2 × seawater, are enriched in ^{18}O by about 2.5‰ above VSMOW. Deuterium is also enriched by evaporation, but because it is a nonequilibrium process, ^{2}H and ^{18}O fractionate differently.

These effects are discussed at length by Gonfiantini (1986). With increasing salinity (after 20 to 50% water loss for seawater), the decreased activity of water decreases the saturated water content in the boundary layer (h/a_w) and reduces the humidity contrast with the adjacent dry air.

At higher salinities, ion hydration imparts an isotope depletion on the water. The hydration sheath, particularly for polyvalent ions, is enriched over free water. For example, the hydration water for $CaCl_2$ in solution is 26‰ enriched over free water for ^{18}O and 341‰ enriched for ^{2}H (Sofer and Gat, 1975). As salts precipitate, the incorporation of crystallization water adds a further effect. These effects are evident in Fig. 2-25 by a reversal in the evaporation trend. O'Neil and Truesdell (1991) have examined the fractionation of ^{18}O between CO_2 and concentrated aqueous solutions for insights into solute-water interactions. They provide measurements of fractionation between pure water and a variety of important aqueous solutes.

Ion hydration causes a discrepancy in the measurements of ^{18}O and ^{2}H in brines. Deuterium is measured on the H_2 gas produced on the water distilled from the solution. This includes both the free water and water in the hydration shells, and so is a measurement of concentration. Oxygen-18, on the other hand, is measured on CO_2 that has equilibrated with the water. This is a measurement of the ^{18}O activity. The isotopic activity of brines decreases with salinity while concentration remains the same. For consistency, $\delta^{18}O$ measurements of activity must be corrected and expressed as concentration before comparison with $\delta^{2}H$. Sofer and Gat (1972) developed the following equation to correct $\delta^{18}O$ measurements for salinity:

$$\delta^{18}O_{corrected} = \left(\frac{1.11\, m\text{Mg} + 0.47\, m\text{Ca} - 0.16\, m\text{K}}{1000}\right) \cdot \left(\delta^{18}O_{meas} + 1000\right) + \delta^{18}O_{meas}$$

Molality m is calculated from ion concentrations (mg/L) as (using Mg as an example):

$$m_{Mg} = \left[\frac{mg/L \, / \, 24.3 \times 10^{-3}}{\text{solution density (kg/L)}}\right]$$

After correction for salinity effects, $\delta^{18}O$ data for brines can be used together with $\delta^{2}H$ measurements for comparison with other data sets and for interpretation of isotope hydrology.

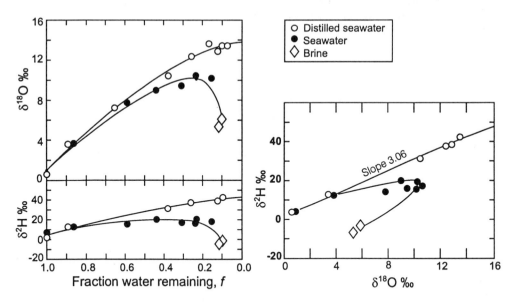

Fig. 2-25 The isotopic evolution of seawater (and distilled seawater) during evaporation under the same temperature and humidity conditions. The effect of solutes on the stable isotope evolution is seen in the brine, where a reversal occurs due to the decreased activity of water which reduces the humidity contrast between the boundary layer and the open atmosphere (from Gonfiantini, 1965).

Problems

1. Is it possible to distinguish on the basis of ^{18}O and ^{2}H analyses the original isotopic composition of surface water which has undergone evaporation? Explain.

2. What controls the ^{18}O and ^2H content in precipitation — kinetic or equilibrium fractionation?

3. The fractionation between vapour and precipitation is greater at lower temperatures. Why then is precipitation at lower temperatures more depleted than at warmer temperatures?

4. What is the variation in the $\delta^{18}O$ value for ocean water between the glacial maximum (ca. 21,500 years ago) and today? Why the change?

5. The data set found at <www.science.uottawa.ca/~eih/ch2/problem5> comprises rainfall data for three meteorological stations at different altitudes from the summit of Ajloun Mountain (Ras Munif) to the Dead Sea in the Jordan rift valley (Der Alla) (Bajjali et al., 1997). From these data establish a local meteoric water line for northern Jordan. The confidence limits for this meteoric water line are calculated on the spreadsheet using standard statistical methods. This spreadsheet can be downloaded and used as a template for other precipitation data.

   Is any amount effect in these data?

   Produce a second plot and local meteoric water line using the >20 mm rains for only the two upland stations at Irbed and Ras Munif, which are situated in the regional groundwater recharge area. Compare this line with that for all rain at all three stations, and comment on the differences in slope, δ^2H intercept and confidence limits.

   Which is a more reasonable baseline for comparing with regional groundwaters?

   Now calculate the average value for deuterium excess d in the Irbid/Ras Munif subset of rainfall data. Why is it so high?

6. What conditions have to be fulfilled if a system is to show isotope effects during a Rayleigh distillation reaction, from which a fractionation factor can be calculated.

7. Graph the progressive Rayleigh enrichment of ^{18}O and ^2H for evaporating water (initial $\delta^{18}O = -6‰$ and $\delta^2H = -38‰$) at 5°C and at 50°C for $h = 0$ and 50% and compare these curves to Fig. 2-8. How do yours compare?

8. The following data for ice in a frost blister are presented in Fig. 2-22 (Michel, 1986), and show a Rayleigh distillation of the residual water, f, as freezing progressed. Determine the average enrichment factor ($\varepsilon^{18}O_{ice-water}$) for this example. Assume an initial $\delta^{18}O_{ice} = 19.9‰$, by extrapolation of these data to $f = 1$. How does your calculated value for $\varepsilon^{18}O_{ice-water}$ compare with values from Table 1-4?

Chapter 3
Precipitation

If we are to use ^{18}O and ^{2}H to trace groundwater recharge, then it is necessary that their concentrations in precipitation provide a characteristic input signal that varies regionally and over time. The rainout process in clouds is driven by decreasing temperature, a parameter with both regional and temporal variability. It is temperature that controls the partitioning of isotopes in precipitation, and provides the variable input function used to trace groundwater recharge.

Characterizing the stable isotope distributions in meteoric waters is essential in determining this input function. The local meteoric water line provides a baseline for groundwaters. The position of meteoric waters on this line is controlled by a series of temperature-based mechanisms that drive the rainout process. These include vapour mass trajectories over continents, rising over topographic features, moving to high latitudes, and seasonal effects. Each has a characteristic effect on the stable isotopic composition of precipitation.

The T–δ^{18}O Correlation in Precipitation

From calculations in Chapter 2, we see that as decreasing temperature drives the rainout process, the precipitation becomes increasingly depleted in ^{18}O and ^{2}H. Weather is of course not so simple, and this evolution is complicated by re-evaporation and atmospheric mixing. Most weather systems acquire new sources of vapour along their paths that can mask evolutionary trends in a evolving vapour mass. Nonetheless, a strong correlation exists between temperature and isotopes in precipitation. Accordingly, where temperature gradients exist, gradients in δ^{18}O and δ^2H should be observed.

In the following discussion, the temperature data used in the T–δ^{18}O correlations are *surface measured temperatures*, and not *in-cloud* temperatures. It is the in-cloud temperatures that control condensation and isotope fractionation. For obvious reasons, these cannot be routinely measured, and so correlation with surface air temperatures or mean annual air temperatures (MAAT) are made.

$\delta^{18}O$ on the global scale

Dansgaard, in 1964, established a linear relationship between surface air temperatures and δ^{18}O for mean annual precipitation on a global basis (Fig. 3-1):

$$\delta^{18}O = 0.695\, T_{annual} - 13.6‰ \text{ SMOW}$$

and $\quad \delta^2H = 5.6\, T_{annual} - 100‰ \text{ SMOW}$

If monthly average temperatures are used, the global relationship for δ^{18}O becomes:

$$\delta^{18}O = (0.338 \pm 0.028)\, T_{monthly} - 11.99\ ‰ \text{ VSMOW} \qquad \text{Yurtsever and Gat (1981)}$$

On average, a 1‰ decrease in average annual δ^{18}O corresponds to a decrease of about 1.1 to 1.7°C in the average annual temperature. Corresponding variations occur for deuterium, and this covariance is the principal reason for the linear relationship or GMWL defined by Craig.

The global map of δ^{18}O values for precipitation makes a nice illustration of the partitioning of isotopes between cold and warm regions. Fig. 3-2 was created from mean annual precipitation data collected within the IAEA-World Meterological Organization survey of precipitation, using a geographical information system (GIS) for contouring. On this global scale, the partitioning of ^{18}O into warmer, low-latitude precipitation is clear.

However, the global T–$\delta^{18}O$ relationship is only an approximation, and on a regional basis it is far from linear. The extensive data base collected from IAEA stations over the past 30 years has been rigorously evaluated by Rozanski et al. (1993). These data are available at the IAEA website <**www.iaea.or.at:80/programs/ri/gnip/gnipmain.htm**>.

The extensive monitoring network of the IAEA (right diagram in Fig. 3-1) shows that the T–$\delta^{18}O$ relationship for worldwide precipitation comprises different curves for specific geographic regions. The distinctions between marine and continental stations in this figure show the importance of geographic effects. Marine stations correlate poorly with global data due to the damping of seasonal variations in temperature and precipitation. The Canadian interior stations depart from the global relationship due to continental effects and the seasonality of precipitation.

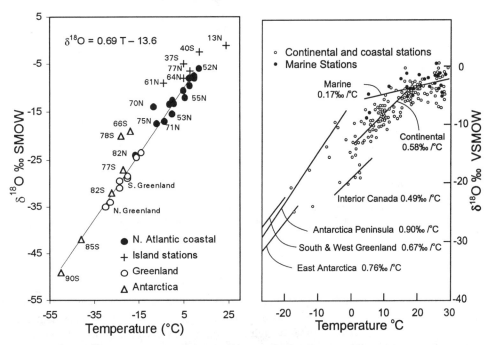

Fig. 3-1 The global T–$\delta^{18}O$ relationship for precipitation, modified from Dansgaard, 1964 (left). Temperature is mean annual air temperature (MAAT) at the station. Data from the extensive IAEA Global Network for Isotopes in Precipitation (GNIP) shows this relationship to be a combination of regional T–$\delta^{18}O$ lines, with strong differences between marine, continental and interior stations, from Rozanski et al., 1993 (diagram on right).

Departures from the global T–$\delta^{18}O$ relationship occur at the regional to local scale, due to physiographic variations. Departures also occur when monitoring data are examined for only short time periods. The correlation of $\delta^{18}O$ with temperature at the event-scale is very poor, and demonstrates that individual weather patterns, storm tracks and air mass mixing are far too chaotic to develop a clear T–$\delta^{18}O$ relationship at the local or event scale. The stochastic nature of weather essentially precludes the use of $\delta^{18}O$ as a proxy for temperature at anything less than seasonal to multi-annual scale. Global climate data sets are available at the NOAA website found at <**http://ncdc.noaa.gov**>.

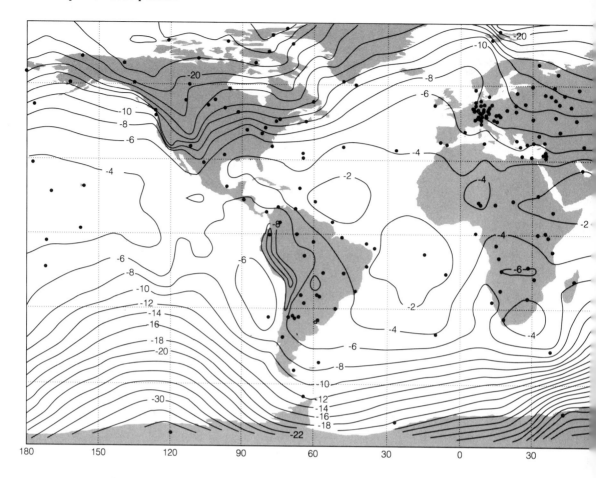

Fig. 3-2 Mean $\delta^{18}O$ distribution in precipitation on a global basis, for stations with at least 24 months of records. (based on IAEA World Meteorological Precipitation monitoring data summarized by Rozanski et al., 1993).

Latitude effect

From the strong δ–T relationship observed by Dansgaard (Fig. 3-1), one can expect that precipitation at higher latitudes tends to have more negative $\delta^{18}O$ values. Furthermore, polar regions are situated at the end of the Rayleigh rainout process (recall from Fig. 2-13), and so the $\delta^{18}O$-gradient should become increasingly steep. The $\delta^{18}O$–T gradients in Fig. 3-1 are on the order of –0.6‰ for $\delta^{18}O$ per degree of latitude for continental stations of the North America and Europe, and a considerably steeper gradient of about –2‰ $\delta^{18}O$ per degree latitude for the colder Antarctic stations. Very shallow gradients are apparent in the low latitudes where over 60% of atmospheric water vapour originates.

The global map of ^{18}O in precipitation in Fig. 3-2 illustrates this point. Here, the depletion in $\delta^{18}O$ values with increasing latitude is clear. Relatively flat gradients are found in the tropics, particularly over the oceans. Gradients increase poleward, consistent with the data in Fig. 3-1. Distortions to these latitudinal gradients are due to continental effects, discussed next, and ocean currents. The North Atlantic Drift, a warm surface water current moving from the Gulf of Mexico and Caribbean Sea to the western coast of Europe, is evident from the $\delta^{18}O$ contours drawn northward off-shore of the British Isles and Scandinavia. Similarly, contours off the

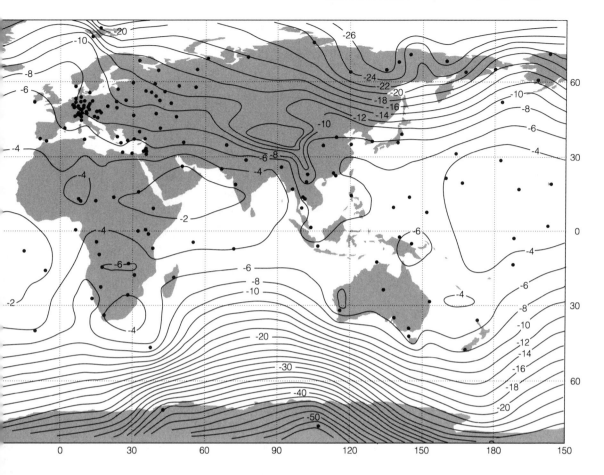

eastern coast of South America are drawn southward by a warm ocean current moving poleward along the western Atlantic.

It is interesting that no continental gradient is observed over the Amazon Basin, although the dominant weather regimes move from the Atlantic towards the high mountains of the Andes. Little if any Pacific rain enters the basin. Recycling of water vapour in the Amazon Basin by evapotranspiration in the tropical forests is responsible for these very flat gradients (Salati et al, 1979). A consequence of deforestation might be that this recycling is interrupted and that locally the daily rain distribution typical for a rain forest will be drastically altered.

Continental effects

Land masses have the effect of forcing rainout from vapour masses. As a vapour mass moves from its source region across a continent, its isotopic composition evolves more rapidly due to topographic effects and the temperature extremes that characterize continental climates. Continental stations are characterized by strong seasonal variations in T, which is a reflection of distance from moderating marine influences and latitude. Coastal precipitations are isotopically enriched, while the colder inner continental regions receive isotopically depleted precipitation with strong seasonal differences. Continentality k is expressed by Conrad's index, an empirical relationship relating the average annual range in temperature (ΔT in °C) and latitude angle φ (Barry and Chorley, 1987):

$$k = \frac{1.7\Delta T}{\sin(\varphi + 10)} - 14$$

The index provides a measure of the regional geographic parameters which influence rainout. For North America as an example, isopleths of continentality have a similar pattern and gradient to those of $\delta^{18}O$ in precipitation (Fig. 3-3). The contours for both follow continental margins, and the steepest gradients for both occur along the mountainous continental margins. The continentality index reaches extreme values in north-central North America where seasonal temperature contrasts are high, while the lowest $\delta^{18}O$ values are shifted westward due to seasonal biases in precipitation and influence of the Rocky Mountains of western Canada.

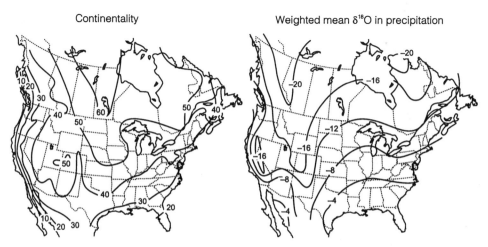

Fig. 3-3 Isopleths of continentality for North America calculated using Conrad's index (modified from Barry and Chorley, 1987) compared with the weighted mean-annual $\delta^{18}O$ composition of precipitation (IAEA GNIP data). The index of continentality is based on seasonal extremes in temperature and latitude. Patterns of $\delta^{18}O$ reflect continentality at this scale, although deviations occur in the north central regions due to seasonal biases in precipitation.

Continentality is observed along inland trajectories across continental margins. An example is $\delta^{18}O$ measured at stations in Europe from the Weathership station in the Atlantic ocean, through Valentia on southeastern Ireland, to Perm (Fig. 3-4). Vapour arriving from the Atlantic Ocean follows an eastward progressing rain-out trajectory over the low-relief European land mass. In this case, mean annual $\delta^{18}O$ follows a near-linear evolution of some 7‰.

By contrast, the evolution of $\delta^{18}O$ in precipitation is considerably more dramatic across a high-relief continental margin such as the Cordillera of western Canada (Fig. 3-5). Yonge et al. (1989) show the evolution for $\delta^{18}O$ in precipitation along a trajectory from the Pacific Ocean to the east of the continental divide over the Rocky Mountains. An exponential Rayleigh distillation is observed in precipitation to the west of the Coast Range Mountains, with a second distillation over the Rocky Mountains (Fig. 3-5). East of the divide, the $\delta^{18}O$ gradient becomes positive due to mixing with weather systems from the east. In this case, rainout is driven not only by the continental effect, but also by the effect of altitude, as discussed below.

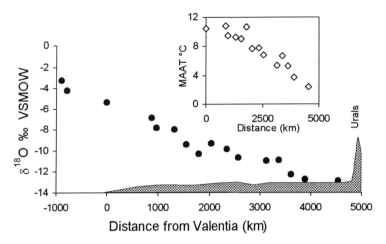

Fig. 3-4 The evolution of $\delta^{18}O$ in precipitation over the low-relief European continental margin from the Atlantic Weathership station, through Valencia on the southwest coast of Ireland to Perm near the Ural Mountains. Data are long-term average annual ^{18}O content of precipitation for European stations (modified from Rozanski et al., 1993). Inset shows the near-linear correlation of mean annual air temperature (MAAT) along this trajectory.

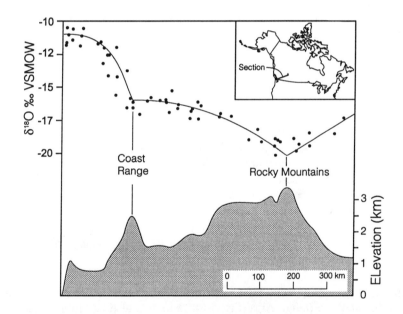

Fig. 3-5 The evolution of $\delta^{18}O$ in precipitation across the high-relief continental margin of the Canadian Cordillera, from the Pacific Ocean to the interior plains of Alberta (profile line shown on inset) (modified from Yonge et al., 1989). Decreasing T with distance and altitude along the trajectory drives rainout and depletion of ^{18}O.

Continentality can also impart significant departures in the global T–$\delta^{18}O$ relationship of $\delta^{18}O$ = 0.7 T – 13.6. For meteorological stations across Canada, Fritz et al. (1987a) find considerable variation in the T–$\delta^{18}O$ relationships for interior continental stations:

West Coast (Victoria)	$\delta^{18}O_{annual} = 0.18\, T_{annual} - 11.2$
Western Interior Canada	$\delta^{18}O_{annual} = 0.49\, T_{annual} - 17.3$
Eastern Canada	$\delta^{18}O_{annual} = 0.43\, T_{annual} - 13.6$

The generally lower slopes noted in these regional temperature relationships are typical for most continental mid-latitude stations for which data of this type have been collected. This appears true for both average annual and average monthly isotope and temperature values. For example, extensive sampling of local precipitation in northern Switzerland and the German Black Forest area produced the following relationships:

$$\delta^{18}O_{monthly} = 0.38\ T_{monthly} - 12.6$$
$$\delta^{2}H_{monthly} = 2.13\ T_{monthly} - 85.7$$

whereas in central Switzerland slopes are somewhat steeper:

$$\delta^{18}O_{monthly} = 0.56\ T_{monthly} - 14.6$$
$$\delta^{2}H_{monthly} = 3.72\ T_{monthly} - 102.7 \qquad \text{(Pearson et al. 1991)}$$

Local effects on T–δ^{18}O

Most studies of groundwater recharge rely on local rather than continental scale variations in the isotopic composition of precipitation. Variations in T and δ^{18}O imparted by the local physiographic setting, i.e. local topography, proximity to surface water bodies, seasonal changes, etc. can provide characteristics that are preserved in the groundwater and provide insights into recharge.

Altitude effect

In any region with even minor relief, orographic precipitation will occur as a vapour mass rises over the landscape and cools adiabatically (by expansion), thus driving rainout. At higher altitudes where the average temperatures are lower, precipitation will be isotopically depleted. For ^{18}O, the depletion varies between about –0.15 and –0.5‰ per 100-m rise in altitude, with a corresponding decrease of about –1 to –4‰ for ^{2}H. This altitude effect (also called the alpine or elevation effect) is useful in hydrogeological studies, as it distinguishes groundwaters recharged at high altitudes from those recharged at low altitude. The effect is observed even in watersheds with elevation contrasts of less than a few hundred metres, provided that sufficient data are collected to resolve seasonal effects.

One of the nicest examples is presented by Bortolami et al. (1979) for a catchment in the maritime piedmont of the Italian Alps (Fig. 3-6). Here, two distinct altitude-δ correlations were observed, each with almost identical gradients (~ –0.31‰ per 100 m rise), but differing slightly in their intercepts. The differences are seasonal: fall precipitation in this region originates over the Atlantic, whereas spring weather comes from the Mediterranean Sea. The meteoric water lines they calculate also reflect these seasonal patterns: for the October line, $\delta^{2}H = 8\ \delta^{18}O + 12$ (close to the GMWL) and April, $\delta^{2}H = 7.9\ \delta^{18}O + 13.4$ (closer to the EMWL for arid Mediterranean climates).

In a study of recharge to a geothermal system at Mount Meager, a Quaternary volcano in the Coast Range of western British Columbia, precipitation collected from 11 sites between 250 m and 3250 m altitude shows an altitude effect of –0.25‰ per 100-m rise (Clark et al., 1982) which provided evidence for the recharge environment of the thermal groundwaters (discussed in Chapter 9). Table 3-1 gives the altitude gradient found in a variety of locations.

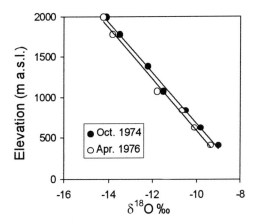

Fig. 3-6 The relationship between altitude and $\delta^{18}O$ in precipitation in Val Corsaglia, maritime piedmont of the Italian Alps (Bortolami, 1979). Samples were collected in October 1974 and April 1976, representing months of the fall and spring seasons with similar mean monthly temperatures. The mean gradient for these data is –0.31 ‰ $\delta^{18}O$ per 100-m rise.

Table 3-1 Range of values for the $\delta^{18}O$-altitude gradient in different studies

Site	Region	Altitude (m asl)	Gradient (‰ per 100 m) $\delta^{18}O$	δ^2H	Reference
Jura Mountains	Switzerland	500-1200	–0.2		Siegenthaler et al., 1983
Black Forest	Switzerland	250-1250	–0.19		Dubois and Flück, 1984
Mont Blanc	France	2000-5000	–0.5*	–4	Moser and Stichler, 1970
Coast Mountains	British Columbia	250-3250	–0.25		Clark et al., 1982
Piedmont	Western Italy	500-2000	–0.31	–2.5	Bortolami, 1979
Dhofar Monsoon	Southern Oman	0-800	–0.10		Clark, 1987
Saiq Plateau	Northern Oman	400-2000	–0.20		Stanger, 1986
Mount Cameroun	West Africa	0-4095	–0.155		Fontes et al., 1977

* Calculated from δ^2H using a slope of 8.

A very detailed analysis of altitude effects was undertaken on Mount Cameroun on the Atlantic coast of Western Equatorial Africa. There, J.-Ch. Fontes and co-workers monitored precipitation during a 4-year period at 20 stations between sea level and 4095 m (Fontes and Olivry, 1977). A rather low gradient of –0.155 ± 0.005‰ per 100-m rise was obtained, since the temperature gradient is not very steep. The isotopic evolution of the vapour reservoir and the resulting precipitation can be described by a modified Rayleigh process where only partial removal of the liquid phase from the vapour reservoir occurs. This allows in-cloud re-equilibration between the liquid and vapour. Gonfiantini (1996) showed that with increasing altitude an increasing amount of liquid is retained in the cloud, and by 4000 m a liquid:total-water ratio of 0.45 was reached. Such may also be the case in other situations and can at least partially explain deviations from a simple Rayleigh distillation.

Seasonal effects

The amplitude of seasonal variations in temperature increases with the continentality of the site. Greater seasonal extremes in temperature generate strong seasonal variation in isotopes of precipitation. These variations in $\delta^{18}O$ and δ^2H give us an important tool to determine rates of

72 Chapter 3 Precipitation

groundwater circulation, watershed response to precipitation, and the time during the year when most recharge occurs.

Rozanski et al. (1993) show seasonal correlations between T and $\delta^{18}O$ for a variety of stations (Fig. 3-7). The highly continental station at The Pas has a strong seasonal variation in both T and $\delta^{18}O$, whereas subtropical Addis Ababa has only minor variation in T and $\delta^{18}O$. The southern hemisphere station at Stanley in the Falkland Islands shows a clear but low-amplitude seasonal variation in $\delta^{18}O$–T, subdued by its maritime setting. The effect of latitude on seasonal variations in isotopes is apparent for average monthly $\delta^{18}O$ data for stations along a north-south transect through North America (Fig. 3-8).

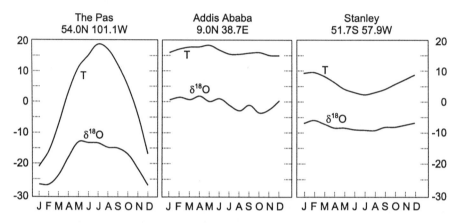

Fig. 3-7 The seasonal variations of ^{18}O and T (long-term averages) for a continental station in the northern hemisphere (The Pas), the equatorial region (Addis Ababa) and a southern hemisphere maritime station (Stanley on the Falkland Islands) (modified from Rozanski et al., 1993).

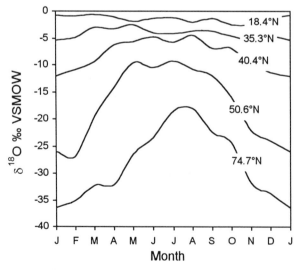

Fig. 3-8 The seasonal variation in $\delta^{18}O$ in precipitation at stations from low to high latitude in North America. Data are for monthly averages at San Juan, Puerto Rico (18.4°N), Cape Hatteras, North Carolina (35.3°N), Coshocton, Ohio (40.4°N), Gimli, Manitoba (50.6°N), and Resolute, NWT (74.7°N).

Unlike the continental stations, the seasonal variation $\delta^{18}O$ for tropical marine stations correlates poorly with temperature, owing to the strong seasonality of monsoon precipitation. Monsoon rains occur along continental margins in low latitude regions due to the landward displacement the inter-tropical convergence zone (ITCZ). This is the equatorial zone along which southerly and northerly trade winds converge and rise. During the northern hemisphere summer, the ITCZ migrates inland in west Africa, the Indian subcontinent and parts of Asia and moist southern trades bring up to 2000 to 4000 mm of rain over the summer months. Here seasonal ΔT is low and so there is less variation in $\delta^{18}O$ and δ^2H. There is also a stronger influence of Dansgaard's amount effect (Chapter 2) in these regions where precipitation during the low-rainfall months experiences evaporation in the low humidity air column.

The influence of the amount effect on the $T-\delta^{18}O$ relationship can be observed for the monsoon stations of Taguac (Guam) and New Delhi (Fig. 3-9). During the dry months, the amount effect enriches ^{18}O and imparts a negative $T-\delta^{18}O$ correlation. Clearly evident for New Delhi are the ^{18}O-enriched light rains during the cooler winter months compared to the summer monsoon rains.

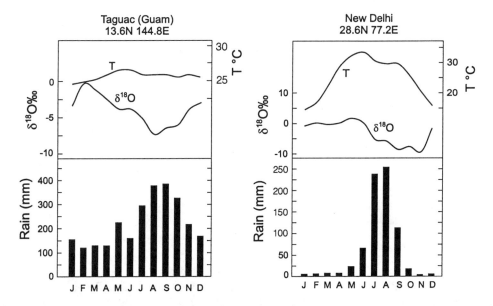

Fig. 3-9 Seasonal variations in $\delta^{18}O$, temperature and the amount of precipitation for a tropical marine station (Guam) and a monsoon station (New Delhi). $\delta^{18}O$ is enriched in seasons with less rain due to evaporation of drops during rainfall. Precipitation during humid high-rainfall months are not affected by this amount effect (from Rozanski et al., 1993).

Long-term, non-seasonal variations in temperature (decadal drifts in climate patterns) also affect the stable isotope contents of precipitation. Direct monitoring of such trends over the past 30 years shows strong $T-\delta^{18}O$ correlation but with amplitudes on the order of 1 to 2‰ and 1 to 2°C (Rozanski et al., 1993). From ice core records we know that temperature and $\delta^{18}O$ variations correlate well with changes in climate (see page 75).

Condensation of coastal fog

Precipitation may occur by condensation of fog on the landscape (occult precipitation), and can be an important source of recharge in some arid regions. It is an extreme example of coastal precipitation, where vapour from the ocean condenses directly onto vegetation as it rises over the coastal margin. The behaviour of isotopes during such a process is useful in tracing this important type of recharge (e.g. Ingraham and Matthews, 1990).

When condensation of a vapour mass directly follows primary evaporation, with little mixing in the upper troposphere, minimal evolution of the primary isotope signature occurs (i.e. fractionation effects during evaporation and condensation virtually cancel one another). As a result, the ^{2}H and ^{18}O in such precipitation is close to seawater. This is the case for the "occult" precipitation in the coastal mountains of southern Oman (Clark et al., 1987). Here, precipitation has an isotopic signature very close to that of the original seawater (Fig. 3-10), reflecting the first rain in closed evaporation-condensation cycle.

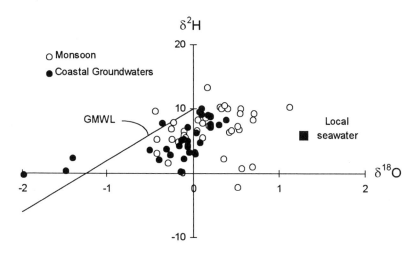

Fig. 3-10 Isotopic composition of monsoon precipitation (occult) in Southern Oman. Local groundwaters have almost the same composition, indicating their recharge source.

In the Cordillera de la Costa region of northern Chile and southern Peru, annual rainfall is extremely low to nil. Only condensation of the dense clouds that meet the landscape at an altitude of about 600 m sustains the vegetation. Below and above this altitude, virtually no indigenous vegetation exists. Aravena et al. (1989) investigated this "Camanchaca" or fog precipitation in central Chile and compared it with local groundwater recharge (Fig. 3-11). From the isotopic composition of transpiration water extracted from Eucalyptus trees, it is clearly the Camanchaca that sustains the vegetation. Groundwaters, on the other hand, are recharged not by the Camanchaca but by the very infrequent rains. They were found to be tritium-free, and so are older than several decades (tritium dating is discussed in Chapter 7).

Kinetic effects of secondary evaporation

Most meteoric and subsurface processes shift the $\delta^{18}O$–δ^2H signature of waters to a position below the local meteoric water line. It is rare to find precipitation or groundwater that plots above the line, i.e. showing a deuterium excess or ^{18}O depletion. However, in low-humidity regions, re-evaporation of precipitation from local surface waters creates vapour masses with isotopic contents that plot above the local meteoric water line. If such vapour is re-condensed in any significant quantity before mixing with the larger tropospheric reservoir, the resulting water will also plot above the LMWL, along a condensation line with slope 8.

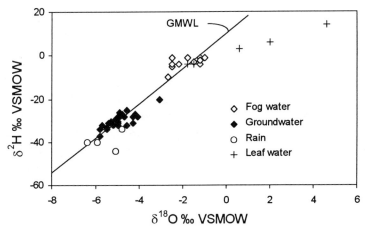

Fig. 3-11 $\delta^{18}O$ vs. δ^2H diagram for fog, rain, groundwaters and water extracted from leaves, in El Tofo, Chile (modified from Aravena et al., 1989).

There are only a few meteorological systems that cause such shifts. Ingraham and Matthews (1988) show the effect for mountain fog in northern Kenya (Fig. 3-12A). Here, vapour evaporated from the hydrologically closed Chalbi desert basin rises into surrounding mountains where it condenses on local vegetation. In an Arctic environment, Lauriol and Clark (1993) show nonequilibrium evaporation from local surface waters as the vapour source for annual ice formations in Arctic caves (Fig. 3-12B). Condensation of this kinetically depleted vapour on the cold cave walls forms water and ice that also plot above the LMWL.

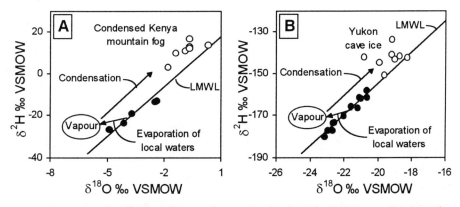

Fig. 3-12 Secondary evaporation effects causing meteoric waters to plot above the LMWL. A — Re-evaporation of local groundwaters in Kenya. (modified from Ingraham and Matthews, 1988). B — Evaporation of local surface waters in the northern Yukon condensing as cave ice in fossil karst terrain. In both cases, the condensed phase is in equilibrium with the vapour (equilibrium fractionation during condensation) and so plots on a line with slope ~ 8.

Ice Cores and Paleotemperature

The T–$\delta^{18}O$ correlation observed in precipitation is not only recognized in groundwater studies, but is also preserved in glacier ice. In many ice sheets, the input variations in isotope contents can be recognized in isotopically distinct layers where enriched summer layers alternate with isotopically depleted winter horizons. This stratification can be preserved for thousands of years. Inter-annual climatic changes are also recorded in great detail, and so the $\delta^{18}O$ in ice cores is a proxy for temperature that provides a record of climate change.

The isotope stratigraphy of the major ice caps in the northern and southern hemispheres has been examined by various groups in an effort to reconstruct Holocene and late Pleistocene climate. The variations in $\delta^{18}O$ for the ice core records in Fig. 3-13 document the changes in temperature as climate deteriorated from the last interglacial into the recent glaciation. Milder interstadial periods are evident during the Wisconsin Glaciation. Temperatures rise rapidly following the last glacial maximum at 21.5 ka and global climate enters the Holocene Interglacial period. Altitude effects on temperature are evident in comparing the Agassiz record, (similar latitude but low altitude) to the higher altitude and more depleted $\delta^{18}O$ record from Camp Century. Climate at Vostok has consistently been the coldest, due to both its high latitude and high altitude.

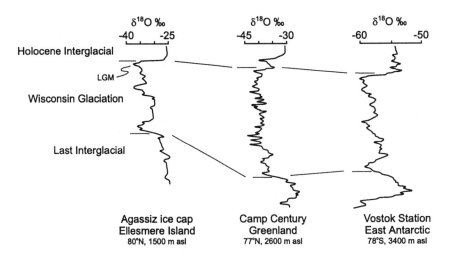

Fig. 3-13 $\delta^{18}O$ values for ice cores from the eastern Canadian Arctic (Agassiz Ice Cap), Greenland (Camp Century) and Antarctic (Vostok) (modified from Koerner, 1989). Vertical scale is depth in ice core. Sections include the late Pleistocene from last (Sangamon) interglacial *ca* 125 ka to the early Holocene *ca* 10 ka.

Most recently, the European Greenland Ice Core Program (GRIP) and American Greenland Ice Sheet Program (GISP) have collaborated in a detailed examination of the record from the Greenland Summit (Dansgaard et al., 1993). The high resolution from these two records suggest rapid and extreme climatic fluctuations characterized the last interglacial maximum (*ca.* 125 ka). Changes in temperature of 5 to 10°C may have occurred over only a few decades due to reorganizations of atmospheric-oceanic circulation patterns. This contrasts with the Holocene climate, which has been remarkably stable up to present, and gives reason to think that it may not always be so stable. Paleoclimate and ice core data sets are available at the NOAA website <**http://ncdc.noaa.gov**>.

The climatic shift from Pleistocene glacial to the Holocene interglacial is recorded not only by old ice, but also by the $\delta^{18}O$ in old groundwaters. Identification of "Pleistocene" recharge to aquifers serves to constrain groundwater age and recharge. This is a valuable paleo-hydrological tool, and will be discussed in Chapter 8.

Problems

1. On the global $\delta^{18}O$ map in Fig. 3-3, account for the equatorial belt of very flat gradients. Identify regions where steeper gradients provide examples of latitude, continental, and altitude effects.

2. We saw earlier that the T–$\delta^{18}O$ relationship varies regionally. Various regression lines for IAEA precipitation data sets are shown in Fig. 3-1 (right). How does the slope for these lines change with latitude? How does this relate to rain-out and Rayleigh distillation?

3. The mean annual $\delta^{18}O$ in precipitation for a given station decreases with decreasing average annual temperature. What changes in $\delta^{18}O$ would you expect in Ontario (Eastern Canada) for a drop in mean annual temperature of 5°C? What about a 3°C increase?

4. Use the results of your regression equation for Greenland and for Antarctic data from Dansgaard (1964) to calculate a mean temperature for the last glacial maximum (LGM) at the end of the Pleistocene, and the early Holocene at the Camp Century site and at Vostok Station. What was the change in mean annual temperature for each site during this period of climate warming?

5. Rozanski et al. (1993) showed that the T–$\delta^{18}O$ varies globally (Fig. 3-1). Other T–$\delta^{18}O$ relationships are given earlier in this chapter. Use these relationships to calculate an altitude effect for these regions (hint: you will need to use the lapse rate discussed in Chapter 2). How do your calculated altitude effects correlate with those measured for various areas (Table 3-1)?

6. What is the maximum resolution that one could determine for the recharge altitude of groundwaters, assuming an altitude effect of –0.25‰/100m, and that the standard deviation for a series of measurements of $\delta^{18}O$ in groundwaters is ± 0.1‰?

7. From the precipitation data used in problem 4 of Chapter 2, determine the altitude effect for northern Jordan. If there is a correlation, what would be the maximum resolution that one could determine for the recharge altitude of groundwaters in this region? Assume that the standard deviation for repeat measurements of $\delta^{18}O$ in groundwaters is 0.25‰, or calculate the actual standard deviation from the data.

8. Download the precipitation data sets for four weather stations found on the EIH web site under Chapter 2 <www.science.uottawa.ca/~eih/chapter3/problem3>, or transcribe the single-year data provided here onto a spreadsheet. These are mean monthly values. Recalling some of the parameters of meteoric waters from Chapter 2 (LMWL, s, d and h), make the following interpretations.

    • Determine the meteoric water line for each station, and the mean monthly values for deuterium excess d. From these parameters, determine the average humidity in the source region for the precipitation at each station, and any effects of secondary evaporation (post-condensation evaporation).

    • Repeat this calculation for Eureka, dividing the data into two sets: Jun–Oct and Nov–Apr. Comment on the seasonal differences in the source of water vapour at this station.

- Are the Waco 1976 data biased by an amount effect, and if so, how does it affect the LMWL? What about the precipitation data for Ottawa?

- Calculate the annual $\delta^{18}O$ value and monthly T–$\delta^{18}O$ correlation for each station. Which stations have the strongest correlations (based on r^2 value).

- Calculate a value for continentality (using Conrad's index) and compare this to the mean $\delta^{18}O$ value, and to the weighted mean $\delta^{18}O$ value at each station. How do these four stations compare (or contrast)?

BERMUDA
32.37N/64.68W, 6 m a.s.l.

month	$\delta^{18}O$	$\delta^{2}H$	T°C	precip (mm)
Jan–63	–3.1	–9	17.2	40
Feb–63	–2.7	–16	17.2	20
Mar–63	0.9	–3	17.7	90
Apr–63	–4.8	–26	17.5	50
May–63	–2.6	–12	21.3	30
Jun–63	–2.7	–12	24.4	70
Jul–63	–3.5	–18	26.9	20
Aug–63	–1.3	–3	27.1	70
Sep–63	–5.4	–35	26.1	5
Oct–63	–5.5	–34	23.3	94
Nov–63	–5.5	–37	20.6	60
Dec–63	–3.7	–16	17.2	15

OTTAWA, ONTARIO, CANADA
45.32N/75.67W, 114 m a.s.l.

month	$\delta^{18}O$	$\delta^{2}H$	T°C	precip (mm)
Jan–88	–15.4	–112	–9	37
Feb–88	–15.6	–114	–9.3	80
Mar–88	–11.3	–78	–3.3	27
Apr–88	–11.6	–82	6	92
May–88	–5.7	–44	14.9	32
Jun–88	–7.3	–50	17.6	94
Jul–88	–9.0	–63	22.7	78
Aug–88	–7.8	–54	20.3	21
Sep–88	–8.2	–61	14.1	68
Oct–88	–13.4	–99	5.9	13
Nov–88	–12.5	–90	2.7	83
Dec–88	–14.2	–95	–8.3	45

EUREKA, N.W.T., CANADA
80.00N/85.56W, 10 m a.s.l.

month	$\delta^{18}O$	$\delta^{2}H$	T°C	precip (mm)
Jan–89	–32.7	–263	–42.5	2
Feb–89	–34.9	–278	–36.6	14
Mar–89	–34.4	–260	–38.7	2
Apr–89	–31.7	–233	–26.3	2
May–89	–37.6	–277	–13.7	2
Jun–89	–22.1	–169	2.2	13
Jul–89	–16.4	–162	6.7	40
Aug–89	–21.0	–163	5.1	17
Sep–89	–26.7	–201	–5.4	11
Oct–89	–33.6	–256	–19.9	9
Nov–89	–34.9	–267	–28.3	13
Dec–89	–39.4	–305	–36.5	2

WACO, TEXAS, 1975
31.62N/97.22 W, 156 m a.s.l.

month	$\delta^{18}O$	$\delta^{2}H$	T°C	precip (mm)
Jan–75	–8.7	–57.1	9.2	36
Feb–75	–7.12	–47.1	8.3	75
Mar–75	–3.98	–15.9	12.4	28
Apr–75	–3.11	–7.5	17.8	16
May–75	–4.44	–23.9	22.2	48
Jun–75	–3.84	–20.8	26.5	72
Jul–75	–2.66	–8.6	27.8	70
Aug–75	0.32	10.8	29.2	18
Sep–75	–5.61	–27.3	25.1	58
Oct–75	–3.25	–8.2	21.4	61
Nov–75	–0.73	6	15.4	10
Dec–75	–7.78	–46.2	10.4	47

VICTORIA, CANADA
48.65N/123.43W, 20 m a.s.l.

month	$\delta^{18}O$	$\delta^{2}H$	T°C	precip (mm)
Jan–76	–11.3	–87	8	94
Feb–76	–11.9	–87	14	47
Mar–76	–9.2	–70	12	57
Apr–76	–10.9	–91	36	42
May–76	–8.2	–63	34	43
Jun–76	–6.3	–49	37	24
Jul–76	–7.9	–71	26	18
Aug–76	–9.4	–67	17	46
Sep–76	–6.7	–52	15	16
Oct–76	–9.3	–67	17	46
Nov–76	–5.8	–40	13	34
Dec–76	–7.6	–54	8	67

WACO, TEXAS, 1976
31.62N/97.22 W, 156 m a.s.l.

month	$\delta^{18}O$	$\delta^{2}H$	T°C	precip (mm)
Jan–76	–4.12	–24.3	8.2	44
Feb–76	0.79	13	15	8
Mar–76	–0.73	–2	15.8	39
Apr–76	–3.45	–22.9	19	66
May–76	–0.17	–5	21.5	27
Jun–76	–3.21	–23.4	26.9	82
Jul–76	–0.76	–11.9	27.9	88
Aug–76	–3.08	–17.8	29.8	6
Sep–76	–5.51	–27	25.5	44
Oct–76	–7.05	–43.9	15.8	32
Nov–76	–6.53	–38.5	10.1	17
Dec–76	–7.28	–45.2	7.5	64

Chapter 4
Groundwater

How is the meteoric ^{18}O–^{2}H signal transferred to groundwater during recharge? For many groundwaters, their isotopic composition will equal the mean weighted annual composition of precipitation. For others, important deviations from precipitation are found. This transfer function must be understood for groundwater provenance studies. It also sheds light on the mechanisms of recharge. Between the surface and the water table, infiltration must traverse the enigmatic "recharge environment" where groundwater infiltration is complicated by differing types of soils and vegetation, non-saturated flow through heterogeneities porosity, losses to evaporation and transpiration, seasonal variations in recharge, not to mention inter-annual and long-term changes in climate. The residence time within the recharge environment can vary from days to years.

Considering that only a small percentage of precipitation actually reaches the water table in most landscapes, the meteoric signal in groundwater can be significantly modified. Isotope variations in precipitation are attenuated and seasonal biases in recharge can be imparted to the newly formed groundwater. Evaporation and other processes can also modify the isotopic composition of groundwaters. This bifurcation of the hydrological cycle closes where groundwater discharges and rejoins surface runoff in streams and rivers. Environmental isotopes play a role in quantifying their relative contributions to stream flow following storm events, and in understanding the hydraulic functioning of a catchment area.

Recharge in Temperate Climates

In temperate climates, generally less than 5 to 25% of precipitation infiltrates to the water table. The rest is lost to runoff, evaporation from soils and transpiration by vegetation. Runoff has little effect on the isotopic composition of groundwater. Transpiration is also a nonfractionating process (Zimmermann et al., 1967; Förstel, 1982). Soil water taken up by roots is quantitatively released by the leaves and so no partitioning can occur. Evaporation, on the other hand, imparts a systematic enrichment on the isotope content of water.

Recharge rates are generally highest during spring runoff when soils are saturated, temperatures are low and vegetation is inactive. Recharge is minimal during summer when most precipitation is transpired back to the atmosphere. In the fall, recharge rates increase again as photosynthesis shuts down. Frost during the winter months precludes recharge. Given this seasonal bias in recharge and the small percentage of precipitation that becomes groundwater, it is surprising to find that groundwaters sampled below the water table generally have an isotopic value that is close to the weighted average of annual precipitation. This implies that that seasonal variations in isotopes are attenuated during movement through the unsaturated zone, which is an important zone of mixing. Students should review the mechanics of groundwater movement through the unsaturated and saturated zone in one of many hydrogeology texts (e.g. Freeze and Cherry, 1979; Domenico and Schwartz, 1990) as a basis for this and subsequent sections. Another important source of groundwater information is the National Ground Water Association (NGWA) <**www.h2o-ngwa.org**>.

Attenuation of seasonal variations

Infiltration through the unsaturated zone occurs via the porous matrix and/or through open fissures and other "fast" pathways. The geometry of this porous network and the degree of water saturation provide a continuum of slow to fast pathways for water to follow and along which mixing occurs. These differential flowpaths are relevant to the isotopic composition of water as it enters the saturated zone, since they are responsible for the smoothing out of seasonal variations in precipitation.

Darling and Bath (1988) describe isotopic variations in lysimeter seepages in the unsaturated zone of the English chalk over a 4-year period. At the lysimeter depth of about 5 m, seasonal variations have been attenuated to less than 5% of that observed in the precipitation (Fig. 4-1). The isotopic composition of the lysimeter waters is also very close to the long-term weighted average for rainfall and local groundwaters.

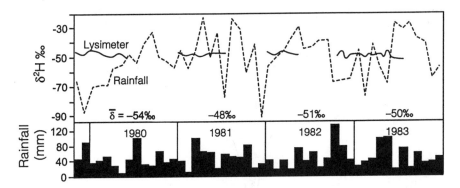

Fig. 4-1 Seasonal variations for δ^2H in precipitation and attenuation during movement to 5-m depth in unsaturated chalk, at Fleam Dyke, eastern England (modified from Darling and Bath, 1988).

Similar observations were made by Eichinger et al. (1984) who looked at the seepage velocity of water in unsaturated Quaternary gravel near Munich (Fig. 4-2). They note that at a depth of 9 m the amplitude of the variations of the precipitation input had been reduced to less than 10%.

Other studies of groundwaters in the Canadian Shield (Douglas, 1997, Fritz, unpublished data), in carbonate aquifers (e.g. Rank et al., 1992) or glacial sediments (Hamid et al., 1989) also show that seasonal variations in isotopes are largely attenuated above the water table, and that the shallow groundwaters closely represent mean annual precipitation.

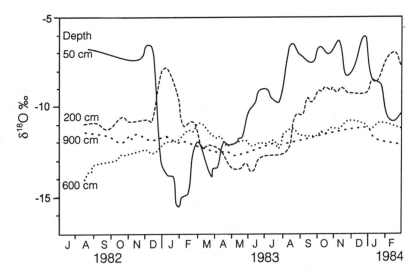

Fig. 4-2 Attenuation of seasonal $\delta^{18}O$ signal at various depths during infiltration through Quaternary gravels near Munich (from Eichinger et al., 1984).

The loss of seasonal variations during infiltration through the unsaturated zone is a function of the physical characteristics of the unsaturated zone, the length of the flowpath and residence time. A "critical depth" can be defined, where the isotopic variation is less than the 2σ error of the $\delta^{18}O$ analysis (Fig. 4-3). In a fine-grained soil with no fast flowpaths, the critical depth may be reached at 3 to 5 m (Zimmermann et al., 1967). In fractured rock, it may be tens of metres or greater. Yonge et al. (1985) note that seepage water entering caves after traversing a >10 to 15-m thick unsaturated zone in fractured limestone retains less than 5% of seasonal variations.

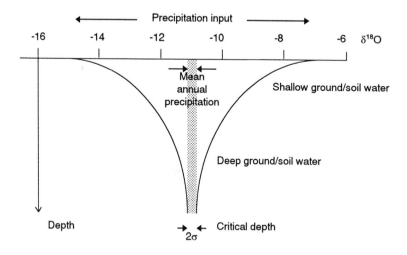

Fig. 4-3 Schematic of the attenuation of seasonal isotope variations (^{18}O or ^{2}H) in recharge waters during infiltration through the unsaturated zone and movement within the saturated zone, and the critical depth below which isotopic variability is less than the analytical precision (2σ).

The critical depth is often situated below the water table. In such aquifers, minor seasonal variations are preserved in shallow groundwaters. These too are eventually lost due to advective dispersion during flow under confined conditions. In advection-dominated systems with a single type of porosity, piston-flow models may be used to determine the critical depth. Where diffusion between fast and slow moving water occurs, exponential models that consider a continuum of pore and fissure sizes are required.

An example from the carbonate bedrock of eastern Ontario shows the attenuation of seasonal variations in ^{18}O in groundwaters (Fig. 4-4). Monitoring was carried out for groundwaters in the sand overburden, the unconfined bedrock outcrop area, and in the artesian aquifer where the bedrock is confined by marine clay (Velderman, 1993). Seasonal variations are increasingly damped in these successively deeper aquifers. The confined zone is beyond the critical depth, and no seasonal variation exists.

The isotopic variability below the critical depth and/or along a single flowpath in a confined aquifer generally does not exceed the 2σ analytical precision. When it does, this signifies preferential pathways (i.e. a bimodal hydraulic conductivity distribution such as channelling in a porous medium or a porous bedrock with fractures) or mixing of different recharge waters. In exceptional cases, variability may signify a change in climate. This is discussed in Chapter 8 in the context of paleogroundwaters. Variations in confined aquifers are usually due to groundwater mixing — not an uncommon process — and is expanded upon later in this chapter.

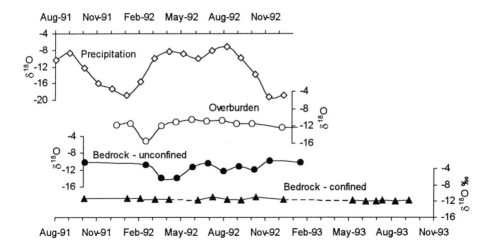

Fig. 4-4 Attenuation of seasonal $\delta^{18}O$ variations below the water table. Unconfined aquifers here retain seasonal variations, which are removed by advective mixing in the carbonate aquifer confined by clay (from Velderman, 1993).

Comparing shallow groundwaters with precipitation

Comparison of $\delta^{18}O$ values for groundwaters with precipitation on a regional scale shows the connection between groundwater and the meteoric input signal. In certain circumstances, the average isotopic composition of groundwater will differ from that of annual precipitation. This occurs when there are seasonal biases to recharge. Spring recharge is the most common example. Infiltration of snowmelt and cool (isotopically depleted) spring rains replenish aquifers when transpiration losses are low (no leaves on the vegetation) and water tables are high (steeper gradients). If recharge rates are greater than at other times of the year, then the groundwater will have slightly less ^{18}O and ^{2}H than the weighted mean for precipitation. Minor deviations from the weighted average in precipitation can also be attributed to land use practices. For example, Darling and Bath, (1988) note that "recharge beneath permanent grass cover is somewhat isotopically depleted relative to the arable plots," reflecting evaporative losses.

Climate may also be responsible for biases between weighted average annual precipitation and groundwater. Fig. 4-5 shows the distribution of ^{18}O in locally recharged shallow groundwater across Canada. The mean weighted isotopic composition of precipitation from stations within the Canadian Network for Isotopes in Precipitation (CNIP) are also shown. The map illustrates the regionally-averaged trajectories of vapour masses, showing the evolution of precipitation to low $\delta^{18}O$ values near the continental interior.

Precipitation and groundwater data correspond well in many areas. For the eastern region and the west coast, the weighted mean values for precipitation are close to that of local groundwaters. In these regions the $\delta^{18}O$ content of precipitation is largely retained by groundwaters. However, deviations of more than 3‰ are apparent in the dry Prairies. Precipitation at Edmonton, Wynyard, The Pas and Gimli is more enriched than local groundwater. This difference increases westward such that precipitation in Edmonton is very different that at Gimli (−17‰ vs. −14‰) despite the fact that their average annual temperatures are very similar (~3°C).

84   *Chapter 4 Groundwater*

This depletion in groundwater reflects a bias in the seasonality of recharge and the nature of summer precipitation. During cooler seasons from October to May, moisture arrives with major storms tracking mainly from the Gulf of Mexico that collide with southward-moving Arctic air (Fig 2-19). However, most precipitation occurs during the summer when shallow groundwater is recycled by evapotranspiration. The result is that mean weighted precipitation is biased towards summer rain while shallow groundwaters retain the cooler weather precipitation of the fall and spring when evapotranspiration is shut down and recharge occurs. The non-weighted averages for these stations are 1 to 3‰ lower.

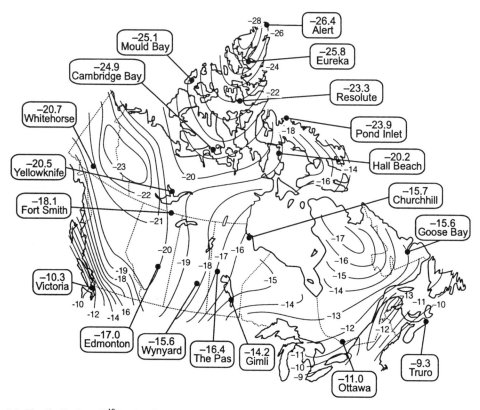

Fig. 4-5 The distribution of $\delta^{18}O$ values in locally recharged groundwater across Canada, and the weighted mean values for $\delta^{18}O$ in precipitation at monitoring stations (modified from Fritz et al., 1987a with additional regional groundwater data; Arctic precipitation data from Moorman et al., 1996).

The seasonal effects of evapotranspiration and frost on the seasonality of recharge in temperate regions cannot be underestimated. Summer precipitation contributes little to recharge due to evapotranspiration, and much of winter precipitation is stored as snow and lost during spring runoff. The isotopic composition of groundwater is largely biased towards the intermediate values of spring and fall rain, which will be close to the mean of annual precipitation.

Other processes that can be observed from the regional distribution of ^{18}O in groundwaters across Canada include:

1. The warm Pacific air masses rise over the western Cordillera where orographic rainout over the Coast Mountains produces a steep $\delta^{18}O$ gradient (Yonge et al., 1989) and a regional alpine effect.

2. A southward gradient extends across the northern Yukon from the north coast, indicating vapour originating from the Beaufort Sea during summer.

3. Low values are found in the vicinity of the high latitude Mackenzie Mountains along the Yukon - Northwest Territories border.

4. The gradient across eastern maritime Canada indicating vapour arriving from the Atlantic, with the contouring of isopleths around the Appalachian Mountains due to the local alpine effect.

Recharge by snowmelt

Snow melt imparts an isotopic depletion on groundwater recharge although its $\delta^{18}O$ is modified by the melting process. During storage and spring thaw, two main processes modify the stable isotope distribution in the melting snowpack. One is sublimation and vapour exchange within the snowpack. The other is exchange between the snow and meltwater as it infiltrates from the melting upper surface through to the base of the snowpack.

The isotopic enrichment of evaporating snow surfaces at below-zero temperatures (–10 °C) was investigated experimentally by Moser and Stichler (1975). Their data document a kinetic isotopic enrichment similar to that of evaporating water, which indicates mass exchange between the vapour and snow (or ice) crystals (Fig. 4-6). Unlike evaporation of water, the high humidity within the snowpack permits a greater degree of equilibrium exchange between solid and vapour. The slope of the "evaporation" trend in Fig. 4-6 (s = 5.75) is therefore steeper than for evaporation of water from an open surface which is generally closer to 5 (Fig. 2-8). Friedman et al. (1991) show the importance of diffusion along the temperature gradient between the base of the snowpack and exterior, and the role of soil moisture diffusing into the snowpack.

When melting of the snowpack occurs, isotope exchange between meltwater infiltrating on snow surfaces and the snow itself will also cause isotopic enrichments. Data for a snow surface in southern Ontario (Fritz, unpublished) showed that $\delta^{18}O$ evolved from about –23‰ to –8‰ with a slope of close to 8 on the $\delta^{18}O$–δ^2H diagram. Runoff from this snow pack had a $\delta^{18}O$ of ~ –11‰, indicating that snowmelt is a mixture of melt from the original snow and an isotopically enriched snow surface. Similar results were found in an experiment undertaken by Bùason (1972), who melted a snow column from the top and noted a continuous enrichment in deuterium in the meltwater draining from the bottom.

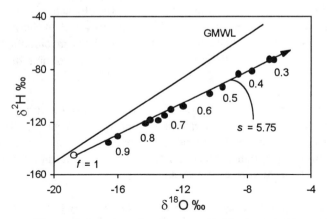

Fig. 4-6 Evolution of $\delta^{18}O$ and δ^2H in snow during evaporation under controlled conditions, with the fraction f of snow remaining during sublimation (Moser and Stichler, 1975).

In this process, continual exchange between the melt water and snow causes a Rayleigh-like enrichment on the meltwater. Using the equilibrium fractionation factors for ice-water exchange ($\varepsilon^{18}O_{i-w}$ = 2.8‰ and ε^2H_{i-w} = 20.6‰; Table 1-4), the theoretical trend can be calculated. Although the slope of this trend line is similar to that of sublimation (Fig. 4-6), the two processes are unrelated. Lauriol et al. (1995) use this characteristic signature of drainage from snowpacks to examine the formation of ice wedges in the northern Yukon. Note that the first water to exit from the snowpack is highly depleted and plots above the meteoric water line.

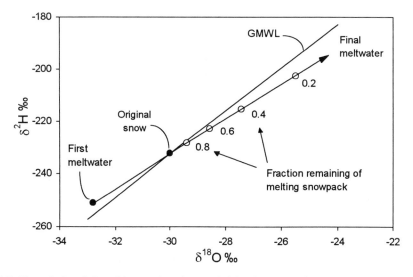

Fig. 4-7 Theoretical evolution of isotopes in meltwater draining from base of snowpack, based on complete equilibrium between the water and the snowpack.

Recharge in Arid Regions

Unlike temperate regions, the isotopic composition of groundwater in arid regions can be considerably modified from that of local precipitation. The cause is the strong isotopic enrichment in water during evaporation. Unlike transpiration, evaporation is a highly fractionating process. This complicates the use of isotopes as tracers of recharge origin, as well as the calculation of surface runoff vs. infiltration. However, the characteristic trends imparted by evaporation can be useful in understanding the mechanisms of recharge, as well as determining recharge rates, which can be as low or lower than 1% of precipitation. In Chapter 2, the "evaporation slope" for $\delta^{18}O$ and δ^2H was shown to be a function of humidity, and to vary between about 3 and 5.

Water can be lost by evaporation from surface waters during runoff prior to infiltration, from the unsaturated zone or from the water table. Evaporation during runoff and infiltration in arid landscapes is generally associated with groundwaters in alluvial aquifers along drainage networks (wadis or arroyos). For groundwaters recharged by direct infiltration through the soil or sand, evaporation from the unsaturated zone occurs.

Despite the strong evaporation in arid regions, it is possible that newly formed groundwater can have isotope contents close to the mean composition of precipitation. A case for this is made in a study of groundwater recharge in Central Africa (Mathieu and Bariac, 1996) where soil profiles show the typical evaporative isotope enrichments (described in the following section of this chapter) yet little to no enrichment is observed in the groundwater. In such cases, macropores and preferential flow channels in the unsaturated zone permit the fast movement of mobile water to the water table with very limited mixing with the enriched water found in the upper parts of the profile.

Evaporative enrichment in alluvial groundwaters

An example of evaporation during recharge of alluvial groundwaters is found in the Sultanate of Oman (Clark, 1987). Samples of precipitation and wadi runoff were collected on the rare occasions that it rained, including two summer storms and a major winter event (Fig. 4-8). A major difference is observed between summer rain (convective) and winter rain (major low pressure systems from the west). The strong evaporation observed in the summer recharge is due to runoff during flow over the hot landscape. The slope of these evaporative enrichments for individual storms is less than 5, typical for evaporation from open surfaces (Fig. 2-8). However, little evidence of evaporation is seen in the winter rains, which plot close to the local meteoric water line. This rain occurs under cooler conditions with greater intensity and duration, which limit evaporation. The high deuterium excess in the winter precipitation is typical of the Middle East (Table 2-1).

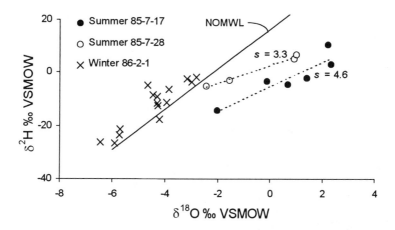

Fig. 4-8 The isotopic composition of wadi runoff for three rainfall events in northern Oman. The regression lines for the summer rains (slopes indicated) show strong evaporation trends at humidities less than 50%. The local water line for northern Oman (NOMWL) is defined as $\delta^2H = 7.5\ \delta^{18}O + 16.1$.

Groundwater in northern Oman reflects these differences in precipitation patterns (Fig. 4-9). Groundwaters in the carbonate bedrock aquifers show minimal evaporative enrichment, and so are dominated by recharge from winter rains. On the other hand, shallow groundwaters in the alluvial aquifers along the wadi channels have an evaporative enrichment, and indicate recharge by summer rain. The displacement of the data from the meteoric water line is a reflection of the evaporative loss, although these groundwaters are a mixture of recharge events. To calculate an average evaporative loss, the local "evaporation slope" must be known or assumed. This slope is a function of humidity, as well as temperature, wind speed etc., and can be estimated from pan evaporation experiments (discussed below).

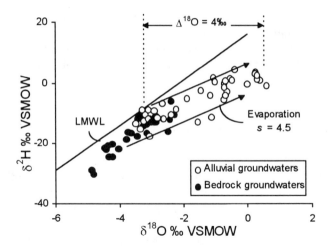

Fig. 4-9 Deep groundwaters from fractured carbonate aquifers and shallow alluvial groundwaters in northern Oman. Alluvial groundwaters have experienced greater evaporative enrichment. Also shown is the average evaporation slope ($s = 4.5$) for the region, with $h = 0.5$.

In this example, a slope of say 4.5 can be adopted from runoff data in Fig. 4-8, which gives a humidity $h = 0.5$ (Fig. 2.8). We can then determine the kinetic fractionation factors using Gonfiantini's equations in Chapter 2, giving $\Delta\varepsilon^{18}O_{v\text{-}bl} = -7.1‰$ and $\Delta\varepsilon^2H_{v\text{-}bl} = -6.3‰$. The overall enrichments ($\varepsilon_{total} = \varepsilon_{v\text{-}l} + \Delta\varepsilon_{v\text{-}bl}$) for evaporation under these conditions, for the mean annual temperature of 30°C (and $\varepsilon_{v\text{-}l}$ Table 1-4), are then $-16.0‰$ for ^{18}O and $-78‰$ for ^2H. The fractional water loss from evaporation can then be modelled according to a Rayleigh distillation. For $\delta^{18}O$, the evaporative enrichment is up to 4‰. According to:

$$\delta^{18}O_{gw} - \delta^{18}O_{prec} = \varepsilon^{18}O_{total} \cdot \ln f = 4‰$$

yielding a residual water fraction f of 0.78 and so an average evaporative loss of 22%.

Recharge by direct infiltration

In sand deserts or inter-channel zones in arid regions where runoff is negligible, evaporative losses may approach 100%. Evaporation from the unsaturated zone and water table is more commonly associated with aquifers recharged by direct infiltration rather than from the surface runoff network (wadis or arroyos), and often is characterized by a stronger evaporative enrichment with a lower slope. Additional data from Oman provide an example (Fig. 4-10). Groundwaters in the Ramlat A'Sharqiya, a 15,000-km² sand desert in northern Oman, are recharged by direct infiltration of rainfall traversing an unsaturated zone of up to 30 m thickness.

Fig. 4-10 shows that these groundwaters have experienced greater evaporative loss than groundwaters in adjacent alluvial aquifers. They are displaced further from the local meteoric water line. Why? During extensive evaporation from the unsaturated zone, kinetic effects by vapour diffusion are greater than those associated with evaporation from open surfaces. Residual soil moisture evolves along an evaporation slope that is shallower, and so displacement from the local meteoric water line is greater.

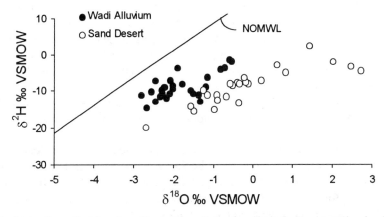

Fig. 4-10 Groundwaters from alluvial wadi aquifers and from the sand aquifer in the Ramlat A' Sharqiya desert in northern Oman.

In Chapter 2 we saw that evaporation from an open surface causes a nonequilibrium enrichment in the residual water. This is due to the difference in gaseous diffusion rates for ^{18}O and ^2H through the thin boundary layer of 100% humidity above the water surface. For open water surfaces this layer is microns thin. In the unsaturated zone it is much thicker and can dramatically increase kinetic evaporation effects. Dinçer et al. (1974) showed that soil moisture between 0.9 and 6 m depth in dune sands from Saudi Arabia was strongly evaporated. The slope of the $\delta^{18}O$–δ^2H relationship was only ~2 (Fig. 4-11), much lower than the range for evaporation from open water surfaces.

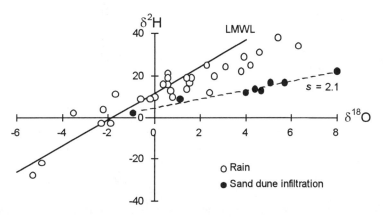

Fig. 4-11 Infiltration through sand dunes in Saudi Arabia (modified from Dinçer et al., 1974). The local meteoric water line is based on monitoring at Bahrain by the IAEA. The evaporation trend for infiltrating waters from the unsaturated zone originates from rains with $\delta^{18}O \sim -2‰$, which are more intense and unaffected by the amount effect. Light (enriched) rains contribute little to recharge.

Soil profiles and recharge rates

Column experiments by Allison (1982) provide an explanation for this enhanced evaporation effect. Using unsaturated soil columns being evaporated from the surface, he found that kinetic isotope effects are augmented as the soil moisture decreases (Fig. 4-12). After addition of water (rain), the soil surface begins to dry. With continued evaporation, the drying front moves down through the soil. As the tension head increases with water loss, water moves up through the soil column by capillary forces. However, at the drying front outward movement of water is by vapour diffusion. The zone of vapour diffusion is equivalent to the boundary layer in Fig. 2-7, in which humidity is 100%. As this zone widens, the kinetic effects of diffusion are augmented. Thus, the soil moisture being evaporated becomes isotopically enriched and the slope of the $\delta^{18}O$–δ^2H relationship for this soil moisture decreases (Fig. 4-11). This characteristic enrichment in the liquid phase moves to the saturated zone during periods of high infiltration. The observed shift away from the LMWL provides an estimate of the evaporative loss.

The maximum enrichment at the drying front creates a concentration gradient along the soil column. The accumulating $H_2^{18}O$ and ^2HHO begin to move downward by aqueous diffusion, giving a characteristic exponential profile (Fig. 4-12). The shape of this profile is controlled by the relative rates of isotopic diffusion downward and the capillary flow of water upward. The trend to low $\delta^{18}O$ near the top of the profile is due to exchange between the evaporation-depleted vapour and residual pore water above the zone of liquid flow. This exponential diffusion profile was first described by Zimmermann et al. in 1967. It has since been used to quantify recharge by direct infiltration under arid conditions (e.g. Aranyossy, 1992).

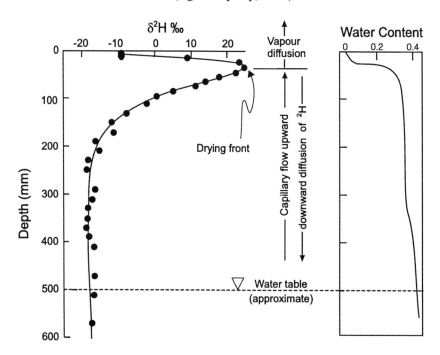

Fig. 4-12 Profile of ^2H in soil column undergoing evaporation. The depth of maximum enrichment marks the boundary between a lower zone of liquid flow and an upper zone of vapour movement by diffusion to the surface (modified from Allison et al., 1983).

The profile in Fig. 4-12 represents steady-state conditions with constant evaporation and flux from the water table. In reality, such profiles are generally disturbed by individual rainfall events that migrate down through the profile to recharge the groundwater. The isotopic composition of such recharge waters will then be a mixture of the pre-existing soil moisture and the rainfall itself. As this process repeats itself, the result is a groundwater with a range of isotopic compositions controlled by the isotopic variation in the rains and the loss of soil moisture by evaporation.

Allison et al. (1984) developed this concept, showing that groundwater recharged under such conditions of direct infiltration often have an $^{18}O-^2H$ composition that plots below but parallel to the local meteoric water line. Fig. 4-13 shows the mixing that occurs between the evaporated soil moisture and a subsequent rain that infiltrates and displaces the residual soil water downward. Ultimately, this mixed parcel of water will reach the water table. If recharge conditions remain relatively uniform over time, groundwaters should follow a line parallel to but displaced from the local meteoric water line.

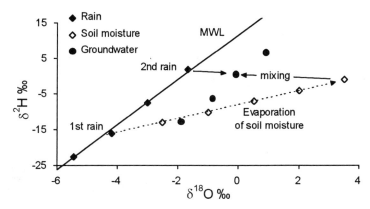

Fig. 4-13 Mixing of evaporated soil water and infiltrating rain in a soil profile. Significant rainfall events will displace the evaporated soil moisture below the influence of evaporation and produce a parcel of water which will eventually reach the water table with little subsequent modification (after Allison et al., 1983).

The displacement of groundwaters from the meteoric water line offers a crude estimate of recharge. For high rates of recharge, evaporative enrichment is minimal, whereas for low recharge rates, a large displacement for groundwaters will be seen. For recharge through thick sands, Allison et al. (1983) give the empirical relationships:

$$\delta^2H \text{ shift} = 22/ \sqrt{\text{recharge (mm / yr)}}$$

$$\delta^{18}O \text{ shift} = 3/ \sqrt{\text{recharge (mm / yr)}} \quad \text{(derived from the } \delta^2H \text{ equation)}$$

where the $\delta^{18}O$ and δ^2H shifts in Fig. 4-13 are on the order of 15‰ and 2‰, respectively, for the evaporated groundwaters, giving a recharge rate of less than 3 mm per year. Applying these relationships to the sand aquifer in Fig. 4-10 gives a recharge rate of 1.5 to 2 mm of rain per year.

In areas of extreme evaporation and net groundwater discharge, evaporative losses from the water table can also be estimated. Aranyossy et al. (1992) used experimental data to calibrate a model which determines evaporative losses from the water table as a function of unsaturated zone thickness. In eastern Niger, annual precipitation is <20 mm and infiltration to the water table is negligible. Isotopic enrichment of groundwater is due only to evaporative discharge. They use the isotopic enrichment profiles in the unsaturated zone to determine the rate of groundwater evaporation from dune sands. Evaporative losses are found to increase exponentially from less than a few mm per year for water table depths > 1.5 m to greater than 1000 mm/year when the water table is within 0.5 m below surface.

Estimating recharge with ^{36}Cl and chloride

Another approach to estimating rates of infiltration where evaporation is important is through mass balance using chloride or isotopic tracers. Increased concentrations of either in groundwater can give quantitative indications of rainfall lost to evaporation vs. infiltration to the water table. An essential parameter is the concentration of the tracer in rainfall. The radioactive isotope of chlorine, ^{36}Cl, can be helpful in such calculations, which are based on the simple mass balance:

$$C_{gw} = C_o/R$$

where R is the fraction of total annual precipitation that becomes groundwater, C_{gw} is the tracer concentration in groundwater, and C_o is the initial concentration in rain. No decay function is used if the tracer is conservative or has a long half-life (300,000 years for ^{36}Cl).

The use of ^{36}Cl is not without complications. Sources of ^{36}Cl in groundwater can originate from atmospheric production, production at the earth's surface (epigenic), thermonuclear weapons testing (anthropogenic) and subsurface (hypogenic) production. The reactions are described in Chapter 8. In studies of groundwater recharge under evaporative conditions, it is atmospheric and epigenic sources that are important to the input function. In semi-arid and arid regions, ^{36}Cl will accumulate in the unsaturated zone due to evaporation. Its concentration in groundwater can then be used to estimate evaporative losses prior to recharge.

Recharge calculations require an estimate of ^{36}Cl fallout, which is a function of latitude (see Chapter 8). An estimate of epigenic production is also necessary. Shallow groundwaters in temperate climates can be several times higher in ^{36}Cl than that attributable to atmospheric fallout, due to epigenic ^{36}Cl production by cosmic radiation on Ca or Cl at the surface (G. Milton, AECL, Chalk River, pers. comm.). Thus, the use of chloride and ^{36}Cl mass balances in recharge studies is limited by how well their input concentrations are defined.

Cook et al. (1994) compared rates of recharge through unsaturated sands using ^3H, ^{36}Cl and stable chloride. They show that reasonable estimates of recharge rates can be obtained by all three techniques although differences do exist. One additional problem identified by the authors are processes that take place in the active root zone where ^3H is lost by evapotranspiration while ^{36}Cl and chloride accumulate.

A comparative case study comes includes the interior-draining limestone aquifers of southern Oman and the unconfined sand aquifer of the A'Sharqiyah desert in eastern Oman. Chloride and ^{36}Cl data are given in Table 4-1. The average annual ^{36}Cl fallout in Oman is between 10 and 15 atoms $m^{-2}s^{-1}$ (Chapter 8) or about $3.7 \cdot 10^8$ atoms m^{-2} per year. Following marine thermonuclear tests in 1964, levels went up to 2000 atoms $m^{-2}s^{-1}$, or about $6 \cdot 10^{10}$ atoms m^{-2} per year.

Table 4-1 Chlorine concentrations and isotope ratios in groundwaters from Oman

Location	Aquifer	^3H (TU)	Cl^- (ppm)	$^{36}Cl/Cl$ ($\times 10^{-15}$)	$^{36}Cl/L$ ($\times 10^5$)
A'Sharqiyah	Desert sand - phreatic	0	1821	20 ± 4	6150
A'Sharqiyah	Desert sand - phreatic	0	1303	16 ± 4	3540
Najd	Carbonate - phreatic	4	138	59 ±10	1350
Najd	Carbonate - artesian	0	213	32 ± 4	1160
Najd	Carbonate - artesian	0	223	56 ± 6	2120

To calculate the average concentration in recharge waters, three evapotranspiration (ET) rates are considered for both pre- and post-test scenarios (Table 4-2). The two cases are made using: (a) the average values for natural (pre-nuclear) fallout, and, for case (b) assuming a component of fallout from thermonuclear weapons testing. One pre-thermonuclear case using 5× rainfall is presented for comparison.

Table 4-2 Calculated concentrations of ^{36}Cl in groundwaters in Oman for given rates of evaporative losses

Scenario	Rainfall mm	ET = 0% $^{36}Cl \cdot L^{-1} \cdot 10^5$	ET = 90% $^{36}Cl \cdot L^{-1} \cdot 10^5$	ET = 99% $^{36}Cl \cdot L^{-1} \cdot 10^5$
Pre-thermonuclear	100	40	400	4000
Pre-thermonuclear	500	8	80	800
Thermonuclear	100	6000	60,000	600,000

A'Sharqiyah — sand aquifer: This desert region receives less than about 100 mm of rain per year on a long-term average, but can have years with no significant rain. Recharge is by infiltration through about 30 m of unsaturated sand. Results from Table 4-1 show that the $6150 \cdot 10^5$ and $3540 \cdot 10^5$ atoms of ^{36}Cl per litre found in these samples from the sand desert, without a contribution of bomb ^{36}Cl (no ^3H in this groundwater), requires infiltration of only about 1% of rainfall, the rest being lost to evaporation. Additional ^{36}Cl from epigenic production would raise the concentration in recharge. Our calculated recharge rate would then be greater.

Chloride mass balance can verify these calculations. Using 20 ppm (aerosols from the ocean) measured in rainwater, for:

ET = 0%, we get 20 ppm in groundwater
ET = 90%, we get 200 ppm in groundwater
ET = 99%, we get 2000 ppm in groundwater

Thus, with the 1303 and 1821 ppm Cl measured in these groundwaters, the Cl⁻ mass balance is compatible with the ^{36}Cl ratio calculations showing an infiltration rate on the order of 1 to 3% of the rainfall and no significant recharge has reached the water table since *ca.* 1955. If thermonuclear ^{36}Cl were present in these waters, one would certainly expect to see ^3H from bomb testing as well (see Chapter 7). These calculations fit well with the estimate of 1 to 2 mm annual recharge determined above from the strong evaporation identified by $^{18}O-^2H$ data (Fig. 4-10). The high salinity in these groundwaters is due to this strong evaporative loss.

Carbonate aquifers of southern Oman: Two paleogroundwaters and a modern phreatic groundwater were sampled from carbonate strata of southern Oman. For the phreatic aquifer, the Cl⁻ mass balance gives an infiltration ratio of about 12% based on 10 ppm Cl⁻ in rainwater. The ^{36}Cl content of $1380 \cdot 10^5$ L^{-1} is high and suggests evaporation rates >90%. However, this groundwater contains tritium (^3H = 4 TU) and probably some thermonuclear ^{36}Cl. Taking this and surface production into account, infiltration estimates should be well above 10%. No evaporation is evident in the $^{18}O-^2H$ data (Clark et al., 1987). Clearly, the karstic terrain of this region allows rapid infiltration.

Calculating infiltration rates for paleogroundwaters is more complicated. Atmospheric and epigenic production rates in the past are not well known. Subsurface production for old groundwaters must also be considered, although this is more important in crystalline bedrock.

The paleogroundwaters of the Najd have ^{36}Cl contents much higher than expected from the calculated levels for pre-bomb infiltration. If increased rainfall is considered for these "pluvial" groundwaters, then even less ^{36}Cl would be expected. All that can be said for these waters is that atmospheric production rates and surface production rates in the past must have been higher. Certainly ^{14}C production during the late Pleistocene was over 20% greater than today.

Water loss by evaporation vs. transpiration

Infiltration of irrigation waters (return flow) can be an important source of groundwater recharge. For example, irrigation canals in the Indus Plain, Pakistan led to increases of up to 90 m in water table elevations. This led to chronic soil logging (soils saturated to surface) and soil salinization due to the high losses of water from evaporation and transpiration. The contrast in isotopic composition between the natural local groundwater and imported irrigation water can be used to determine the relative contribution of irrigation return flow to recharge and the relative importance of evaporation vs. transpiration in salinity build-up.

Another aspect that can be assessed with stable isotopes is the efficiency of the irrigation network, i.e. the relative loss of water through transpiration by vegetation vs. evaporation losses. While both cost water and contribute to salinization, evaporation can be controlled. Recall from above that the two mechanisms differ in their effects on stable isotopes. Evaporation losses can be from surface waters during irrigation and from the unsaturated zone by diffusion. The enrichment of ^{18}O and ^{2}H in return flow water will follow a slope of about 2.5 to 5.5, depending on the relative humidity. Transpiration also concentrates salts in the soil, but the residual water remains isotopically unchanged (see page 80).

A study of this nature was undertaken for shallow groundwaters (irrigation return flow) in the Nile Delta, where dissolved salt concentrations are between 2 and 10 times greater than that of the irrigation waters from the Nile River due to water loss (Simpson et al., 1987). The question was whether agricultural practices should be modified to reduce evaporative water loss. The authors take great pains in determining the relative losses from evaporation and transpiration. They do this on the basis of *pan evaporation* experiments, which merit some explanation.

In studies where evaporation is an important process in water loss and/or modification of the isotopic composition of water, it is important to understand the effects of humidity and to establish evaporation-enrichment relationships for ^{18}O and ^{2}H. Pan evaporation experiments involve monitoring the evolution of δ^{18}O and/or δ^2H in residual water in a pan left to evaporate at the study site (Welhan and Fritz, 1977). As local humidity, temperature and wind shear effects prevail, the results should be comparable to the local surface waters. Controversy exists as to how representative pan experiments results are, considering that wind and temperature effects, and even humidity can vary considerably in micro-environments and for different surface types. Nonetheless, they are often used for baseline considerations.

From the pan experiments in the Nile Delta study, it was determined that evaporation imparts for δ^2H an enrichment of 0.65‰ per 1% water loss (Fig. 4-14) and 0.185‰ increase in δ^{18}O per 1% water loss. Their pan evaporation experiments using Nile River water (δ^{18}O = 4.4‰, δ^2H = 28.3‰) determined the slope of the local evaporation trend line to be 3.8 (Fig. 4-15), which is consistent with evaporation under humidity conditions of 40%. The isotopic composition of irrigation drainage waters show only minor enrichment (Fig. 4-15). Both ^{2}H and chloride increase in the pan evaporation data, while in the groundwaters waters only chloride becomes more concentrated. This non-fractionating water loss can only be due to transpiration.

The minor evaporative loss can be calculated from the relationship:

$$\% \text{ evaporation} = \frac{\delta^2H_{gw} - \delta^2H_{Nile}}{0.65}$$

According to δ^2H, losses are only 5%, and using $\delta^{18}O$ the authors get 3%. Thus, in this case, transpiration by the crops is responsible for the majority of the salt build-up in the groundwaters, and altering irrigation practices to reduce evaporative losses will have little effect on the accumulation of salts.

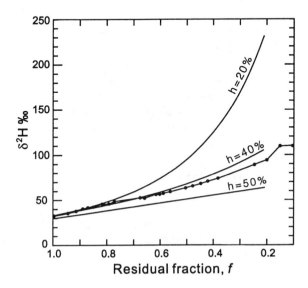

Fig. 4-14 Cairo evaporation pan data for δ^2H with turbulent boundary layer model calculations using $\delta^2H_v = 100‰$ and three cases for humidity: $h = 50\%$, 40% and 20% (Simpson et al., 1987). The low enrichment for the case of high humidity reflects the process of re-equilibration with the ambient vapour in the air.

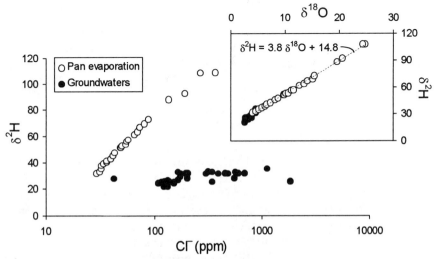

Fig. 4-15 δ^2H vs. Cl⁻ for Nile Delta agricultural drainage (groundwaters) and Cairo evaporation pan experiments. Inset plot shows the slope ($s = 3.8$) of the evaporation trend for the pan data (Simpson et al., 1987).

Recharge from River-Connected Aquifers

Infiltration from rivers or other surface water bodies is a critical component of recharge to many groundwaters, particularly where heavy extraction occurs. Recharge by infiltration from rivers occurs for many of the world's great alluvial aquifers such as along the Nile, the Tigris and Euphrates, or the Indus River in Pakistan. Water resource engineers rely on such river-connected aquifers to sustain supplies and for purification. One would hesitate to drink water from the Danube River, which drains one of the more heavily industrialized basins in Europe. But once treated by infiltration through an extensive alluvial aquifer, the Danube slakes the thirst of over 2 million Bratislavans.

The relative contributions of infiltrated river water and groundwater to river-connected aquifers are needed for water management studies. The potential contamination of river-connected aquifers by infrequent spills or leaks of contaminants is of great concern, particularly in industrialized basins. In the case of reservoirs and artificial recharge schemes, the efficiency of the recharge mechanism must be established. While the use of artificial tracers in water supply aquifers is at best controversial and often not possible, environmental isotopes (^{18}O, ^{2}H and ^{3}H) are natural, conservative tracers of surface water-groundwater dynamics.

Time series monitoring in a river-connected aquifer

The seasonal variation in stable isotopes in precipitation and runoff is highly attenuated in groundwaters. This contrast allows a separation of well water extracted from a river-connected aquifer into both the river water and groundwater components. The approach is based on the measurement of $\delta^{18}O$ and/or $\delta^{2}H$ in the river and well water on a monthly, weekly or even on a daily basis. The isotopic contrast needed for such a separation can also be provided by significant rainfalls.

This approach was applied to water supply wells located near the confluence of the Iller and Weihung rivers in Germany. The seasonal variations in the rivers (Fig. 4-16) were used to determine relative contributions to the wells and to derive aquifer characteristics. These were the basis for hydraulic dispersion modelling to determine velocity and dispersion of contaminants in the aquifer in the case of river pollution (Maloszewski et al., 1987). Monthly monitoring of the two wells showed H I and H II to be largely recharged from the Iller River, although H II showed strong seasonal contributions from the Weihung River. Dispersive model results provided contaminant concentrations and arrival times for given river pollution scenarios.

The Swiss tritium tracer "experiment"

An almost ideal, albeit accidental tracer experiment was initiated by an accidental spill of 500 Ci of tritium water (^{3}HHO) into a small river in Switzerland (Hoehn and Santschi, 1987). Environmental tritium as a tracer is discussed in Chapter 7. As a part of the water molecule, it is a very conservative tracer (with the exception of loss by decay over longer time periods). The tritiated river water passed an experimental site instrumented with piezometers to monitor contaminant movement from the river to the water supply aquifer. The site was close to a 2600 m^{3}/day production well located about 1000 m downstream and at a distance of 500 m from the river.

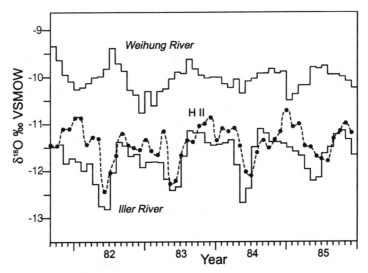

Fig. 4-16 Weekly variation in ^{18}O for the Iller and Weihung Rivers and from monthly samples of well H II during the period 1981-1985 and (B) modelling results (Maloszewski et al., 1987).

As the tritum plume passed, all piezometers and the production well responded. From the ^{3}H concentration profiles at the piezometers (all less than 30 m from the river) and the well (Fig. 4-17), the hydraulic parameters and average flow velocity for the two layers of this sand and gravel aquifer were derived. The early ^{3}H peak in the well resulted from rapid flow through the high permeability layer, whereas flow through the low permeability zone attenuated and slowed the arrival of ^{3}H. Similar studies could be undertaken using heavy water (^{2}H$_2$O), which unlike tritium, is benign and poses no water quality concerns.

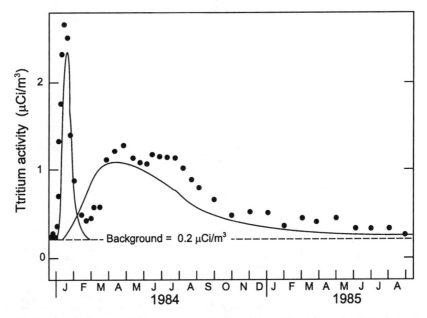

Fig. 4-17 Tritium breakthrough in the production well during an "accidental" tracer experiment in Switzerland. Lines represent a computer simulation using a two-layer model. The early peak shows tritium arrival through more permeable aquifer. The attenuated and delayed arrival peak indicates flow through the low permeability aquifer intersected by the well (from Hoehn and Santschi, 1987).

Water balance with ^{14}C

In studies where long-term monitoring of ^{18}O in river water and well water is not feasible, other species may be examined to quantify the relative contributions of local groundwater and induced river water infiltration in the water supply well. In one such case, the authors used radiocarbon. The Grand River in southern Ontario carries in its dissolved inorganic carbon (DIC) a modern ^{14}C signal that at the time of the study was close to 125 pmC. The local groundwater had $^{14}C_{DIC}$ values of about 70 pmC. The ^{14}C activity of the well water DIC was between 96 and 101 pmC. A simple mass balance shows that only about 50% of the water recovered in the wells was river water. This calculation assumes that the DIC of the river water was not modified during the short underground passage.

Recharge from the Nile River

A case study from the Nile River (El Bakri et al., 1992) provides an excellent demonstration of alluvial aquifer recharge from a surface water source. The alluvial aquifer in the Nile River valley in central Egypt is used for irrigation, and has a clear link with discharge in the Nile. Understanding the temporal and spatial variations in recharge of the alluvial aquifers by the Nile is important for aquifer management and for the control of salinity build-up due to irrigation.

In the early 1960s, the construction of major reservoirs in the upstream Nile changed the hydrologic regime, with significant losses due to evaporation. Fig. 4-18 shows two principal groups of groundwaters: Group 1 groundwaters are depleted and similar to the mean values of precipitation in the Nile headwaters (Addis Ababa), and Group 2 groundwaters plot with samples from the modern Nile. The mixing line is considerably steeper ($s = 6$) than evaporation lines for this region, which are closer to 3.5 (Simpson et al., 1987), and which rules out local evaporative enrichment related to irrigation as a salinization process.

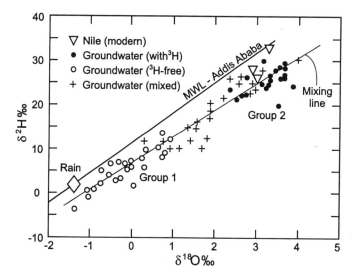

Fig. 4-18 The isotopic composition of groundwaters and surface waters from the alluvial aquifer along the Nile in central Egypt. Group 1 waters (open circles) are pre-Nile reservoir (pre-1953 with low ^3H) while Group 2 waters (filled circles) are more recently infiltrated. Group 2 waters experienced evaporation during storage in reservoirs (enriched in ^3H). Modern Nile waters have an enriched signature due to evaporation in reservoirs upstream. Meteoric water line is for Addis Ababa: $\delta^2H = 6.95\ \delta^{18}O + 11.51$, weighted mean of rain is $\delta^{18}O = -1.31$ and $\delta^2H = 1.8$ (from El Bakri et al., 1992).

The Group 1 waters have low ^{3}H contents and predate reservoirs on the Nile (early 1950s), whereas Group 2 waters are modern (elevated ^{3}H). Thus, Group 2 waters represent infiltration of modern Nile water, whereas Group 1 groundwaters were recharged in the 1950s or earlier, before the reservoirs on the upper Nile were constructed. The groundwaters falling between these end-members are mixtures between the two. The distribution of these two groundwater types reflects flowpaths and circulation times in this aquifer.

Recharge by desert dams

The construction of dams across dry wadi channels is gaining a lot of attention in the Middle East and Magreb countries. The idea is to capture storm runoff for infiltration to shallow aquifers. Underground storage of runoff water offers incredible benefits over surface water storage, with sharply reduced losses from evaporation (losses of >50% reduced to only a few percent) and protection from contamination and water-borne disease. Over 200 desert dams have been built in Saudi Arabia alone and several are now built or under construction in Oman and other arid countries. Stable isotopes offer a way to monitor the efficiency of recharge, and the relative contribution of a given runoff event to groundwater reserves. The input function is defined by the variations in the isotopic composition of discrete storm events. Evaporative losses are determined from the isotope shift that is measured in the groundwater downstream of the dam, according to calculations presented in Chapter 2 and the example on page 88.

Hydrograph Separation in Catchment Studies

The recharge, storage and discharge characteristics of a watershed is reflected by its behaviour during rainfall and base flow. Surface runoff resulting from storm events is of major concern to engineers responsible for the routing of water through channels, drainage systems and natural waterways. To assess the volume of runoff from a storm or continuous drainage from a given basin, empirical models have been developed that satisfy engineering needs but do not provide an actual understanding of the processes involved. Traditional models for storm runoff often assume that baseflow from groundwater to the river decreases during high water stages. Isotope techniques show that such is not the case and provide some basic insights into the runoff process.

In basin budget studies, it is important to assess the proportion of precipitation that actually recharges groundwater, i.e. the percentage of precipitation that actually recharges groundwater and that which is lost by surface runoff. The rate at which baseflow runoff changes over time is a measure of the actively circulating groundwater in the basin. Stream hydrograph separation into baseflow and storm runoff components is the basic tool in determining the components of discharge from a catchment.

Attempts were made to use artificial and natural tracers for hydrograph separation, including chemical data from groundwater and runoff but it soon became evident that none of the parameters analysed were conservative or could be assigned to a specific source. Environmental isotopes provided a breakthrough. One of the earliest applications of stable isotopes to define the baseflow component in runoff from a discrete watershed is described by Fritz et al. (1976), with various subsequent applications (e.g. Sklash et al., 1976, Rodhe, 1984; Turner, 1992). Buttle (1994) provides an excellent overview of the methodology and field applications.

The principle of hydrograph separation using stable isotopes is based on a contrast in the isotopic composition of the basin groundwater and that of a given storm. The groundwater in the basin will have an isotopic composition which reflects the long-term averaged input value whereas the storm will have a discrete δ-value falling somewhere within the range of the mean annual value (Fig. 4-19). If there is no contrast between the storm water and basin groundwater, then a hydrograph separation using isotopes is not possible. Other components that can be expected to vary with the ratio of baseflow to storm runoff will be dissolved solids, usually represented by electrical conductivity (E.C.) or chloride.

In the early studies, the separation was based on the assumption that only two major components participate in the runoff: the groundwater baseflow and rainfall. The actual storm runoff component has been viewed in the past as overland flow, or interflow (flow in the permeable upper few centimeters of the soil). More recently, it has been recognized that this approach is only approximately correct and that there may be more than one distinct subsurface component (Hinton et al., 1994).

The two-component separation of a storm hydrograph into storm runoff and prestorm baseflow can be described as:

$$Q_t = Q_{gw} + Q_r$$

where Q is a discharge component, and the subscripts represent total stream flow (t), prestorm groundwater (gw), and storm runoff (r). With sampling of the prestorm baseflow, the rain itself, and the total discharge during a storm, the isotope contents of these components can be established. Written then as an isotopic mass balance, it becomes:

$$Q_t \cdot \delta_t = Q_{gw} \cdot \delta_{gw} + Q_r \cdot \delta_r$$

and so by substitution for $Q_r = Q_t - Q_{gw}$ and rearranging:

$$Q_{gw} = Q_t \left(\frac{\delta_t - \delta_r}{\delta_{gw} - \delta_r} \right)$$

Assuming minimal spatial variability of the groundwater isotopic composition, this isotope mass balance will then be valid for any particular gauged point along a creek where the prestorm baseflow has been monitored for stable isotopes. Another parameter such as E.C. (electrical conductivity), Cl⁻ or Si, can be used as well. The requirement is that the parameter is different in the stormwater and groundwater components, and that it behaves conservatively. For example, Cl⁻ would not be conservative if evaporation salts had accumulated on the soil surface prior to the storm event. The separation of a two-component system on the basis of one parameter is illustrated in Fig. 4-19, based on the equation above.

Hinton et al. (1994) and Kendall et al. (1995) summarize the important runoff mechanisms and point out that the simple two-component model has its limits: the isotopic composition of rain, forest throughfall, meltwater, soil water and groundwater have variable and often differing compositions. However, for each additional component, another independent parameter will be required in the mass balance. A number of studies using the isotopic composition of dissolved constituents were undertaken with variable results since a) no source-specific compounds exist, b) the source terms are often poorly known (e.g. composition of throughfall under different climatic conditions) or c) the compounds are not conservative.

For example, it may be possible to recognize groundwater on the basis of its radon content yet radon is a gas which does move predictably through the unsaturated zone. Radiocarbon and ^{13}C may be useful to distinguish soil and groundwater yet it is virtually impossible to define the composition of DIC in different compartments. One could consider the isotopic composition of dissolved strontium compounds, yet their origin in the stream is not easily defined, especially where dry fallout plays a significant role.

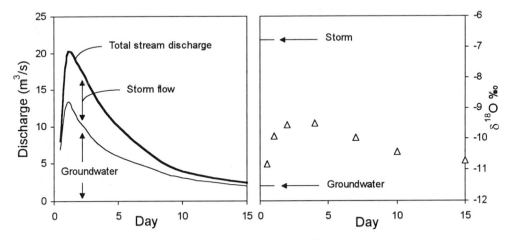

Fig. 4-19 Storm hydrograph separation for a two-component system using $\delta^{18}O$.

Not to despair, it is possible to use chemical parameters with caution. Hinton et al. (1994) used $\delta^{18}O$ and SiO_2 as complementary indicators and presented the results on a three-component diagram in which groundwater, soil water and new water (rain water) was recognized. The approach assumes that all SiO_2 comes from the soil or groundwater although dry fallout and washoff during a rain event can result in a substantial chemical load in the runoff. Nevertheless, the study demonstrates that multiple contributions to stream flow from a basin can be determined.

A three-component approach uses two parameters (for example, $\delta^{18}O$ and dissolved silica, represented as δ and Si) and three equations to resolve the three components of total discharge at a given time during and following the storm. In this case, total stream discharge = Q_t, rainstorm runoff = Q_r, soil water = Q_s and groundwater = Q_{gw}. The equations are based on the basic mass-balance:

$$Q_t = Q_r + Q_s + Q_{gw}$$

Through substitution of this equation into the chemical mass balance equations for the two parameters, δ and Si, we can determine two of the three components:

$$Q_r = \frac{Q_t(\delta_t - \delta_{gw}) - Q_s(\delta_s - \delta_{gw})}{\delta_r - \delta_{gw}}$$

and $$Q_{gw} = \frac{Q_t(Si_t - Si_r) - Q_s(Si_s - Si_r)}{Si_{gw} - Si_r}$$

These three equations can be solved iteratively for the three unknowns Q_r, Q_{gw} and Q_s. The simplest solution is to use a spread sheet and through trial and error or a "goal seek" utility, find a value for (in this case) Q_s that solves the initial Q mass-balance equation. The results of this solution are presented in the two-parameter diagram and the separated hydrograph in Fig. 4-20.

A statistical summary of the relative contribution of pre-event groundwater to streamflow was made by Buttle (1994) for field sites in North America, Europe, Western Australia and New Zealand (Table 4-3). From these data, it can be seen that it is pre-event groundwater, and not storm water that contributes the greater amount to stream flow during precipitation events.

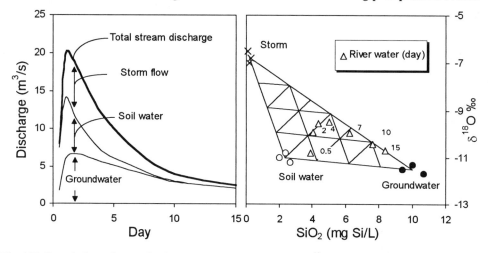

Fig. 4-20 Storm hydrograph separation for a three-component system using $\delta^{18}O$ and dissolved silica (Si).

Table 4-3 Summary of pre-event contribution to streamflow discharge for rainfall and snowmelt events (n) for various landuse drainage basins (from Buttle, 1994)

Basin type	Rainfall (n)	Snowmelt (n)
Forest	0.77 ± 0.17 (32)	0.58 ± 0.18 (32)
Agriculture	0.65 ± 0..02 (10)	0.80 (1)
Urban	0.02 ± 0.04 (3)	0.51 ± 0.28 (2)
Other	0.67 ± 0.24 (9)	0.38 ± 0.23 (3)

Example of the Big Otter Creek Basin, Ontario

Big Otter Creek (Sklash et al., 1976), situated on the north side of the Lake Erie catchment in southern Ontario, is 90 km long and drains about 700 km² of Quaternary deposits. These include sandy tills and glaciofluvial deposits in the upper reaches and a deltaic sand sequence for the remaining 75% of the catchment. Gauging stations in the upper, middle and lower portions of the drainage basin were monitored (discharge, E.C. and $\delta^{18}O$) over a pre-storm period, then during and following a major storm in May 1974. The results are shown in the hydrographs and the $\delta^{18}O$ and E.C. profiles in Fig. 4-21 and Fig. 4-22.

The hydrograph separations for these gauging points show that at peak discharge, the majority of the stream flow comes from pre-storm groundwater ($\delta^{18}O = -10.0‰$), with less from storm rain ($\delta^{18}O = -4.0‰$), and that the groundwater component for this particular catchment increases down-basin.

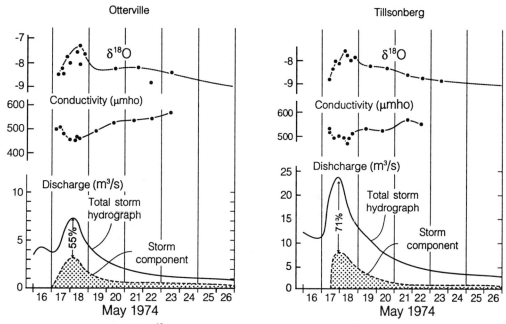

Fig. 4-21 The temporal variations in $\delta^{18}O$, E.C. and discharge at gauging stations on Big Otter Creek in southern Ontario, for a May 1974 storm event. Hydrograph separations are made on the basis of the Q–$\delta^{18}O$ relationship given above (modified from Sklash et al., 1976).

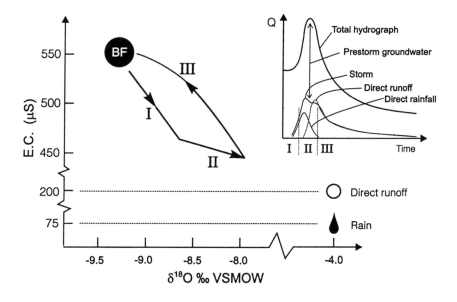

Fig. 4-22 Conceptual version of the $\delta^{18}O$ and E.C. response to rainfall in a watershed, using the example for Big Otter Creek during a May 1974 storm (modified from Sklash et al., 1976).

The relationship between E.C. and $\delta^{18}O$ for these monitoring points was used to generate a generalized three-component mixing model, shown in Fig. 4-22. Their model can be simplified to show three main discharge phases:

Phase I → Simple mixing between pre-storm baseflow (gw) and rain falling on the channel

104 *Chapter 4 Groundwater*

Phase II → Increasing influence of direct runoff (r) which has $\delta^{18}O$ of the rain, but higher E.C. due to flow over the ground from some distance from the channel.

Phase III → Decrease in the direct runoff/baseflow ratio and establishment of a new baseflow, which includes the influence of the storm recharge.

An example from Australia

Turner et al. (1992) have used deuterium and Cl⁻ instead of ^{18}O and E.C. to separate hydrographs from semi-arid catchments near Perth in western Australia. However, in contrast with the glaciofluvial outwash deposits in southern Ontario, the Wattle Retreat basin they describe is underlain by shales with a sandy clay overburden. This example is particularly useful in showing the limitations of solute-based separations as opposed to stable isotopes. In Fig. 4-23 the ^2H separation shows that during this rainfall event, about 75% of the runoff was rain and only 25% was from increased groundwater discharge, which is perhaps what one would expect for rainfall over clayey terrain.

The Cl⁻ separation, however, indicates that only about half of the total flow is attributable to rain. The discrepancy is due to chloride salts which can accumulate at the surface by evaporation under the semi-arid climate. While the ^2H in the soil water could also be affected by evaporation, this would not be incorporated in surface runoff.

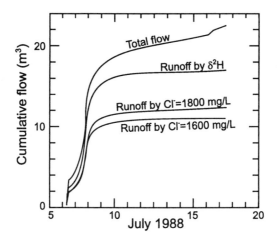

Fig. 4-23 Hydrograph separation using δ^2H and Cl⁻ for the Wattle Retreat catchment, Australia (from Turner et al., 1992). Figure shows surface flow contributions to total runoff.

Groundwater Mixing

Mixing between groundwaters occurs at various scales. We saw at the beginning of this chapter how hydrodynamic dispersion attenuates seasonal variations in groundwater by micro-scale mixing. Isotopic methods can be used to quantify groundwater mixing at the local to basinal scales, where mixing between groundwaters of different recharge origins, from different aquifers and different flow systems can occur.

Mixing of groundwaters can also occur in wells which have intake zones that integrate multiple flow paths and aquifers. Uncased boreholes, for example, can link two or more aquifer zones. Their discharge will then be an artificially mixed groundwater — a note of caution to those proposing convoluted recharge and mixing interpretations for groundwater samples from uncased boreholes.

Binary and ternary groundwater mixing

Mixing between two distinct groundwaters can be quantified by simple linear algebra using $\delta^{18}O$ or δ^2H. Fig. 4-24, for example, shows the proportion of groundwater A in a mixture of A and B. The proportion of mixing for a given sample will then relate directly to its position on the mixing line, according to:

$$\delta_{sample} = \chi \cdot \delta_A + (1-\chi)\delta_B$$

This simple mixing relationship does not indicate where the mixing occurs, which could be within the aquifer or within the well. Unlike most geochemical tracers, ^{18}O and ^2H are conservative in mixing relationships and so will preserve the mixing ratio. By contrast, most solutes are not always conservative due to geochemical reaction.

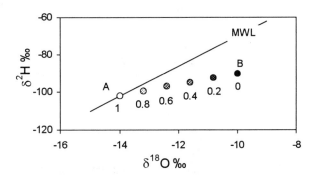

Fig. 4-24 The fractional mixing of two groundwaters quantified on the basis of their stable isotope contents, shown as the fraction of groundwater "A" in the "A-B" mixture.

A three-component mixing system was identified by Douglas (1997) for groundwaters from a crystalline setting in the northern Canadian Shield. In this study, mixing between a high salinity brine and modern meteoric water was demonstrated by the decrease in salinity over time at certain sampling sites. Shallow groundwaters have low salinity and a $\delta^{18}O$ composition that match with the mean of local precipitation (Fig. 4-25). The Shield brine is characterized by Cl⁻ contents of 170 g/L and an enriched $\delta^{18}O$ value of about –13.5‰ (see Chapter 9). However, additional sampling sites showed that a relatively low salinity, ^{18}O-depleted groundwater was present. Identification of this third end-member accounted for the displacement of most groundwaters from the simple brine-modern groundwater mixing line.

Mixing of groundwaters in regional flow systems

Unlike the attenuation of seasonal signals along a single discrete flowpath, mixing within regional groundwater flow systems has the effect of averaging the isotopic composition of various groundwaters from different recharge environments. The isotopic composition of deep or well-mixed groundwater in such cases converges on the mean weighted value of all recharge contributions. Such regional scale mixing can be observed in deep groundwater within a basin as well as in shallow but confined systems with a long subsurface flowpath. In such cases, stable isotopes are used as indicators of regional flow rather than to identify discrete areas of recharge.

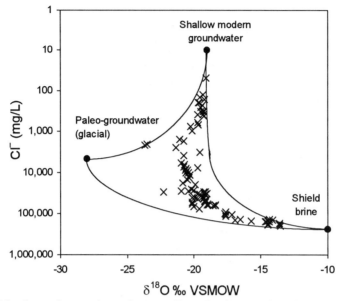

Fig. 4-25 Ternary mixing diagram for groundwaters from crystalline rocks of the Canadian Shield. The glacial meltwater end-member was identified by the isotopic depletion observed in many of the intermediate depth groundwaters, and its ^{18}O–Cl$^-$ composition determined by extrapolation to a ^{3}H-free water (Douglas, 1997).

An example comes from a study in the Altiplano of central Mexico where thermal groundwaters in a Tertiary volcanic basin are heavily exploited for irrigation and thermoelectric power generation (Fig. 4-26). The δ^{18}O for groundwaters from both the shallow unconfined aquifer and deeper (to 450 m depth) zones shows a dramatic decrease in variance with increased circulation depth as indicated by temperature. The δ^{18}O contents of the warmest (deepest) groundwaters converge on the mean value for the data set. The mean δ^{18}O value for the deep groundwaters (–9.8‰) falls within the standard deviation for the shallow groundwaters, signifying a common origin.

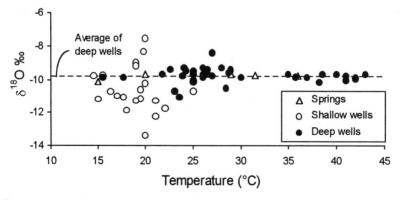

Fig. 4-26 δ^{18}O vs temperature for groundwaters, Villa de Reyes volcanic basin, Mexico. Variability in the shallow, cold groundwaters is attenuated by mixing at greater depths (higher T) (from Carrillo-Rivera et al., 1993).

Groundwater mixing in karst systems

Karst systems are an interesting example of groundwater mixing where a dual porosity generally exists. Connected porosity occurs in small-scale fissures and the porous matrix that provide much of the storage in the system and which contribute to baseflow. A second porosity can be identified in the high-velocity conduits through the system, which are most active during storm events. Stable isotopes can serve to distinguish flow through these two systems.

Rank et al. (1992) have measured the ^{18}O contents of local precipitation and the discharge from two karst springs (Fig. 4-27). Tritium contents in the groundwaters are used in a dispersion model to show that mean circulation times in the fissure/matrix porosity are on the order of 4.5 years (Wasseralmquelle spring) and 2.5 years (Siebenquellen spring). However, estimating the transit time and contribution to discharge for flow in the high-velocity conduits requires high-frequency sampling program. This can be provided by the seasonal variations in $\delta^{18}O$. Fig. 4-27 shows the 8-year $\delta^{18}O$ monitoring record for these two springs. The authors use these data in a tracer mass balance to examine the contribution of high-velocity conduit flow to discharge from the karst. They find that this flowpath has a mean transit time of 2 months.

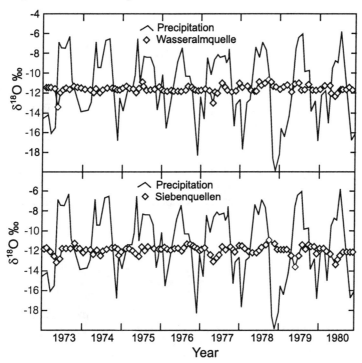

Fig. 4-27 Variations in ^{18}O for precipitation and discharge from the alpine karst in southeastern Austria, due to mixing within a dual porosity drainage network. Long-term circulation in the fissure/porous matrix network (4.5 years for Wasseralmquelle; 2.5 years for Siebenquellen) attenuates the precipitation-input signal. Rapid flow in conduits (mean circulation of ~ 2 months) imparts sharp deviations from the attenuated signal (from Rank et al., 1992).

Despite the attenuation of seasonal variations in these two groundwater systems, inputs from storm flow through channels and conduits can have a short-term but significant effect on isotope contents. This is typical of groundwaters in karst and other unconfined aquifers. As a groundwater enters confined conditions, it is isolated from further seasonal and storm contribution and its isotopic composition is attenuated to a value representing the weighted mean of meteoric water inputs.

Where variations occur, a principal component separation is possible. Such analyses (based on chemical and isotopic data) were undertaken for a perennial karst spring in Indiana, USA (Lakey and Krothe, 1996). Surprisingly, it turned out that only 20 to 25% of the discharge could be attributed to the storm event and that the bulk discharged from groundwater storage (fissure/porous matrix network). Typical phreatic and vadose storage may have contributed to discharge with the onset of the storm flow, while water from soil moisture and epikarst may have contributed at the onset of the discharge recession.

Problems

1. The following data were collected during a storm event on a river in central Kentucky in the summer of 1995:

Day	SiO_2	$\delta^{18}O$	Q_t (m^3/s)
0.1	4.30	−4.7	3
0.5	4.95	−4.4	10
1	5.63	−4.24	22
2	6.60	−4.28	15
3	7.83	−4.7	8
5	9.11	−5.1	4
10	11.2	−5.8	2
groundwater	12	−6	
rain	0.1	−2	
soil water	3.5	−4.8	

    Perform a hydrograph separation with these data using the single parameter method ($\delta^{18}O$) to determine the proportion of groundwater and runoff in discharge for each day. Now, using the two component method with $\delta^{18}O$ and Si, determine the percentage of groundwater, soil water and runoff in discharge. Compare the results for the peak discharge (t = 1 day). What is the discrepancy in estimation of the groundwater component of discharge between the two methods?

2. The following data were measured for groundwater in a sand aquifer beneath an agricultural field in a region that receives 460 mm of precipitation annually. Precipitation has an average of 8 mg/L Cl⁻. The enrichment of Cl⁻ in the groundwaters indicates some loss of recharge water during infiltration. Is this due to *evaporation* or *transpiration* from the soil by crops? Calculate the fraction of water lost, and the annual recharge rate (mm/yr). The local meteoric water line is $\delta^2H = 8.0\ \delta^{18}O +12$ ‰. Comment on the seasonality of any water loss.

$\delta^{18}O$	δ^2H	Cl⁻	$\delta^{18}O$	δ^2H	Cl⁻
−13.9	−92	11	−13.7	−95	8
−6.6	−40	45	−10.7	−74	42
−14.2	−108	11	−14.3	−99	9
−9.5	−61	45	−13.7	−99	7
−7.1	−48	46	−6.6	−38	53
−7.4	−46	52	−6.7	−42	52
−8.1	−52	44	−10.3	−74	41
−8.4	−58	48	−11.3	−82	22
−8.4	−53	44	−14.8	−105	8
−8.9	−52	38	−11.9	−83	26
−14.3	−101	12	−12.0	−89	22
−10.5	−70	41	−11.4	−85	30

3. Plot the following data (or download from <www.science.uottawa.ca/~eih>) for the stable isotope composition of groundwaters sampled from deep bedrock (BR) and shallow alluvial (AL) aquifers in a region with mean annual air temperature of 28°C and humidity of 40%.

Well ID	Cl⁻ (mg/L)	$\delta^{18}O$	δ^2H	Well ID	Cl⁻ (mg/L)	$\delta^{18}O$	δ^2H	Well ID	Cl⁻ (mg/L)	$\delta^{18}O$	δ^2H
AL12	770	−0.36	3.0	AL38	12470	0.44	3.5	BR5	484	−2.64	−5.3
AL13	495	0.01	4.6	AL39	574	−0.01	5.8	BR6	835	−3.28	−14.1
AL14	538	0.10	5.2	AL40	363	−7.60	2.6	BR7	1607	−3.18	−13.5
AL15	460	−0.31	5.1	AL41	363	−0.69	2.4	BR8	417	−3.14	−8.0
AL16	2450	−0.76	0.2	AL42	9620	0.07	1.9	BR9	468	−3.26	−11.7
AL17	11150	0.46	3.7	AL43	930	−1.66	−4.3	BR10	1365	−3.67	−16.6
AL18	4170	−0.55	1.2	AL44	406	−1.66	−8.0	BR11	546	−2.68	−8.1
AL19	335	−0.44	4.4	AL45	406	−1.42	−9.0	BR12	339	−1.91	−3.4
AL20	335	−0.40	4.0	AL46	380	−1.30	−4.5	BR13	252	−2.89	−7.1
AL21	6130	−0.20	2.8	AL47	380	−1.19	−4.1	BR14	259	−2.39	−5.8
AL22	337	−0.51	1.2	AL48	382	−1.00	−1.0	BR15	343	−2.42	−4.8
AL23	337	−0.46	1.1	AL49	382	−0.91	−0.9	BR16	304	−2.32	−2.7
AL24	404	−0.59	1.5	AL50	528	−1.45	−6.0	BR17	555	−3.80	−14.3
AL25	404	−0.54	1.4	AL51	644	−1.29	−4.1	BR18	343	−2.70	−7.6
AL26	577	−0.07	3.5	AL52	3620	−0.68	0.1	BR19	555	−2.22	−4.7
AL27	18540	1.01	5.1	AL53	14760	0.84	4.1	BR20	468	−3.41	−10.6
AL28	7840	−0.06	2.6	AL54	740	−0.70	−0.5	BR21	410	−3.45	−11.7
AL29	437	−0.68	0.8	AL55	620	−0.95	−3.3	BR22	1170	−3.36	−11.4
AL30	14050	0.70	4.4	AL56	11660	0.26	4.4	BR23	794	−2.62	−8.1
AL31	554	−0.11	3.0	AL57	5480	−0.26	1.3	BR24	302	−2.73	−5.7
AL32	374	−1.10	0.6	AL58	19680	1.26	6.1	BR25	306	−3.44	−14.0
AL33	374	−1.00	0.5					BR26	357	−3.76	−16.2
AL34	9000	0.17	3.4	BR1	257	−3.66	−12.4	BR27	264	−3.52	−14.7
AL35	4950	−0.68	2.2	BR2	293	−2.89	−9.5	BR28	293	−3.04	−11.6
AL36	650	−1.27	−2.8	BR3	295	−2.13	−4.3	BR29	519	−2.90	−12.7
AL37	523	−0.77	3.2	BR4	278	−3.09	−9.7	BR30	335	−2.26	−5.7

- Compare these data with the local meteoric water line ($\delta^2H = 7.6\,\delta^{18}O + 14‰$). and comment on the relative degree of evaporation that occurred during recharge for alluvial and bedrock groundwaters. What can you say about the isotopic composition of the recharge waters to both aquifers?

- Many of the alluvial aquifer groundwaters are situated in a coastal zone of heavy exploitation, and intrusion of seawater into this coastal aquifer is suspected. Using a combination of $\delta^{18}O$, δ^2H and Cl⁻ data, determine whether the observed chloride enrichments could be produced by evaporation, or by saline intrusion. In the case of seawater intrusion, determine the percent mixing with seawater ($\delta^{18}O = 1‰$, $\delta^2H = 5‰$ and Cl⁻ = 19500 mg/L).

- For the alluvial groundwaters that are unaffected by seawater intrusion, use a Rayleigh model and the T–h conditions given above to determine an average evaporative loss during recharge. Assuming the same T–h conditions, what percentage of precipitation was loss to evaporation during recharge of the bedrock groundwaters?

4. Sketch a schematic graph of $\delta^{18}O$ in runoff vs. time for the meltwater from a winter snowpack during spring melt. Assume a mean value of −20‰ for the snow prior to any melting. Revise your sketch for the condition that melting was largely due to spring rains, with $\delta^{18}O = -14‰$.

5. The following precipitation and groundwater data were collected from a heavily irrigated agricultural region. A buildup of salinity has been noted over the past decade which threatens the viability of certain crops. What is the average water loss experienced by these groundwaters? Can irrigation practices be improved to reduce evaporative losses?

Precipitation			Groundwater		
$\delta^{18}O$	$\delta^{2}H$	Cl^-	$\delta^{18}O$	$\delta^{2}H$	Cl^-
−1.6	−9	6	−4.2	−31	78
−3.8	−27	10	−4.5	−32	150
−4.2	−25	9	−4.2	−29	171
−5.6	−42	14	−4.9	−37	227
−2.6	−19	12	−5.2	−39	65
−3.8	−25	12	−4.5	−33	201
−1.9	−14	4	−3.9	−28	141
−1.7	−12	15	−4.8	−35	114
−5.3	−39	3	−4.1	−31	50
−3.4	−26	8	−4.1	−33	196
−4.7	−37	6			
−5.4	−31	9			
−6.6	−43	9			
−7.6	−56	14			
−1.1	−3.8	12			
−3.5	−25	9			
−2.7	−14	15			
−8.9	−63	3			

Chapter 5
Tracing the Carbon Cycle

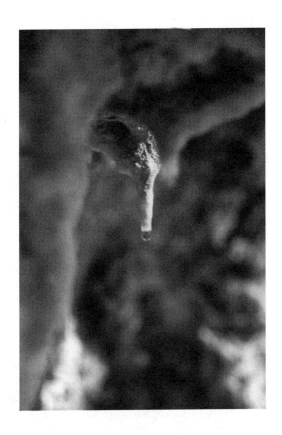

The carbon cycle in groundwaters begins with weathering reactions in the recharge area driven by CO_2 dissolved from the soil. Carbonate reactions dominate the geochemical evolution in shallow groundwaters, and so bicarbonate is generally the dominant anion in fresh water resources. The carbonate system evolves with subsequent organic and inorganic reactions in soils and aquifers. Microbiological activity plays a key role in the degradation of organic compounds and evolution of redox conditions. The chemistry and isotopes of carbon species provide insights into carbonate evolution and carbon cycling in groundwaters: insights required for an understanding of groundwater quality, fate of contaminants and for a correct interpretation of groundwater age. This review of the all-important carbon system provides a basis for the topics of groundwater quality, groundwater dating, and water-rock interaction, which are discussed in subsequent chapters.

Evolution of Carbon in Groundwaters

Fresh groundwaters invariably originate as meteoric waters, in most cases infiltrating through soils and into the geosphere. Along the way, dissolved inorganic carbon (DIC) is gained by dissolution of CO_2 and evolves through the weathering of carbonate and silicate parent material. As carbonate acidity is "consumed" by weathering, the pH rises and the distribution of dissolved inorganic carbonate species shifts towards bicarbonate (HCO_3^-) and carbonate (CO_3^{2-}). The groundwater generally approaches equilibrium with calcite, whose solubility will control pH and the equilibrium of carbonate species.

At the same time, labile organic matter from the soil can be dissolved. Oxidation of the dissolved organic matter (DOC) will take initially place by aerobic bacteria that use O_2. If the supply of DOC is exhausted before the O_2 is consumed, then redox conditions will not evolve much further unless another electron donor in the system is found (ferrous iron minerals, sulphides, etc.). If an excess of DOC accompanies the groundwater below the water table and beyond the influence of atmospheric O_2, then anaerobic bacteria will consume it using electron acceptors such as NO_3^-, Fe^{3+}-oxyhydroxides or SO_4^{2-}. Ultimately, methanogenic reactions can take place. Redox evolution is accompanied by mineral dissolution and precipitation reactions that affect the mass balance of dissolved solids and the distribution of isotopes.

All sources of carbon are linked through these acid-base and redox reactions, which are most often mediated by bacteria. Bacterial involvement is important for two reasons. They derive their energy from redox reactions (usually oxidation of organics), and so act as catalysts, speeding up reactions that are otherwise kinetically impeded. Secondly, bacteria are isotopically selective, preferring to break the weaker, light-isotope bonds. Bacterially mediated reactions are then accompanied by large isotope fractionations. The huge range for $\delta^{13}C$ in various carbon reservoirs is a demonstration of isotope selectivity by bacteria (Fig. 5-1). Biologically mediated redox reactions for carbon are kinetic, proceeding in one direction only. However, for relatively stable environmental conditions, fractionation is constrained to definable ranges. Understanding the distribution of isotopes in the carbon cycle begins with a look at carbonate geochemistry.

Carbonate Geochemistry

Activity, concentration and mineral solubility relationships

Carbonate geochemistry involves acid-base reactions and the determination of carbonate equilibria and state of mineral saturations. Calculations are based on low-temperature aqueous geochemistry, which is the subject of several textbooks. References include Garrels and Christ (1965), Freeze and Cherry (1979), Drever (1997), Domenico and Schwartz (1990) and Stumm and Morgan (1996). Here we review of the essential terminology and methods. The solved problems for this chapter serve as examples for those unfamiliar with aqueous geochemistry.

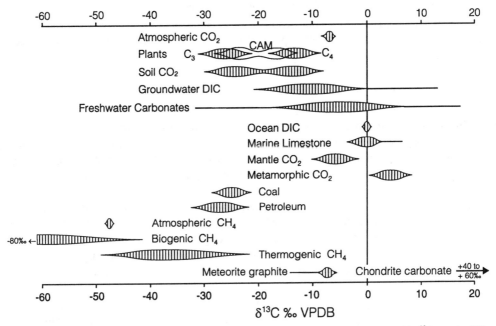

Fig. 5-1 Ranges for $\delta^{13}C$ values in selected natural compounds. Especially noteworthy is the spread in ^{13}C seen in different plant groups and the resulting soil CO_2.

The concentration of a solute is generally analysed as ppm (mg/kg solution) or mg/L. These must usually be converted to molality (moles/kg or m) or molarity (moles/L or M) for geochemical calculations:

$$m = \frac{\text{ppm}}{\text{gfw} \times 1000} \qquad M = \frac{\text{mg/L}}{\text{gfw} \times 1000}$$

In most groundwaters, salinity is low and so $m \cong M$ and ppm $\cong$ mg/L. Solute concentration can also be expressed in terms of equivalents, which is their concentration multiplied by their charge, z, and divided by their weight:

$$\text{epm} = \frac{\text{ppm}}{\text{gfw}} \times z \qquad \text{meq/L} = \frac{\text{mg/L}}{\text{gfw}} \times z$$

In geochemical equations, round brackets indicate concentration as molality m, i.e. $(Ca^{2+}) = mCa^{2+}$. However, the total concentration m of a solute differs from its thermodynamic concentration or activity a. It is activities that must be used in thermodynamic calculations of aqueous speciation and mineral solubility. Ion activities are also expressed as moles per kg or L solution, and can be denoted by square brackets, i.e. $aCa^{2+} = [Ca^{2+}]$. A solute's activity is lower than its concentration due to electrostatic interferences in solution, and is related to m by an activity coefficient γ:

$$a_{Ca} = m_{Ca} \cdot \gamma_{Ca}$$

The activity coefficient is affected by temperature, but mainly by the salinity or ionic strength I of the water. Ionic strength is calculated from the sum of the molalities of all anions and cations:

$$I = \tfrac{1}{2} \sum m_i \cdot z_i^2$$

where m_i is the molality of the ith ion in solution and z its valence. An ion's activity coefficient γ is calculated from the solution's ionic strength by some form of the Debye-Hückel equation. In dilute groundwaters at temperatures up to about 50°C, the Debye-Hückel equation can be simplified to:

$$\log \gamma_i = -0.5\, z_i^2 \sqrt{I}$$

Note that measurements with ion-specific electrodes, such as pH, are measurements of ion activity a, not m, and so are not corrected with γ before use in thermodynamic equations. Gas concentrations in a gas phase can be expressed as parts per million by volume (ppmv) but in thermodynamic equations, are expressed as a fraction or partial pressures (i.e. $P_{CO_2} = 10^{-3.5}$).

The expression of solute concentrations as activities allows us to calculate solubility index of a given mineral. This index is a measure of whether a given mineral, e.g. calcite, can be dissolved or is being precipitated. Calcite dissolution and precipitation is an important control on the evolution of DIC and $\delta^{13}C_{DIC}$ in groundwaters, particularly in carbonate aquifers or aquifers where DOC supports processes such as sulphate reduction or methanogenesis.

The solubility index for a mineral is the ratio of the activity product of the relevant ions in solution to the equilibrium mineral solubility product (thermodynamic reaction constant, K_T). Solubility products represent the activity product of mineral-forming ions under equilibrium conditions. For example, the solubility product for calcite is:

$$K_{calcite} = [Ca^{2+}] \cdot [CO_3^{2-}] = 10^{-8.48}$$

The solubility product is derived from thermodynamic data and varies with temperature. Table 5-1 provides values for some common minerals in the carbonate system.

Table 5-1 Mineral solubility products for some low temperature minerals (source: Drever, 1997)

Mineral	Activity product	$\log K_T$ @ 25°C
Calcite	$[Ca^{2+}] \cdot [CO_3^{2-}]$	−8.48
Dolomite	$[Ca^{2+}] \cdot [Mg^{2+}] \cdot [CO_3^{2-}]^2$	−16.54
Gypsum	$[Ca^{2+}] \cdot [SO_4^{2-}]$	−4.36

The solubility index for calcite is then calculated from $K_{calcite}$ and the activities of Ca^{2+} and CO_3^{2-} in the groundwater:

$$SI_{calcite} = \frac{[Ca^{2+}] \cdot [CO_3^{2-}]}{K_{calcite}}$$

When $\log SI_{calcite} > 0$, calcite is oversaturated in the water and is likely being precipitated. If $\log SI_{calcite} < 0$, the waters will dissolve calcite (if present in the aquifer). If $\log SI_{calcite}$ is = 0, then the groundwaters are in equilibrium with calcite.

A review of these basic elements of low-temperature aqueous geochemistry in any of the above mentioned textbooks is highly recommended. These simplified calculations can be easily made using a spreadsheet that includes the relevant equations for I, a_i, γ_i, and SI. More complicated geochemical systems (i.e. including more than carbonate and sulphate minerals, higher salinity or higher temperature) should be analysed using a computer programmed geochemical model. Langmuir (1997) and the supporting web <www.igginc.com/iggi/langmuir/don.htm> site provide a good review of available models and where to find them. United States Geological Survey (USGS) programs are available at <http://h2o.usgs.gov/software>, and the US Environmental Protection Agency (EPA) geochemical models are available at <ftp://ftp.epa.gov/epa_ceam/wwwhtml/software.htm>.

Atmospheric and soil CO_2

The atmosphere is the smallest global reservoir of carbon, with an average concentration of some 360 ppmv or a partial pressure of $10^{-3.5}$. Its short-term natural concentration is regulated largely by biological activity (photosynthesis and respiration) in the oceans and on the temperate land masses. Long-term variations are controlled by tectonism and weathering. The $\delta^{13}C$ of atmospheric CO_2 was about −6.4‰ (Craig and Keeling, 1963; Friedli et al., 1986), but is now decreasing due to the burning of fossil fuels.

When waters infiltrate to the subsurface, they equilibrate with soil CO_2. Bacterial oxidation of vegetation in soils and respiration of CO_2 in the root zone maintains CO_2 levels between 1000 and 100,000 ppmv (P_{CO_2} ~10^{-3} to 10^{-1}). These levels are considerably higher than the ~360 ppmv of CO_2 in air today. Dissolution of soil CO_2 produces carbonic acid, which lowers pH and increases the weathering capacity of groundwater. The amount of carbon dioxide that can dissolve will depend on the geochemistry of the recharge environment: the temperature and initial pH of the water, and the partial pressure of soil CO_2.

Dissolution of soil CO_2 and carbonate speciation

When $CO_{2(g)}$ diffuses into water, it forms four main species of dissolved inorganic carbon (DIC).

$CO_{2(aq)}$	Dissolved or aqueous CO_2
H_2CO_3	Carbonic acid or hydrated CO_2
HCO_3^-	Bicarbonate or dissociated carbonic acid
+ CO_3^{2-}	Carbonate or the second dissociation species of carbonic acid
DIC	

Their distribution, or relative concentration, is a function of pH. Dissolution of $CO_{2(g)}$ in water takes place according to the reactions:

CO_2 diffusion into water	$CO_{2(g)} \leftrightarrow CO_{2(aq)}$
CO_2 hydration	$CO_{2(aq)} + H_2O \leftrightarrow H_2CO_3$
1st dissociation of carbonic acid	$H_2CO_3 \leftrightarrow H^+ + HCO_3^-$
2nd dissociation of carbonic acid	$HCO_3^- \leftrightarrow H^+ + CO_3^{2-}$

giving the net reaction:

$$CO_{2(g)} + H_2O \xleftrightarrow{K_{CO_2}} H_2CO_3 \xleftrightarrow{K_1} H^+ + HCO_3^- \xleftrightarrow{K_2} 2H^+ + CO_3^{2-}$$

The concentrations of HCO_3^- and CO_3^{2-} together comprise carbonate alkalinity. Alkalinity is an essential measurement that can be carried out in the field. The formal definition of alkalinity is the concentration of dissolved species which act as proton (H^+) acceptors and buffer pH. For example, groundwaters with high HCO_3^- concentrations will consume acidity generated through, say, oxidation of sulphides or acid rain, by reversing the two dissociation reactions above. Alkalinity is then the total of such species as HCO_3^-, CO_3^{2-}, HS^-, S^{2-}, OH^-, $H_3SiO_4^-$, $H_2BO_3^-$, although species other than HCO_3^- and CO_3^{2-} rarely contribute significantly to alkalinity. Alkalinity measurements are discussed in the field procedures in Chapter 10.

At chemical equilibrium, the activities of the reactants and products are determined from the reaction constants (K_T) according to the following equations (at 25°C):

$$K_{CO_2} = \frac{[H_2CO_3]}{P_{CO_2} \cdot [H_2O]} = 10^{-1.47}$$

$$K_1 = \frac{[H^+][HCO_3^-]}{[H_2CO_3]} = 10^{-6.35}$$

$$K_2 = \frac{[H^+][CO_3^{2-}]}{[HCO_3^-]} = 10^{-10.33}$$

These equations define the distribution of DIC species according to pH (Fig. 5-2). At low pH, it is H_2CO_3 that dominates the DIC, whereas HCO_3^- is the major species between pH 6.4 and 10.3. CO_3^{2-} dominates only under very alkaline conditions. The second dissociation of carbonic acid (K_2) is a minor reaction at pH values less than neutral. Note that K_{CO_2} defines the equilibrium between gas phase CO_2 and dissolved CO_2. Dissolved $CO_{2(aq)}$ by convention is expressed in its hydrated form as carbonic acid, H_2CO_3, although at low pH, it exists mainly as unhydrated CO_2.

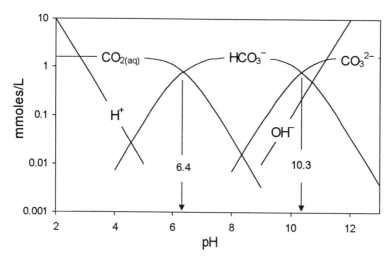

Fig. 5-2 Distribution of carbonate species in pure water as a function of pH at 25°C, calculated for a total dissolved inorganic carbon content, DIC, of 1.6 mmoles/L (or 100 mg/L as HCO_3^-).

Thermodynamic reaction constants (K_T) vary with temperature. The pK values (negative log of K_T) are given for a range of temperatures in Table 5-2. DIC speciation must be calculated for the temperature of the groundwater. The following equations were regressed from data in Drever (1997) and can be used to calculate reaction constants for temperatures up to 50°C.

$$pK_{CO_2} = -7 \cdot 10^{-5} T^2 + 0.016\, T + 1.11$$

$$pK_1 = 1.1 \cdot 10^{-4} T^2 - 0.012\, T + 6.58$$

$$pK_2 = 9 \cdot 10^{-5} T^2 - 0.0137\, T + 10.62$$

$$pK_{CaCO_3} = 6 \cdot 10^{-5} T^2 + 0.0025\, T + 8.38$$

Table 5-2 Equilibrium constants for the carbonate system (Drever, 1997)

$T\,°C$	pK_{CO_2}	pK_1	pK_2	pK_{CaCO_3}
0	1.11	6.58	10.62	8.38
5	1.19	6.52	10.55	8.39
10	1.27	6.46	10.49	8.41
15	1.34	6.42	10.43	8.43
20	1.41	6.38	10.38	8.45
25	1.47	6.35	10.33	8.48
50	1.72	6.28	10.16	8.65

These reaction constants also vary with the ionic strength of the solution (He and Morse, 1993), which must be taken into account for groundwaters with high salinity. In carbonate-free soils with little acid buffering potential, the major DIC species are $CO_{2(aq)}$, H_2CO_3 and HCO_3^-. The equilibrium pH for groundwater in a soil with a fixed P_{CO_2} (say 10^{-2}) can be calculated (note that the $[H_2O] = 1$ for dilute solutions):

$$[H_2CO_3] = 10^{-1.46} \cdot P_{CO_2} = 10^{-1.46} \cdot 10^{-2} = 10^{-3.46}$$

and $$[H^+] = 10^{-6.35} \cdot [H_2CO_3] / [HCO_3^-] = 10^{-9.81}/[HCO_3^-]$$

Because this is a dissociation reaction, we can assume that $[H^+]$ will be equal to $[HCO_3^-]$:

$$[H^+]^2 = 10^{-9.81}$$

and the pH will be 4.9 and $[HCO_3^-] = 10^{-4.90}$. The distribution of DIC species will be:

$$DIC = [H_2CO_3] + [HCO_3^-] = 10^{-3.46} + 10^{-4.9} = 10^{-3.44} \text{ mole/L}.$$

pH buffering and mineral weathering

The higher the CO_2 concentration in the soil atmosphere, the lower will be the initial groundwater pH. The low pH is then buffered by mineral weathering in the soil and upper bedrock. Calcite dissolution is perhaps the most common and effective buffering reaction. The alteration of silicate minerals to clays also consumes H^+, although these reactions proceed more slowly. Calcite dissolution can be written as the simple dissociation reaction:

$$CaCO_3 \rightarrow Ca^{2+} + CO_3^{2-}$$

The solubility constant for this reaction is low in pure water ($K_{CaCO_3} = 10^{-8.48}$) and does not account for the high bicarbonate content of most natural groundwaters. In most groundwater systems, calcite dissolution is enhanced by carbonic acid from soil CO_2 according to the net reaction:

$$CO_{2(g)} + H_2O + CaCO_3 \rightarrow Ca^{2+} + 2HCO_3^- \qquad K_T = 10^{-6.41}$$

Carbonate dissolution according to this reaction is sensitive to the partial pressure of CO_2. The higher the CO_2 concentration in the soil, the greater will be the amount of calcite dissolved, and the higher will be the groundwater's DIC. However, another condition must be considered in determining the DIC gained during infiltration: that is the degree of "openness" between the groundwater and soil atmosphere.

Under *open system* conditions, calcite dissolution proceeds with a constant supply of soil CO_2. Open system conditions are typical of the unsaturated zone where gas and aqueous phases coexist. Much more calcite will be dissolved due to the continual replenishment of CO_2, and the final equilibrium concentration of DIC will be high. Under *closed system* conditions, the groundwater infiltrates through the soil and reaches the saturated zone before it begins dissolving calcite. Here the groundwater is closed off from the source of soil CO_2. The fixed concentration of CO_2 gained in the soil is not replenished as carbonate dissolution proceeds, and so the amount of dissolution and the final DIC concentration will be lower. Closed system conditions are typical of recharge areas where infiltration to the water table is fast (e.g. karst) or in saturated soils with little or no carbonate content.

In reality, partially open systems are most typical, where dissolution of calcite begins under open system conditions and is completed below the water table under closed system conditions. The pathway followed by groundwater during the incorporation of soil CO_2 and calcite dissolution for given pH and P_{CO_2} conditions is presented in Fig. 5-3. This figure shows the strong increase in DIC when calcite saturation is reached under open system conditions, and the minimal increase in DIC for closed system conditions.

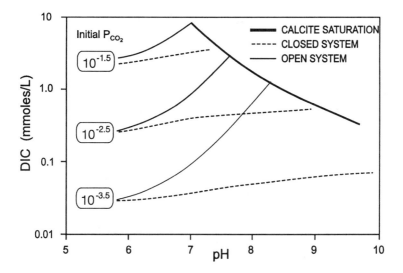

Fig. 5-3 The evolution of pH and DIC as groundwater dissolves calcite under open and closed system conditions, for varying initial P_{CO_2} conditions. Replenishment of CO_2 in open system conditions allows considerably more calcite to be dissolved and produces much higher DIC than for closed system conditions.

Weathering of silicate minerals has a different effect on the carbonate system. The DIC is derived solely from the CO_2 consumed by the alteration of feldspars such as albite [$NaAlSi_3O_8$] or anorthite [$CaAl_2Si_2O_8$] to kaolinite [$Al_2Si_2O_5(OH)_4$]:

$$NaAlSi_3O_8 + CO_2 + \tfrac{11}{2}H_2O \rightarrow Na^+ + \tfrac{1}{2}Al_2Si_2O_5(OH)_4 + 2H_4SiO_4° + HCO_3^-$$

$$CaAl_2Si_2O_8 + 2CO_2 + 3H_2O \rightarrow Ca^{2+} + Al_2Si_2O_5(OH)_4 + 2HCO_3^-$$

In such reactions, the only change to the carbonate system is the associated increase in pH, which shifts the distribution of DIC species to the HCO_3^- field. If this occurs under non-saturated conditions (open system), additional CO_2 will be dissolved from the soil zone. Below the water table (closed system), no additional soil CO_2 is available and so the extent of the reaction is limited. In either case, the DIC is derived solely from the soil CO_2. For anorthite weathering, released Ca^{2+} will participate in carbonate reactions and calcite precipitation.

Carbon-13 in the Carbonate System

Vegetation and soil CO_2

Carbon-13 is an excellent tracer of carbonate evolution in groundwaters because of the large variations in the various carbon reservoirs. The evolution of DIC and $\delta^{13}C_{DIC}$ begins with atmospheric CO_2 with $\delta^{13}C \sim -7‰$ VPDB. Photosynthetic uptake of $CO_{2(atm)}$ is accompanied by significant depletion in ^{13}C. This occurs during CO_2 diffusion into the leaf stomata and dissolution in the cell sap, and during carboxylation (carbon fixation) by the leaf's chloroplast, where CO_2 is converted to carbohydrate (CH_2O). The combination of these fractionating steps results in a 5 to 25‰ depletion in ^{13}C (Fig. 5-1). The amount of fractionation depends on the pathway followed. Three principal photosynthetic cycles are recognized: the Calvin or C_3 cycle, the Hatch-Slack or C_4 cycle, and the Crassulacean acid metabolism (CAM) cycle.

The C_3 pathway operates in about 85% of plant species and dominates in most terrestrial ecosystems. C_3 plants fix CO_2 with the Rubisco enzyme, which also catalyses CO_2 respiration through reaction with oxygen. CO_2 respiration is an inefficiency that remains as an artifact of evolution in an atmosphere with high CO_2 and increasing P_{O_2} (Ehleringer et al., 1991). The diffusion and dissolution of CO_2 has a net enrichment in ^{13}C, whereas carboxylation imparts a 29‰ depletion on the fixed carbon (O'Leary, 1988). The result is an overall ^{13}C depletion of about 22‰. Most C_3 plants have $\delta^{13}C$ values that range from –24 to –30‰ with an average value of about –27‰ (Vogel, 1993). The natural vegetation in temperate and high latitude regions is almost exclusively C_3. They also dominate in tropical forests. Most major crops are C_3, including wheat, rye, barley, legumes, cotton, tobacco, tubers (including sugar beats) and fallow grasses.

The more efficient C_4 pathway evolved as atmospheric CO_2 concentrations began dropping in the early Tertiary. Under low $CO_2:O_2$ conditions and at higher temperatures, increased respiration in C_3 plants interferes with their ability to fix CO_2. C_4 plants add an initial step where the PEP carboxylase enzyme acts to deliver more carbon to Rubisco for fixation. The result is a reduction in ^{13}C fractionation during carboxylation. C_4 plants have $\delta^{13}C$ values that range from –10 to –16‰, with a mean value of about –12.5‰ (Vogel, 1993). C_4 species represent less than 5% of floral species, but dominate in hot open ecosystems such as tropical and temperate grasslands (Ehleringer et al., 1991). Common agricultural C_4 plants include sugar cane, corn and sorghum.

120 Chapter 5 Tracing the Carbon Cycle

This difference provides a tool to monitor food products marketed as "100% natural", such as fruit juices and maple syrup that can be cut with inexpensive cane sugar. Carbon-13 and other isotopes are now used routinely by customs and excise departments to check the origin of these and other imports (Hillaire-Marcel, 1986).

CAM photosynthesis is favoured by about 10% of plants and dominates in desert ecosystems with plant species such as cacti. They have the ability to switch from C_3 photosynthesis during the day to the C_4 pathway for fixing CO_2 during the night. Their isotopic composition can span the full range of both C_3 and C_4 plants, but usually is intermediate (Fig. 5-1).

As vegetation dies and accumulates within the soil, decay by aerobic bacteria converts much of it back to CO_2. Soils have CO_2 concentrations 10 to 100 times higher than the atmosphere. Microbially-respired CO_2 has much the same $\delta^{13}C$ as the vegetation itself. However, outgassing of CO_2 along this steep concentration gradient imparts a diffusive fractionation on the soil CO_2. Cerling et al. (1991) show that measured fractionations are over 4‰, and very close to the theoretical fractionation of 4.4‰ for CO_2 diffusion through air (see page 24). Aravena et al. (1992) provide addition field evidence for enrichment of soil CO_2 by outward diffusion. For this reason, the $\delta^{13}C$ of soil CO_2 in most C_3 landscapes is generally about –23‰ (Fig. 5-1). In soils hosting C_4 vegetation, it would be closer to about –9‰.

^{13}C fractionation in CO_2 – DIC reactions

As infiltrating waters dissolve soil $CO_{2(g)}$, some of it hydrates and dissociates into HCO_3^- (bicarbonate) and CO_3^{2-} (carbonate), The distribution of DIC species is set by pH (Fig. 5-2). A different isotope fractionation factor comes into play between each aqueous species and the soil gas (Fig. 5-4). The largest occurs during CO_2 hydration.

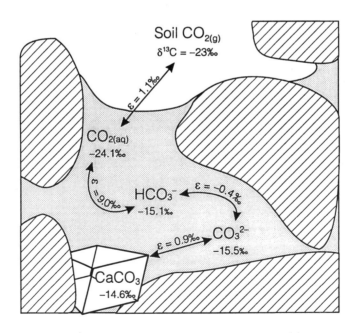

Fig. 5-4 Schematic of fractionation of ^{13}C during equilibrium exchange of carbon between CO_2, DIC and calcite at 25°C. Conditions of geochemical equilibrium and calcite saturation are assumed for isotopic equilibrium.

The fractionation factors for carbonate species at varying temperatures are determined with the following equations. Values for a range of temperatures are given in Table 5-3.

$$10^3 \ln\alpha^{13}C_{CO_{2(aq)}-CO_{2(g)}} = -0.373\,(10^3\,T^{-1}) + 0.19 \quad \text{(Vogel et al., 1970)}$$

$$10^3 \ln\alpha^{13}C_{HCO_3-CO_{2(g)}} = 9.552\,(10^3\,T^{-1}) - 24.10 \quad \text{(Mook et al., 1974)}$$

$$10^3 \ln\alpha^{13}C_{CO_3-CO_{2(g)}} = 0.87\,(10^6\,T^{-2}) - 3.4 \quad \text{(Deines et al., 1974)}$$

$$10^3 \ln\alpha^{13}C_{CO_{2(g)}-CaCO_3} = -2.9880\,(10^6\,T^{-2}) + 7.6663\,(10^3\,T^{-1}) - 2.4612 \quad \text{(Bottinga, 1968)}$$

Table 5-3 ^{13}C fractionation in the system $CO_{2(g)}$, $CO_{2(aq)}$, HCO_3^-, CO_3^{2-} and $CaCO_3$.

T°C	$\varepsilon^{13}C_{CO_{2(aq)}-CO_{2(g)}}$	$\varepsilon^{13}C_{HCO_3-CO_{2(g)}}$	$\varepsilon^{13}C_{CO_3-CO_{2(g)}}$	$\varepsilon^{13}C_{CaCO_3-CO_2}$	$\varepsilon^{13}C_{CaCO_3-HCO_3}$
0	−1.2	10.9	11.35	14.4	3.6
5	−1.2	10.2	9.8	13.5	3.3
10	−1.1	9.6	9.2	12.7	3.0
15	−1.1	9.0	8.6	11.8	2.8
20	−1.1	8.5	8.1	11.1	2.6
25	−1.1	7.9	7.6	10.4	2.4
35	−1.0	6.9	6.6	9.0	2.2
50	−1.0	5.5	5.2	7.4	1.9
75	−0.9	3.3	3.2	5.1	1.8

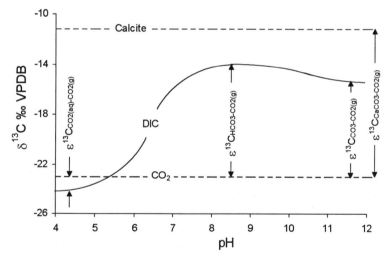

Fig. 5-5 The ^{13}C composition of dissolved inorganic carbon (DIC) in equilibrium with soil CO_2 at 25°C. Values are calculated using the relative contribution of individual DIC species (Fig. 5-2) to weight their respective enrichment factor. Equilibrium is assumed with soil CO_2 which has $\delta^{13}C = -23‰$ VPDB.

The weighting of these species-dependent fractionation factors affects the overall fractionation of ^{13}C between CO_2 and DIC over the range of pH found in the soils and groundwater in recharge areas (Fig. 5-5). Accordingly, DIC and $\delta^{13}C_{DIC}$ in groundwater evolve to higher values during weathering reactions in the soil or aquifer. Controls on this evolution include the degree to which it takes place under open or closed system conditions, and whether the parent material is silicate or carbonate. In purely silicate bedrock terrains, the DIC does not evolve much beyond the conditions established in the soil. In carbonate terrains, however, dissolution of calcite and/or dolomite provides an additional source of carbon to the DIC pool. As these carbonates are generally enriched in ^{13}C, this contribution has a large effect on the $\delta^{13}C_{DIC}$.

Evolution of $\delta^{13}C_{DIC}$ during carbonate dissolution

Marine carbonates make great aquifers. Their porosity increases with use, and if you don't mind a little hardness, the groundwater quality is generally excellent. For this reason, much attention has been given to the evolution of $\delta^{13}C$ and carbonate geochemistry in limestone aquifers. However, carbonate dissolution can take place in silicate aquifers as well, where calcite may be present as a fracture mineral (metamorphic or hydrothermal calcite), or as carbonate grains, cobbles or cement in alluvial aquifers. Marine carbonates have ^{13}C contents similar to the reference VPDB (also a marine carbonate) with $\delta^{13}C \approx 0‰$ (Fig. 5-1), although values can deviate from this value by several permil (see Fig. 5-12, below). Carbonate minerals in igneous and metamorphic rocks can also have a range in $\delta^{13}C$ (Fig. 5-1 and 5-12).

The $\delta^{13}C$ value of carbonate minerals is then about 15‰ more enriched than the DIC from the soil zone. As carbonate is dissolved, $\delta^{13}C_{DIC}$ will evolve to more enriched values. How far it evolves and how much carbonate is dissolved depends on the "openness" of the system. If dissolution takes place under fully open system conditions, then the $\delta^{13}C$ will be controlled by the soil CO_2. This is because of the continual exchange of CO_2 between the DIC and soil atmosphere, which is a large reservoir compared to DIC in non-saturated zone groundwaters. If all the dissolution takes place under fully closed conditions, then the $\delta^{13}C$ will simply be diluted by DIC from the carbonate mineral source ($\delta^{13}C \sim 0‰$). The combination of P_{CO_2} mDIC and $\delta^{13}C_{DIC}$ can provide an indication of recharge conditions.

The evolution of $\delta^{13}C_{DIC}$ during calcite dissolution under open and closed system conditions is illustrated by Fig. 5-6. High ($10^{-1.5}$) and low ($10^{-2.5}$) soil P_{CO_2} conditions are considered, and the diagrams have been calculated for recharge environments characterized by both C_3 and C_4 vegetation. In both cases, the initial $\delta^{13}C_{DIC}$ value on the diagram is calculated for dissolved CO_2 under the specified P_{CO_2} conditions. For open system conditions, final $\delta^{13}C_{DIC}$ values are enriched by about 7‰ from the original soil CO_2. This enrichment reflects equilibrium exchange between soil CO_2 and DIC at increasing pH (as observed in Fig. 5-5). As the pH increases, the ~8‰ enrichment between $CO_{2(g)}$ and HCO_3^- becomes increasingly important. For closed system evolution, the final $\delta^{13}C$ values are also enriched by several permil from the initial value of the soil CO_2. In this case, the enrichment reflects a roughly 50:50 mixture between the $\delta^{13}C$ of the CO_2 dissolved during recharge and the calcite dissolved in the subsurface.

Open and closed system dissolution of carbonate can be distinguished by differences in groundwater P_{CO_2} and concentration of DIC. Open system dissolution retains a high P_{CO_2} close to that of the soil, with elevated DIC. In a closed system the P_{CO_2} becomes much lower, and the DIC concentration is much lower. As $\delta^{13}C$ increases during both open and closed system evolution, this parameter is less diagnostic, although ^{14}C measurements can be useful to evaluate the degree of "openness". DIC derived from the soil will have a modern ^{14}C activity whereas aquifer carbonate is ^{14}C-free. Under closed system recharge conditions, the $^{14}C_{DIC}$ activity will be diluted to less than modern values. Under open system conditions it will have the same activity as the soil CO_2 (see Chapter 8). Once the groundwater has reached calcite saturation, additional geochemical processes including matrix exchange and redox reactions can take place, and will continue this evolution of DIC concentration and $\delta^{13}C$.

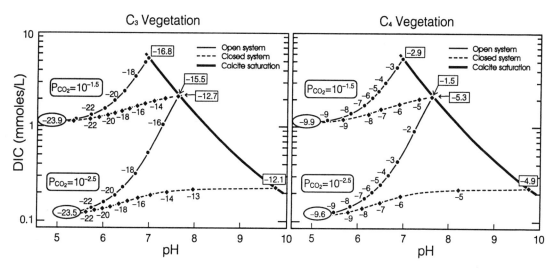

Fig. 5-6 The evolution of DIC and $\delta^{13}C_{DIC}$ in groundwaters as calcite ($\delta^{13}C = 0‰$ VPDB) is dissolved to the point of saturation. Open and closed system conditions are shown for the cases where the water is in equilibrium with soil P_{CO_2} of $10^{-1.5}$ and $10^{-2.5}$. Diagrams are shown for soil CO_2 from C_3 vegetation ($\delta^{13}C_{CO_2} = -23‰$) and from C_4 vegetation ($\delta^{13}C_{CO_2} = -9‰$). Final $\delta^{13}C_{DIC}$ values at calcite saturation are shown in squares. Open system dissolution maintains a fixed P_{CO_2} while under closed system conditions, the P_{CO_2} of the groundwater decreases as calcite dissolves. For open system conditions, the enrichment in ^{13}C reflects continual exchange with soil CO_2 at increasing pH (Fig. 5-5). The greater enrichments observed for closed system dissolution are due to dilution of the initial DIC with marine carbonate.

Incongruent dissolution of dolomite

The magnesium concentration in groundwater is an indication that dolomite and high magnesium calcite may be present in the aquifer. Dissolution of these minerals is slower than that of calcite, and so usually takes place below the water table under closed system conditions. The dissolution of dolomite by calcite-saturated groundwaters generally proceeds incongruently, i.e. calcite is precipitated as dolomite is dissolved:

$$Ca^{2+} + HCO_3^- + CaMg(CO_3)_2 + H_2O \rightarrow Ca^{2+} + Mg^{2+} + 2HCO_3^- + CaCO_3 + OH^-$$

While the loss of calcite has only a minor effect on the $\delta^{13}C_{DIC}$, the gain of dolomite adds carbonate with $\delta^{13}C$ of ~0‰ to the DIC pool. The effect is a secondary dilution of the initial DIC and a further ^{13}C enrichment. The extent of this dilution can be examined on the basis of the Mg^{2+} content in groundwater. Both Mg^{2+} and the dolomite saturation index will increase along the flow path as dolomite is dissolved. The modification of the DIC pool by dolomite dissolution is proportional to the excess Mg^{2+} over groundwaters in the recharge area:

$$mMg_{excess} = mMg_{meas} - mMg_{rech}$$

The effect of dolomite dissolution on carbonate evolution can be tested by comparing the measured $\delta^{13}C_{DIC}$ with the value calculated from the excess Mg^{2+}, according to the isotope mass balance equation:

$$\delta^{13}C_{DIC} = \frac{\delta^{13}C_{rech} \cdot mDIC_{rech} + \delta^{13}C_{dol} \cdot 2mMg_{excess}}{mDIC_{rech} + 2mMg_{excess}}$$

Dissolved Organic Carbon

The geochemical evolution of groundwaters can involve more than the CO_2-based weathering reactions discussed above. Dissolved organic carbon (DOC) plays an important role through reduction-oxidation (redox) reactions. Redox reactions involve the exchange of electrons between two complementary species or redox pairs. One species is oxidized (loss of electrons) and the other is reduced (gain of electrons). Photosynthesis is an example of an endothermic redox reaction, using energy (photons) to reduce carbon from the +IV (oxidized) to the 0 (fixed) redox state:

$$CO_2 + H_2O \rightarrow O_2 + CH_2O$$

Respiration is simply the reverse reaction:

$$O_2 + CH_2O \rightarrow CO_2 + H_2O$$

These two redox reactions are fundamental in the evolution of groundwater geochemistry. Photosynthesis fixes carbon, which is stored as living or dead biomass within the soil. Respiration by bacteria in the soil decomposes this biomass. Much is oxidized to CO_2 that is recycled through photosynthesis. However, decomposition also produces various organic molecules of heterogeneous structure collectively referred to as organic matter. Much of it is soluble in groundwater and represents DOC.

So what constitutes DOC? The operative boundary between DOC and particulate organic carbon is set as what will pass through a commercially available 0.45-μm filter. The decay products of vegetation form a spectrum of molecular sizes and charges. They are composed of C, O, N, H and S in varying proportions. The most common soil-derived organic materials are the humic substances, defined as high molecular weight (up to several hundred thousand mass units), refractory, heterogeneous organic substances. In non-contaminated groundwaters, low molecular weight (LMW) compounds make up the rest. LMW DOC includes cellulose, protein, and organic acids such as carboxylic, acetic and amino acids.

The structure and formation of humic substances have yet to be fully understood, and their characterization has been based largely on methods of separation. They are alkali-soluble acids that give the dark colour to soil and wetland waters. Humic acids (HA) precipitate from solution at pH less than 2, while the fulvic acid (FA) fraction is soluble at all pH values. Insoluble humic substances or *humin* may be that refractory component that is strongly sorbed to the soil mineral component (Stevenson, 1985).

Structurally, humic acids appear to be networks of phenol rings bridged and fringed with carbohydrate, amino acids, fat and protein residues including various O, OH, CH_2, NH and S functional groups (Fig. 5-7). Fulvic acids are low molecular weight compounds (Stevenson, 1985; Orlov, 1995).

Compositionally, humins are C–O–H–N–S compounds, varying according to vegetation and decompositional history. Humic acids are roughly 50 to 60% carbon, with 30 to 40% oxygen (Table 5-4). Hydrogen and nitrogen represent on the order of 5% each. Fulvics tend to have slightly lower carbon contents. HA and FA result from the humification of vegetation (cellulose and other carbohydrates, proteins, lignins, tannins etc.) by bacterial metabolism and oxidation. Their phenolic and amino acid products then polymerize to form humic substances.

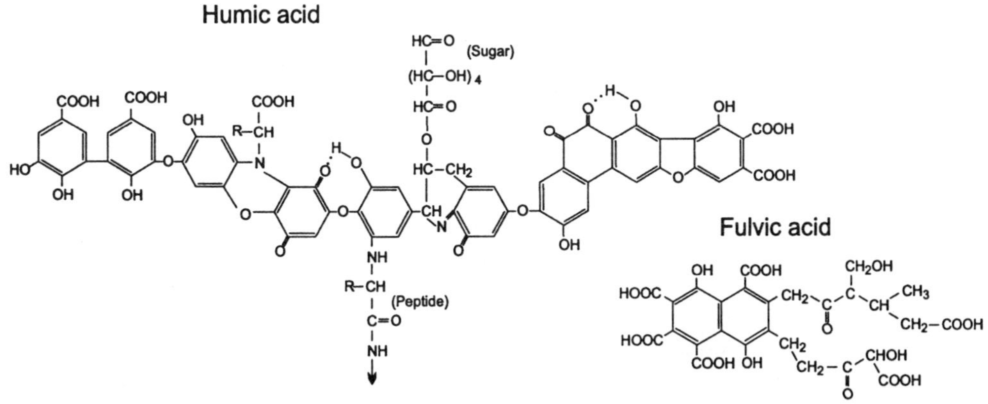

Fig. 5-7 Typical structure of humic and fulvic acids as proposed by Stevenson (1985) and Buffle (1977).

The transport of organic carbon from the soil to groundwater is influenced by soil water conditions and soil structure. DOC in soil moisture reaches a maximum of 10 to 100 mg-C/L in the root zone, and drops off towards the water table (Aiken, 1985). Groundwaters often contain less than 1 to 2 mg-C/L, although groundwaters in certain environments can recharge with much higher DOC concentrations (Wassenaar et al., 1990). Periods of high water table and storm events can flush significant quantities of DOC to the saturated zone, particularly in agricultural areas or in the spring when soil microbial activity is low. Groundwaters recharged through saturated soils, such as in tundra and peat bogs, also typically have high DOC. Groundwaters contaminated with high dissolved organics from landfill sites or septic tanks are another example. In such cases, DOC concentrations can exceed 10 to 100 mg-C/L.

Table 5-4 Mean elemental composition of humic and fulvic acids from vegetation, soils, surface water and groundwater, in weight percent (from Thurman, 1985; Orlov, 1995)

Medium	DOC (mg-C/L)	C	H	O	N	S
Peat	—	58	5.6	33	2.6	0.2–0.5
Plant residue	—	50	6.3	42	2.5	—
Humic Acid						
Soils	—	53–58	3.4–5.2	35–38	2.7–5.0	—
Lakes	1–30	45–52	3–5	37–47	0.7–2	0.5–4.3
Groundwaters[1]	<0.2–10	55–63	5.8–6.3	30–35	0.5–1.8	—
Fulvic Acid						
Soils	—	53–58	3.4–5.2	35–38	2.7–5.0	—
Lakes	1–30	43–53	3–5	35–52	0.5–2.2	0.5–4.3
Groundwaters[1]	<0.2–10	58–62	3–5	24–30	3.2–5.8	—

1 HA and FA in groundwaters from agricultural areas or bogs can exceed 100 mg–C/L (e.g. Aravena et al., 1993a; 1995; Cane, 1996).

DOC can also be gained from organic sources within the aquifer. Buried peat is often a component of Quaternary sediments, and may also be a source of DOC in groundwaters (Aravena and Wassenaar, 1993). Marine carbonates and shales can also host sedimentary organic carbon (SOC), ranging from immature kerogen with elevated O/C ratios to highly reduced, thermally mature hydrocarbons such as bitumen. Coal horizons, besides having aquifer potential, can act as a substrate for redox reactions. In most cases, the $\delta^{13}C$ value of soil or aquifer-derived organic carbon is about -25 ± 2–3‰ (Fig. 5-1).

DOC and redox evolution

The kinetics of sterile reaction of organics with O_2 or other electron acceptors are slow, and of little importance in geochemical evolution. However, the energy released by the oxidation of organic carbon is of interest to bacteria. Bacteria mediate virtually all aerobic and anaerobic oxidation of organics in groundwater. Their metabolism can increase reaction rates by several orders of magnitude. Bacteria are also essential in the oxidation of iron and sulphides, in the reduction of sulphate, and denitrification. The involvement of bacteria in a wide variety of hydrogeological settings up to thousands of metres deep is only now being appreciated.

The sequence of redox reactions depends on the availability of other oxidants, and the energy released from their reaction with DOC, as shown in Fig. 5-8. The presence of a given redox pair will buffer the groundwater Eh, as DOC is consumed and the electron acceptor is reduced. Denitrification and/or sulphate reduction reactions invariably impart a ^{13}C depletion to the DIC, through the addition of oxidized DOC. These reactions are discussed in Chapter 6. The modification of the DIC can be quantified for sulphate reduction, which is important for radiocarbon dating (Chapter 8). The DIC dilution is proportional to the concentration of dissolved sulphide, assuming none has been lost by reaction or diffusion, according to δ^{13}C-DIC mass balance:

$$\delta^{13}C_{DIC} = \frac{\delta^{13}C_{rech} \cdot mDIC_{rech} + \delta^{13}C_{org} \cdot 2mHS^-}{mDIC_{rech} + 2mHS^-}$$

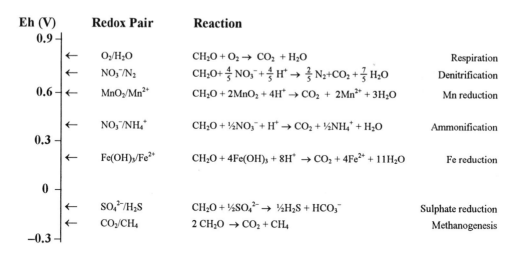

Fig. 5-8 The sequence of principal redox reactions involving fixed organic carbon for neutral pH groundwaters, according to their electromotive potential (Eh) in volts (V).

The absence of a given electron acceptor forces the redox environment to decreasing Eh potentials where lower free-energy electron acceptors can operate. We see in Fig. 5-8 that the persistence of DOC in the absence of NO_3^-, Mn^{4+}, Fe^{3+}, or SO_4^{2-} forces the redox state to very low Eh values where even methane can be produced.

Bacterially mediated redox reactions proceed only in the forward direction and so are not equilibrium reactions. Unlike the isotope fractionations associated with equilibrium inorganic reactions discussed in Chapter 2, these are kinetic fractionations. We have only qualitative values for bacterial fractionation in redox reactions, based on observation. They occur because bacteria prefer to metabolize the isotopically light organics and oxidizers (recall from Chapter 1 that it is easier to break ^{12}C–H bonds than ^{13}C–H or C–^{2}H bonds). Consequently, the fractionation factor is affected by more than temperature. Other factors such as the bacterial species and biochemical pathway, the structure of the organics, and perhaps the availability of nutrients could alter metabolic rates and thus the kinetic fractionation factor.

Methane in Groundwaters

Methane is a ubiquitous gas, found in natural environments ranging from deep crustal settings and sedimentary basins to soils, surface waters and the atmosphere. As a component of carbonate evolution in groundwaters, it participates in the carbon cycle and contributes to greenhouse gases. In geothermal and deep bedrock groundwaters, mantle and crustal methane reflect geologic processes. Methane is also a contaminant generated at landfill sites, and in some water supply aquifers. It becomes hazardous when it collects in basements and water distribution systems.

Methane solubility in water at atmospheric pressure ranges from 2.6 mmoles/L at 0°C to about 1 mmole/L at 50°C, decreasing by roughly 1% per degree increase in temperature. Increased salinity plays a role as well, decreasing solubility by 0.7% for each part per thousand (permil) increase in total dissolved solids (Yamamoto et al., 1976). However, at depth in confined aquifers, increased pressure allows dissolved CH_4 concentrations several times higher before a separate gas phase will develop. Groundwaters that are close to saturation with methane at depth will effervesce at surface. Therein lies the danger of explosion as CH_4 collects in water distribution systems.

There are three principal origins of methane in groundwater: (i) *biogenic methane* is the most common in shallow groundwater systems, forming from the bacterial reduction of organic matter, (ii) *thermocatalytic methane* forms by the breaking down of higher mass hydrocarbons at elevated temperatures and represents the natural gas in sedimentary basins, and (iii) *abiogenic and mantle methane* can be produced without the involvement of bacteria when strongly reducing conditions and inorganic catalysts such as Fe are found.

What insights can ^{13}C, ^{14}C and ^{2}H bring to investigations of the origin of methane? Isotope fractionation during methanogenesis is large, and depends on the reaction pathways followed. Research currently focuses on isotope distribution in the precursor organics and among the final reaction products, in order to sort out both sources and production pathways of methane. The isotopes of interest include $\delta^{13}C_{CH_4}$, $\delta^{13}C_{DIC}$, $\delta^2H_{CH_4}$, $\delta^2H_{H_2}$ and $\delta^2H_{H_2O}$.

Biogenic methane

The generation of methane by bacteria requires a fully saturated environment that excludes atmospheric O_2 and an organic carbon substrate in the absence of other free-energy electron-acceptors such as NO_3^- and SO_4^{2-}. These conditions are most closely matched by wetlands and Arctic tundra, although bovine digestive tracts also meet the requirements. In natural freshwater systems, the pathways for biological methanogenesis have been reviewed by Klass (1984), and generally involve mixed bacterial populations. Fermentive bacteria begin the process by reducing complex organic structures of carbohydrates, proteins and lipids that originate in vegetation and sediments, to simpler molecules including acetate (CH_3COOH), fatty acids, CO_2

and H_2 gas. In these example reactions from Klass (1984), CH_2O is used to represent these complex organic molecules:

$$2CH_2O \rightarrow CH_3COOH$$

$$6CH_2O \rightarrow CH_3CH_2CH_2COOH + 2CO_2 + 2H_2$$

Acetogenic bacteria thrive on fatty acid products to produce acetate, with CO_2 and H_2 by-products in reactions such as:

$$CH_3CH_2CH_2COOH + 2H_2O \rightarrow 2CH_3COOH + 2H_2$$

$$CH_3CH_2COOH + 2H_2O \rightarrow CH_3COOH + CO_2 + 3H_2$$

The products of these reactions support a variety of methanogens. Some use the acetate food source to produce CO_2 and methane:

$$CH_3COOH \rightarrow CH_4 + CO_2 \quad\quad \text{Acetate fermentation}$$

while others use the hydrogen gas to reduce CO_2:

$$CO_2 + 4H_2 \rightarrow CH_4 + 2H_2O$$

or $\quad HCO_3^- + 4H_2 \rightarrow CH_4 + 2H_2O + OH^- \quad\quad CO_2$ reduction

In all these reactions, inorganic carbon is represented as CO_2, although it naturally will hydrate and dissociate to form bicarbonate at the ambient pH in most groundwaters.

The simplified, "pathway-independent" reaction can be written as:

$$2CH_2O \rightarrow CO_2 + CH_4$$

although the full reaction pathway and the relative proportions of reaction products are controlled by the structure of the initial organic substrate, as well as environmental factors such as pH, P_{H_2} and P_{CO_2}.

As one might expect, the ^{13}C fractionation between CO_2 and CH_4 is large ($\varepsilon^{13}C_{CO_2-CH_4}$ = 75 ± 15‰). The metabolic pathway of methanogenic bacteria favours the light isotopes. From Fig. 5-9, biogenic methane in groundwater has $\delta^{13}C$ values depleted from coexisting CO_2 by some 50 to 80‰. This distinguishes biogenic methane from thermocatalytic and abiogenic origins.

The fractionation of ^2H can also help characterize the source of methane. The ^2H contents of methane are established during methanogenesis by the ^2H content of the organic substrate and the water participating in the reaction. However, these are kinetic reactions, and the fractionation factors are generally higher than for equilibrium exchange between CH_4 and H_2O. Deuterium values of –150 to –400‰ can be measured in the CH_4. Subsequent exchange of ^2H with water only occurs at geothermal temperatures, well beyond the range of biogenic systems. The combination of δ^2H and $\delta^{13}C$ can then be used to distinguish biogenic methane from other sources (Fig. 5-10).

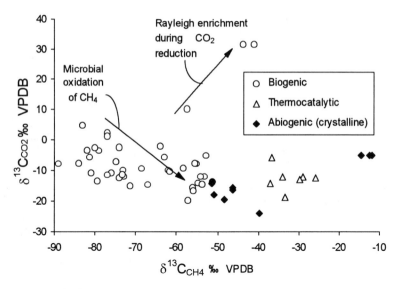

Fig. 5-9 The distribution of ^{13}C between CH$_4$ and coexisting CO$_2$ for fresh and brackish groundwaters with different sources of methane. Data sources: Biogenic — Aravena et al. (1993b, 1995); Grossman et al. (1989), Schoell (1980); Biogenic and thermocatalytic — Barker and Fritz, (1981b); Abiogenic — Fritz et al. (1987b), Fritz et al. (1992) [δ^{13}C$_{CH_4}$ > −20‰].

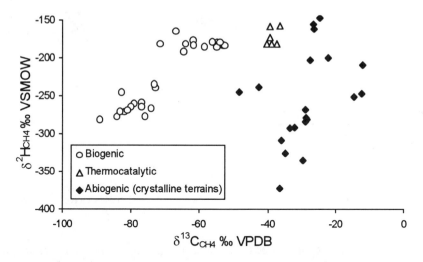

Fig. 5-10 The origin of methane in groundwater according to its δ^{13}C–δ^2H composition. Data sources: Biogenic — Aravena et al. (1995), Grossman et al. (1989), Schoell (1980); Thermocatalytic — Barker and Pollock (1984); Abiogenic — Fritz et al. (1987b), Sherwood Lollar et al. (1993), Fritz et al. (1992) [δ^{13}C$_{CH_4}$ > −20‰].

The reaction pathway, i.e. acetate fermentation or CO$_2$ reduction, will affect differently the isotopic composition of the methane and the evolution of DIC. It was traditionally accepted that CH$_4$ is produced by acetate fermentation in freshwater settings and by CO$_2$ reduction in marine sediments. In fact methanogenesis by CO$_2$ reduction occurs in many freshwater settings. Using ^{14}C-labelled CO$_2$ and acetate at a freshwater bog, Lansdown et al. (1992) show that their observed CH$_4$ production is essentially via CO$_2$ reduction. The most diagnostic tool to identify the pathway appears to be the ^{13}C fractionation between coexisting CH$_4$ and CO$_2$, calculated as:

$$\alpha^{13}C_{CH_4-CO_2} = \frac{\delta^{13}C_{CH_4} + 1000}{\delta^{13}C_{CO_2} + 1000}$$

This fractionation factor is less than about 0.935 for CH_4 production by CO_2 reduction, whereas for acetate fermentation, it is generally greater than about 0.95 (Whiticar et al., 1986; Lansdown et al., 1992). Data from various sources are shown with equal-fractionation lines in Fig. 5-11. Data for which other lines of evidence indicate methanogenesis via CO_2 reduction are dominated by a fraction factor less than 0.935. Other data suggest that acetate fermentation or a combination of the two pathways occurs in various freshwater settings.

The ^{13}C distribution is also affected by several additional factors. One is the isotopic composition of the organic substrate within the aquifer, which can be preserved in the methane (Grossman et al., 1989). Another is the bacterial oxidation of methane. This produces a positive shift in $\delta^{13}C_{CH_4}$ and corresponding depletion in the DIC (Fig. 5-9) (Barker and Fritz, 1981a). A positive shift in the ^2H content of the CH_4 occurs as well.

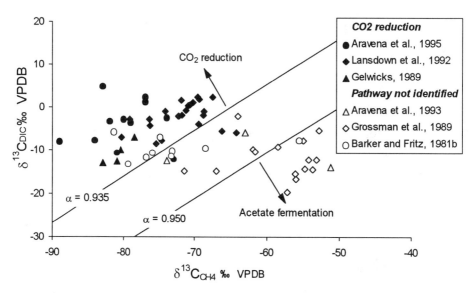

Fig. 5-11 The $\delta^{13}C$ composition of biogenic methane and CO_2 or DIC in groundwater. Filled symbols indicate sites where other evidence indicates CH_4 production by CO_2 reduction. Fractionation lines for CH_4 and CO_2 provide an empirical division between the two methanogenic pathways.

Methane oxidation reactions can proceed according to a number of reactions, although the two principal electron acceptors are O_2 and SO_4^{2-}. Incorporation of atmospheric O_2 generally occurs in the groundwater discharge area such as at the well head, spring vent, or if the groundwaters mix with shallow groundwaters in a phreatic aquifer, oxidizing CH_4 according to:

$$CH_4 + 2O_2 \rightarrow HCO_3^- + H_2O + H^+$$

On the other hand, sulphate incorporated along the flow path can act as an electron acceptor:

$$CH_4 + SO_4^{2-} \rightarrow HCO_3^- + HS^- + H_2O$$

Bacteria can thrive on either reaction, considering the relative position of these redox pairs with respect to CH_4/CO_2 in Fig. 5-8.

Another important factor is whether the reaction products remain with the groundwater, or are lost (e.g. trapped as a separate gas phase within the aquifer), which may affect Rayleigh type reactions. Brackish groundwaters from 850 m depth in dolomites of the St. Lawrence Lowlands provide an example, where measured $\delta^{13}C$ values for DIC reach an astounding +31.6‰ and are ~–41‰ for CH_4 (Fig. 5-9). Such high $\delta^{13}C_{DIC}$ values are rarely observed in nature, and cannot be attributed to reaction with the dolomites ($\delta^{13}C$ = +1‰). In this case, DIC levels are over 600 mg/L (pH = 9.62). DOC reaches 120 mg-C/L and is derived from low-maturity bitumen in the dolomites. TDS reaches 2500 mg/L as NaCl salinity. In this case, reduction of DIC, producing a ~75‰-depleted CH_4 product, imparts a progressive enrichment on the residual DIC pool.

Thermocatalytic methane

Many of the studies referenced in Fig. 5-10 sought to determine whether methane was produced within the aquifer or had migrated there from other sources. The combination of highly depleted ^{13}C and ^2H in methane distinguishes biogenic methane from thermocatalytic or abiogenic (geogenic) methane.

The incorporation of thermocatalytic methane migrating from hydrocarbon-bearing sedimentary basin strata is a plausible source, particularly for deep water wells completed in bedrock. Gas field methane is produced by the thermal cracking of higher mass hydrocarbons. Isotope fractionation is suppressed and $\delta^{13}C$ values are seldom below about –50‰. The relative abundance of methane (C_1) to ethane (C_2) and propane (C_3) is also less than for biogenic methane. The ratio $C_1/(C_2+C_3)$ is generally less than about 10 for thermocatalytic gases, whereas C_2+ hydrocarbons are found at trace levels if at all in groundwaters with biogenic CH_4.

Abiogenic and mantle methane

In groundwater with very low redox potential in the absence of organic matter, methanogenesis can proceed *abiotically* by CO_2 reduction during the alteration of mafic minerals. Apps (1985) found that forsterite-enstatite (Fe-olivine and pyroxene) alteration in the presence of CO_2 from carbonate produced methane and hydrogen according to the reaction:

$$15FeO_{(fo-en)} + 3H_2O + CO_2 \rightarrow 5Fe_3O_4 + H_2 + CH_4$$

Where this occurs, the methane is shown to be far from equilibrium with the groundwater (Sherwood et al., 1988, 1993; Fritz et al., 1992). Further, these abiotic methanes are enriched in ^{13}C with values generally above –40‰ (Fig. 5-10), a prime indicator that bacterial processes are not involved. Such reactions are found almost exclusively in unusual groundwaters such as brines from shield terrains or in hyperalkaline groundwaters, where the interest is geological rather than in water resources.

Deep crustal or mantle methane is typically enriched in $\delta^{13}C$ (–20 to –15‰) due to exchange at high temperatures with mantle carbon (Welhan, 1987). It also has enriched δ^2H values, reflecting high-temperature equilibrium with water. Isotope partitioning among CH_4, H_2, CO_2 and H_2O provides geothermal temperature estimates. Geothermal water may incorporate high-temperature methane, providing insights into the movement of volatile fluids in the crust.

^{14}C and sources of carbon

While ^{13}C systematics is central to unravelling carbon evolution in groundwaters, ^{14}C can play an important role as well. Chapter 8 reviews much of the work with ^{14}C as a groundwater dating tool. It can also be used as an interpretative tool, to constrain the origin of carbon. Various species in the carbon cycle can be characterized by their ^{14}C content. DOC and DIC derived from living biomass and modern soils will have ^{14}C activities close to modern levels. Carbon from old sources such as marine limestone or fossil hydrocarbon (aquifer kerogen, bitumen, thermocatalytic methane) will be ^{14}C free. Buried organic material in aquifers, peatlands and other Quaternary environments can have low ^{14}C activities that reflect decay since time of deposition.

For example, Charman et al. (1994) use this approach to quantify peat accumulation and decay rates. They find that both DIC and methane in the groundwater have higher ^{14}C activities than the associated peat, indicating a downward flux with groundwater recharge. Aravena et al. (1995) use the ^{14}C of DOC and methane in a confined aquifer to show that the source of methane is older than the groundwater, and derived from buried peat within the Quaternary alluvial aquifer.

Isotopic composition of carbonates

The storage and movement of carbon through the global cycle is governed by tectonics and life, which control weathering rates, carbon fixation and respiration. Understanding the sources and sinks of the carbon cycle is the basis for prediction of global warming effects. Carbon-13 contributes much to tracing this cycle. The largest reservoir of carbonate is found in sedimentary basins where limestone and dolomite hold roughly 75% of terrestrial and oceanic carbon. Its isotopic composition is largely established during carbonate growth, sedimentation and diagenesis. Controlling parameters include temperature and reactions with organic carbon, as well as interaction with diagenetic fluids and subsequent recrystallization reactions. For both $\delta^{13}C$ and $\delta^{18}O$, values are generally close to the VPDB standard, although some strong deviations can exist (Fig. 5-12). We have measured $\delta^{13}C$ values up to 3.4‰ in Permian limestones in Oman, and up to 5.7‰ for associated dolomites. By contrast, Tertiary marl units had $\delta^{13}C$ values as low as –8‰ (Clark, 1987). Dolomites are typically more enriched in ^{13}C and ^{18}O than limestone due to dolomite-calcite fractionation controls (Table 1, front cover).

Secondary calcite and other carbonate minerals occur in marine limestones and carbonates as fracture fillings, often precipitated from hydrothermal solutions. Carbonate fracture fillings are also common in metamorphic, granitic and other crystalline rock masses. Their isotopic composition tends to be depleted from that of marine limestones (Fig. 5-12), and reflects the origin of DIC in the fracture fluid. Reduced ^{18}O fractionation at higher temperatures and the involvement of meteoric waters affects the $\delta^{13}C$ of these minerals.

Terrestrial or freshwater carbonates, including speleothem, travertine, calcrete and secondary mineralization in soils and near surface fractures, precipitate from groundwaters at or near the point of discharge. The geochemical situations that provoke calcite precipitation can include depressurization and CO_2 degassing (e.g. speleothem and travertine), evaporation (e.g. soil calcrete), freezing of bicarbonate waters, sulphate reduction and other biotic reactions. Their isotopic composition is a reflection of the geochemical conditions during precipitation, and can be used to interpret their origin. Fig. 5-12 shows examples of typical and extreme isotope values associated with different freshwater carbonates.

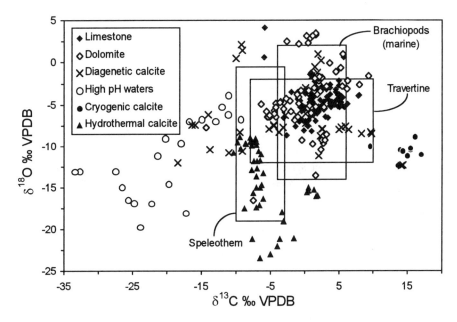

Fig. 5-12 The $\delta^{13}C$ and $\delta^{18}O$ composition of marine carbonates and secondary "terrestrial" carbonates. Sources of data: Limestone, Dolomite — various sources including Al Aasm (pers. comm.), Land (1991), Azmy (1997); Brachiopod aragonite (marine carbonate) — Veizer et al. (1997); Speleothem — Schwarcz (1986), Lauriol et al. (1996); Diagenetic calcite — Irwin et al. (1977), Dimitrakopoulos and Muehlenbachs (1987), Hutcheon et al. (1989), Al-Aasm (1997); High pH waters (kinetic) — Barnes and O'Neil (1969), Clark et al. (1990); Cryogenic calcite — Clark and Lauriol (1992); Hydrothermal calcite — Bottomley (1992), Douglas (1997), McDermott et al. (1996), Shemesh et al. (1992).

$\delta^{18}O$ in secondary calcite and paleotemperatures

As we have seen, the $\delta^{13}C$ composition of various carbon pools is a tracer for the origin of carbon. On the other hand, ^{18}O in the various dissolved carbonate phases ($CO_{2(aq)}$, HCO_3^- and CO_3^{2-}) exchanges rapidly with the water, and so cannot be used to trace pathways. However, when calcite is precipitated from water, it preserves a record of the $\delta^{18}O$ composition of the water. Urey recognized the potential of the ^{18}O content of carbonates to record temperature information (Urey et al., 1951), which led the $\delta^{18}O_{CaCO_3}$ paleotemperature scale (Epstein, 1953).

In fact, there are two temperatures recorded by the $\delta^{18}O$ of freshwater calcite: (1) the water temperature at which precipitation took place, which controls the equilibrium fractionation between water and calcite ($\varepsilon^{18}O_{CaCO_3-H_2O}$ = 28.4‰ @ 25°C), and (2) the climatic temperature (mean annual air temperature, MAAT) recorded by the $\delta^{18}O$ of the water, according to the local T–$\delta^{18}O$ relationship (Chapter 3).

To derive paleotemperature data from calcite, either the temperature during calcite precipitation or the $\delta^{18}O$ value of the water must be known. In the case of marine shells, the $\delta^{18}O$ of seawater is well constrained. Where neither is well established, a qualitative interpretation of paleotemperature can be made. The following example illustrates this point. Data were collected for speleothem (laminated secondary calcite in karst) in the northern Yukon. Their average $\delta^{18}O$ composition is –14.0‰ VPDB. This speleothem was dated using the U/Th disequilibrium method (see Chapter 8) and found to be about 80,000 years old.

Speleothem:	$\delta^{18}O_{CaCO_3}$	$= -15.5‰$ VPDB
		$= +14.9$ VSMOW
Modern groundwater:	$\delta^{18}O_{mod\text{-}gw}$	$= -21.5‰$ VSMOW
	T_{gw}	$= 2$ to $4°C$
Modern MAAT:	MAAT	$= -15°C$
T–δ^{18}O gradient:	$\Delta^{18}O_{paleo\text{-}mod}$	$= \delta^{18}O_{paleo\text{-}gw} - \delta^{18}O_{mod\text{-}gw}$
		$= 0.5 \: \Delta T_{MAAT}$
Change in MAAT:	ΔT_{MAAT}	$= MAAT_{paleo} - MAAT_{modern}$

Making these calculations for a series of estimated paleogroundwater temperatures gives the range for the shift in MAAT as recorded by the $\delta^{18}O$ in the speleothem (Table 5-5).

Table 5-5 Calculated shift in MAAT between paleo- and modern climate based on $\delta^{18}O$ in speleothem

$T_{paleo\text{-}gw}$	$\varepsilon^{18}O_{CaCO_3\text{-}H_2O}$	$\delta^{18}O_{paleo\text{-}gw}$ VSMOW	$\Delta^{18}O_{paleo\text{-}mod}$	ΔT_{MAAT}	$MAAT_{paleo}$
2°C	33.8‰	(14.9 – 33.8) = –18.9	2.6‰	5.2	–9.8°C
6°C	32.8‰	(14.9 – 32.8) = –17.9	3.6‰	7.2	–7.8°C
10°C	31.8‰	(14.9 – 31.8) = –16.9	4.6‰	9.2	–5.8°C

We can see that for a 4°C increase in the assumed paleogroundwater temperature (fractionation temperature), the calculated $\delta^{18}O_{paleo\text{-}gw}$ increases by 1‰, and the MAAT by only 2°C. As groundwater temperature changes would be much less than MAAT during changes in climate, a paleogroundwater temperature of about 6°C would likely be most representative. The $\delta^{18}O$ for the speleothem then shows that it was formed under climatic conditions that were on the order of 5 to 7°C warmer than today.

The $\delta^{18}O$ composition of secondary calcite is a powerful tool to resolve paleohydrogeological processes and events. Secondary calcite in fractures can record influxes of meteoric waters, glacial meltwaters or hydrothermal waters. At Devils Hole, Nevada, a metre thick calcite coating in a fault zone has provided a paleotemperature record of MAAT in the groundwater recharge area that documents the waxing and waning of ice sheets over the past 500 ka (Winograd et al., 1992). In marl and lacustrine shells, $\delta^{18}O$ can be a record of changing water balances, where evaporative loss of lake water is recorded by ^{18}O enrichments in the calcite.

Essential in the use of ^{18}O as a paleotemperature tool is the establishment of a reliable chronology for the material. Calcite can be dated with ^{14}C providing the sources of carbon that have formed the calcite can be established (soil CO_2, DOC, limestone, etc.). Reliable dates can generally be produced with U/Th. These dating methods are discussed in Chapter 8.

Problems

1. What is the pH of a groundwater in equilibrium with a soil atmosphere that has $P_{CO_2} = 10^{-1.8}$? The groundwater temperature is 25°C. Determine the distribution of carbonate species (i.e. mH_2CO_3, $mHCO_3^-$ and mCO_3^{2-}).

2. Dishonest producers of "100% pure" fruit juices and maple syrup, who spike their product with cane sugar water, can be caught with a few isotope analyses. How?

3. Give the reaction pathway for isotope exchange between the following reactants. What conditions are required for isotopic equilibrium?
    ^{13}C: $CO_{2(g)} — HCO_3^-$
    ^{18}O: $H_2O — CO_{2(g)}$
    ^{13}C: $CO_{2(g)} — CaCO_3$
    ^{18}O: $H_2O — CaCO_3$

4. What is the $\delta^{18}O$ of atmospheric CO_2 if it is in equilibrium with ocean water? Assume an ambient temperature of 18°C. How would you expect $\delta^{18}O_{CO_2}$ to change between equatorial and polar seas regions, and between oceans and terrestrial landscapes?

5. The CO_2 equilibration method for measuring $\delta^{18}O$ in waters requires a low (<6) pH to assure rapid exchange in the CO_2. What would be the pH of a sample of pure water after equilibration with the system CO_2 at a working pressure of 0.1 atmospheres? What would it be if the water had an initial pH of 7.3 and 225 mg/L HCO_3^-?

6. What would be the $\delta^{13}C$ of DIC in recharge water that is in isotopic equilibrium with a soil CO_2 having $\delta^{13}C = -23‰$ at 25°C for pH values of 5.8, 7.3 and 9.8?

7. You have sampled groundwater from two springs in differing carbonate terrains. JR1 is from an alluvial aquifer with well drained soils while YK1 has saturated arctic tundra soils in the recharge area. In both regions, the soil P_{CO_2} is $10^{-2.0}$ with $\delta^{13}C_{CO_2} = -23‰$. The bedrock at YK1 has $\delta^{13}C = 1.5‰$ VPDB, and the alluvium at JR1 has $\delta^{13}C = -1.5‰$. From the following data (in mg/L), calculate the equilibrium P_{CO_2} and calcite saturation index for each. Did open or closed system conditions prevail during recharge?

Now determine the distribution of carbonate species, and by means of appropriate isotope mass balance equations, calculate values for $\delta^{13}C_{DIC}$. How do these values compare with the measured $\delta^{13}C_{DIC}$ values?

Parameter	JR1	YK1
pH	7.31	8.35
T	25	5
Ca^{2+}	63.0	25.5
Mg^{2+}	1.3	0.3
Na^+	3.1	2.7
HCO_3^-	205	79.1
SO_4^{2-}	<0.5	<0.5
Cl^-	2.8	4.1
$\delta^{13}C_{DIC}$	–15.9	–11.4

8. A farmer's well, completed in a Quaternary sediment aquifer, is contaminated with CH_4 that has a measured $\delta^{13}C$ of -58‰ VPDB. Can you determine whether this is methane originating within the aquifer itself, migrating from a nearby landfill, or is it leaking from a natural gas storage reservoir in the underlying Paleozoic rocks?

9. Secondary calcite minerals were sampled from a fissure in carbonate rock with high kerogen content (sedimentary organic carbon). The associated groundwater had high

concentrations of H_2S. Describe, with the relevant reactions, the geochemical process taking place. What would you predict for the ^{13}C content of the calcites? What about the DIC?

10. Take a look at the geochemical and isotope data for these four wells sampled along a flow system in a confined basaltic aquifer. No carbonate minerals have been observed, although gypsum from pyrite oxidation has been noted in the outcrop area. In the outcrop region, the soil P_{CO_2} was measured at $10^{-2.6}$ with $\delta^{13}C_{CO_2} = -23‰$. The DOC in groundwater BC1 was identified as humic and fulvic acids with $\delta^{13}C_{CH_2O} = -26‰$ Values are in mg/L.

	BC1	BC2	BC3	BC4
pH	7.00	7.30	7.95	8.30
T	10	15	17	18
Ca^{2+}	15.3	17.0	18.9	21
Mg^{2+}	4.1	2.2	1.3	0.1
Na^+	7.2	12.4	14.2	18.1
K^+	2.1	2.9	3.1	3.5
HCO_3^-	28.1	41.1	55.1	76.1
SO_4^{2-}	32	24	16	0
HS^-	0	2.8	5.5	11
Cl^-	11	12	10	11
DOC (mg-C/L)	8	6	4	0
$\delta^{13}C_{DIC}$	−15.8	−17.6	−18.9	−21.8

- Account for the concentration and $\delta^{13}C$ of DIC in BC1. Is this an open or closed system recharge environment?

- Using the appropriate geochemical reactions, show the process responsible for the evolution of DIC and $\delta^{13}C_{DIC}$ in this groundwater flow system. From your reaction pathway, calculate a final HCO_3^- concentration and $\delta^{13}C_{DIC}$ for BC4 and compare with the measured values.

11. The following analyses (in mg/L) were made for groundwater sampled from successively greater depths in a carbonate aquifer overlain by tundra vegetation. The soil has $P_{CO_2} = 10^{-1.8}$ but are saturated at a shallow depth. The DOC was identified as soil-derived humic material. Write out a series of equations to describe the carbonate evolution in this aquifer.

	FR1	FR2	FR3	FR4
pH	8.05	7.85	7.50	7.42
T	5.1	5.1	5.1	5.1
Eh (mV)	325	−175	−213	−207
Ca^{2+}	31	48	61	68
Mg^{2+}	1.9	1.8	2.1	1.7
Na^+	2.4	3.2	1.8	2.9
HCO_3^-	108	163	199	223
SO_4^{2-}	<0.5	<0.5	<0.5	<0.5
Cl^-	3	2	4	3
DOC (mg-C/L)	44	28	12	2
CH_4	0	21	43	56
$\delta^{13}C_{DIC}$	−13.7	−2.1	3.1	5.2

- How does the state of calcite saturation and the P_{CO_2} of this system evolve?

- Account for the evolution in $\delta^{13}C$ in these groundwaters.

- What would you predict for the $\delta^{13}C$ of the methane observed in these groundwaters?

Chapter 6

Groundwater Quality

Preceding chapters have dealt with the origin and evolution of groundwater. Yet, where water resources are concerned it is not only the source of water that is relevant, but also its quality. What is groundwater quality? Fetter (1988) quotes Aristotle: "When water chokes you, what are you to drink to wash it down?"

Salinity is certainly a component of water quality. In most arid regions, the availability of non-saline groundwater has played an historical role in the development of agriculture, and thereby of civilization. Most sources of salinity in groundwaters are natural, although groundwater salinization can also arise from agricultural, industrial and urban activities. Salinity can encompass a variety of geochemical types or *facies*. The dominant ones are Na-Cl and Ca-SO$_4$ due to the high solubility of chloride and sulphate minerals. More obscure geological environments can host high-salinity groundwaters with some rather unusual geochemical facies such as Ca-OH (e.g. Khoury et al., 1985), Ca-F, Fe-SO$_4$ or Ca-Cl groundwaters. In either natural or polluted groundwaters, it is important to know the mechanism of salinization. Groundwater development strategies depend on understanding where the salinity comes from, and how it can be minimized.

The quality of groundwater is also diminished by organic and inorganic contaminants. Environmental isotopes have much to offer where the contamination of groundwater is of concern. Their analysis can provide information on the origin of contaminants, geochemical reactions and microbiological processes in the subsurface. They are also important in evaluating the sensitivity of groundwater to contamination from surface waters, or determining the integrity of aquitard barriers.

Beyond the context of groundwater as a resource, the quality of groundwater is also a characterization of its geochemical composition, and the geochemical evolution or pathway followed to produce it. Understanding the geochemical evolution of groundwater is the basis of geochemical exploration for metals, geothermal research, groundwater age dating and a host of related hydrogeological studies. In this chapter, we look at the isotope geochemistry of water and solutes important to water quality.

Sulphate, Sulphide and the Sulphur Cycle

Sulphur is an essential nutrient for vegetation. It is also a major element of seawater and marine sediments. Its four main oxidation states ranging from +VI to –II make it both an electron acceptor and donor in redox reactions. Major forms of sulphur in the subsurface include sulphate and sulphide minerals, dissolved sulphate (SO_4^{2-}), dissolved sulphide (HS$^-$) and hydrogen sulphide gas (H$_2$S). Organic sulphur is a component of organic compounds such as humic substances, kerogen and hydrocarbons. Its oxidation and recycling in soils produces the "terrestrial" sulphates found in semi-arid regions. Atmospheric sulphur sources include natural and "technogenic" or industrial SO$_2$, particulate sulphur and aerosols of marine sulphate. The movement of sulphur through these various reservoirs in soils and the hydrosphere constitutes the sulphur cycle.

Sulphur compounds from these various sources participate in the geochemical evolution of groundwater. They also contribute to groundwater salinization. Some of the applications of sulphur isotope geochemistry include the cycling of sulphur in agricultural watersheds, the origin of salinity in coastal aquifers or sedimentary strata, groundwater contamination by landfill leachate plumes, acid mine drainage, and dating sulphate-reducing groundwaters.

Sulphur-34 is highly fractionated between sulphur compounds due to biological cycling. Similarly, the ^{18}O content of sulphate is an important tool to trace the sulphur cycle. A summary of the ranges for ^{34}S in natural materials is shown in Fig. 6-1. Meteorites and magmatic sulphur are close to the standard Cañon Diablo Troilite (CDT). Values exceeding +20‰ are found in association with evaporites and limestones. Negative $\delta^{34}S$ values are typical of diagenetic environments where reduced sulphur compounds are formed (Krouse, 1980). The most common reaction product is pyrite, which is present in many shales or other organic-rich sedimentary rocks and is formed by bacteria reducing seawater sulphate in marine sediments.

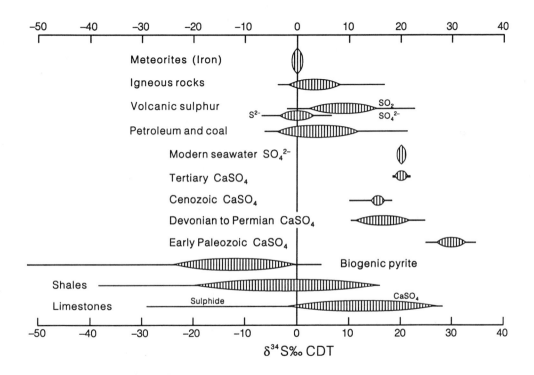

Fig. 6-1 Ranges in $\delta^{34}S$ contents of sulphur and sulphur compounds in different materials and environments (modified from Krouse 1980).

Marine sulphate

The sulphate concentration in groundwater can readily exceed 1 g/L due to the high solubility of gypsum. Gypsum [$CaSO_4 \cdot 2H_2O$] and its unhydrated polymorph, anhydrite [$CaSO_4$], are principal constituents of marine evaporites in sedimentary strata and can be major contributors to groundwater sulphate. Gypsum also accumulates in the soil of arid regions and in sabkha environments. In addition to dissolution of evaporites, marine sulphate in groundwaters comes from mixing with seawater in coastal aquifers and seawater-derived sulphate that accumulates in soils by evaporation. Sulphur isotopes and geochemistry can be used to distinguish between these sources. Other tools include ionic ratios, mineral saturation indices, carbonate isotope geochemistry, and ^{37}Cl to trace chloride salinity. Table 6-1 gives the ionic composition of seawater, along with values for isotopic composition.

The solubility of gypsum in pure water is governed by the equilibrium equation:

$$CaSO_4 \cdot 2H_2O \rightarrow Ca^{2+} + SO_4^{2-} + 2H_2O$$

and $\quad K_{gyp} = [Ca^{2+}][SO_4^{2-}] = 10^{-4.36}$

As equal moles of SO_4^{2-} and Ca^{2+} are produced, the activity of SO_4^{2-} in groundwater which has dissolved gypsum to saturation will be:

$$[SO_4^{2-}] = (10^{-4.36})^{1/2} = 10^{-2.1} = 0.0079 \text{ mole/L} = 763 \text{ mg/L}.$$

This simple calculation overlooks the activity coefficient, γ_{SO_4}, ($m_{SO_4} = a_{SO_4}/\gamma_{SO_4}$) which would be in the order of 0.7 (recall the influence of ionic strength I on γ from Chapter 5), and so concentrations would exceed 1000 mg/L. In more saline groundwaters, γ_{SO_4} increases with ion complexation, allowing even higher concentrations in solution.

Table 6-1 The major ion geochemistry and isotopes of seawater (data from Drever, 1997; Longinelli, 1989; Tan, 1989; Bassett, 1990; Chan et al., 1992)

Species	mg/L	mmole/L	Isotopic composition
Cl⁻	19,350	546	$\delta^{37}Cl = 0‰$ SMOC
SO_4^{2-}	2710	28.3	$\delta^{34}S = 21‰$ CDT
"	"	"	$\delta^{18}O_{SO_4} = 9.5‰$ VSMOW
HCO_3^-	142	2.33	$\delta^{13}C = -1$ to $+2‰$ VPDB
Br⁻	67	0.84	
F⁻	1.3	0.068	
NO_3^-	0.005–2	0.8–$300 \cdot 10^{-4}$	
PO_4^{3-}	0.001–0.05	1–$50 \cdot 10^{-5}$	$\delta^{18}O_{PO_4} = 19.7‰$ VSMOW
DOC	0.3–2 as C	0.02–0.2	$\delta^{13}C \sim -25$ to $-30‰$ VPDB
Na⁺	10,760	468	
Mg^{2+}	1290	53.1	
Ca^{2+}	411	10.3	
K⁺	399	10.2	
Sr^{2+}	8	0.091	$^{87}Sr/^{86}Sr = 0.70924 \pm 0.00003$
SiO_2	0.5–10	<0.35	
B^{3+}	4.5	0.42	$\delta^{11}B = 40.4 \pm 1.7‰$ NBS 951
Li⁺	0.18	0.026	$\delta^6Li = -32.3 \pm 0.5‰$ LSVEC

Gypsum and anhydrite are not restricted to major evaporite sequences in the geological record. Minor sulphate nodules and laminations on bedding planes can occur throughout limestone and dolomite sequences, and can be overlooked during geological mapping. The dissolution of sulphate minerals increases permeability, and so can enhance concentrations in groundwater.

In pure water, the Ca^{2+}/SO_4^{2-} molar ratio equal to 1 would distinguish gypsum dissolution from other sources of sulphate salinity such as seawater (0.36) (Table 6-1). Calcite precipitation also affects this ratio. Groundwaters typically move towards calcite saturation within the soil or along the flow path. Dissolution of sulphate then forces calcite precipitation according to the common ion effect, resulting in a disproportionate increase in SO_4^{2-}:

$$Ca^{2+} + HCO_3^- + CaSO_4 \cdot 2H_2O \rightarrow Ca^{2+} + SO_4^{2-} + CaCO_3 + H^+ + 2H_2O$$

The dissolution of gypsum or anhydrite occurs without measurable isotope effects, and so the isotope contents of SO_4^{2-} can be used as a tracer for sulphate origin. The sulphate of modern seawater has a very homogeneous and well-defined isotopic composition (Table 6-1):

$\delta^{34}S_{SO_4} = 21‰$ CDT

$\delta^{18}O_{SO_4} = 9.5‰$ VSMOW

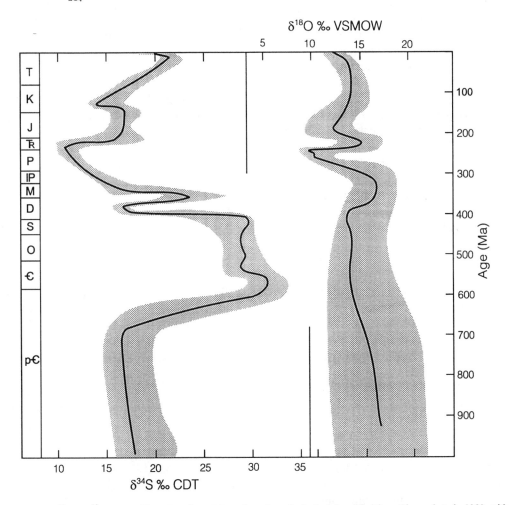

Fig. 6-2 The ^{34}S and ^{18}O composition of marine sulphate through geologic time (modified from Claypool et al., 1980, with data from Fritz et al., 1988).

This composition was not constant in the past. Pronounced variations exist throughout geologic time (Claypool et al., 1980; Fig. 6-2). These variations are found in all major marine evaporite deposits and were most likely controlled by major inputs or removal of sulphide from the oceanic reservoirs during changes in tectonic activity and weathering rates. Simple removal of sulphate (increase in evaporite formation) would not be accompanied by such dramatic isotope effects. The $\delta^{18}O$ content of seawater sulphate has been more stable over geologic time, and is controlled largely by the sulphide weathering reactions that contribute sulphate to seawater. Comparison of the isotope contents of sulphate in groundwater with the appropriate geological period in Fig. 6-2 can distinguish a geogenic source from other sources of sulphate salinity.

Oxidation of sulphide and terrestrial sulphate

In surface and near-surface environments, oxidation of aqueous sulphide can be an important source of sulphate in groundwaters. Fractionation during oxidation of dissolved sulphide is considerably less than sulphate reduction. As with sulphate reduction, ^{32}S reacts more quickly and imparts a ~5‰ depletion on the accumulating sulphate when dissolved sulphide is oxidized chemically by O_2 (Fry et al, 1988). When aerobic oxidizers like *Thiobacillus concretivorous* are involved, sulphate may be depleted by up to 20‰ from the sulphide (Kaplan and Rittenberg, 1964). However, anaerobic oxidation of sulphide by photosynthetic algae can enrich the new sulphate and organic sulphur by about 2‰ (Fry et al., 1986). The ^{18}O content of sulphate from oxidation of dissolved sulphide is dependent on the reaction pathway followed, and is similar to sulphate from sulphide minerals.

Where sulphide-bearing rocks are exposed at the surface, oxidation of pyrite and other sulphide minerals can contribute considerable amounts of sulphate (plus metals and acidity) to surface and groundwaters, or produce gypsum crusts in arid regions. Bacteria catalyse reactions such as:

$$FeS_2 + 3½O_2 + H_2O \rightarrow Fe^{2+} + 2SO_4^{2-} + 2H^+ \qquad \textit{Thiobacillus thiooxidans}$$

Other reactions dominate at low pH:

$$FeS_2 + 14Fe^{3+} + 8H_2O \rightarrow 15Fe^{2+} + 2SO_4^{2-} + 16H^+ \qquad \text{abiological}$$

$$Fe^{2+} + ¼ O_2 + H^+ \rightarrow Fe^{3+} + ½ H_2O \qquad \textit{Thiobacillus ferrooxidans}$$

The potential of sulphide oxidation to generate acidity, sulphate salinity and dissolved metals can be alarming. Nordstrom et al. (1990) report SO_4^{2-} concentrations exceeding 80 g/L in drainage from the Iron Mountain site in California, with high levels for Fe^{3+} and other metals. In the interior plains of North America, the accumulation of sulphate salts in playas and small depressions greatly affect local groundwater and surface water quality (Dowuona et al., 1992).

These "terrestrial" sulphates are distinguished from marine sources by their isotopic composition (Fig. 6-3). The δ^{34}S composition of sulphate from oxidation of sulphides is in most cases marginally depleted from the original sulphide. Toran and Harris (1989) summarize the literature on the subject of isotope effects during sulphide oxidation, and show that SO_4^{2-} is depleted in ^{34}S by 2 to 5.5‰ for biologically-mediated oxidation of base-metal sulphides. The lighter sulphate is accounted for by the lower energy required by the bacteria to break ^{32}S–Fe bonds. However, dissolution reactions on solid-phase surfaces tend to be quantitative, with limited opportunity for fractionation.

The δ^{18}O composition of sulphate is more complicated. Both atmospheric oxygen and oxygen in the water molecules present during oxidation can have an influence. The ^{18}O content of sulphate will be a mixture of these two end-members, plus any fractionation effects (Krouse, 1980; Longinelli, 1989; Toran and Harris, 1989). Experimental results suggest that $\varepsilon^{18}O_{SO_4-O_2}$ during oxidation by Fe^{3+} reduction range between −8.7‰ (Lloyd, 1967; abiological) and −11.4‰ (Taylor et al., 1984). By contrast, $\varepsilon^{18}O_{SO_4-H_2O}$ is closer to 2 to 4‰ (Toran and Harris, 1989). The δ^{18}O of the new sulphate can then be determined from the fraction f of O_2 vs. H_2O and their respective enrichment factors according to (Van Everdingen and Krouse, 1985):

$$\delta^{18}O_{SO_4} = f_{H_2O} (\delta^{18}O_{H_2O} + \varepsilon^{18}O_{SO_4-H_2O}) + 0.825 f_{O_2} (\delta^{18}O_{O_2} + \varepsilon^{18}O_{SO_4-O_2})$$
$$+ 0.125 (\delta^{18}O_{H_2O} + \varepsilon^{18}O_{SO_4-H_2O})$$

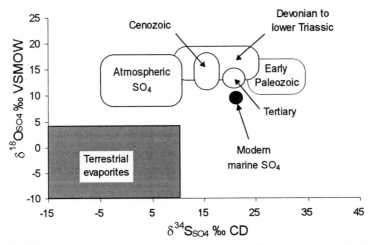

Fig. 6-3 The $\delta^{34}S–\delta^{18}O$ composition of terrestrial sulphates in comparison with marine and atmospheric sulphate.

The proportion of the two oxygen-contributors varies according to the enzymes involved in the biological reaction pathway. Expected $\delta^{18}O$-values for different isotopic compositions of the water are graphically shown in Fig. 6-4 for different O_2/H_2O ratios. Note that the most common case predicts the contribution from water to be between 25% for non-saturated conditions to 75% for anoxic, saturated conditions (Lloyd, 1968; Taylor et al., 1984; Toran and Harris, 1989). Based on this, one would expect $\delta^{18}O$ values of $0 \pm 2‰$ VSMOW for sulphate formed by the oxidation of reduced sulphur in groundwaters typical for central Canada.

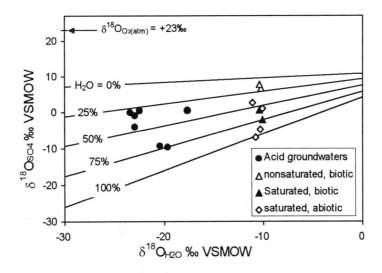

Fig. 6-4 The oxygen isotopic composition of sulphate according to the percent water in the oxidation reaction, determined from mass balance reaction above (modified from van Everdingen and Krouse, 1985). Calculations assume a ^{18}O fractionation of 4.1‰ between SO_4^{2-} and H_2O and –11.4‰ between SO_4^{2-} and O_2. Acidic groundwaters and mine waters — Van Everdingen and Krouse (1985); triangles and diamonds — experimental oxidation reactions, with biotic reactions mediated by *T. ferrooxidans* (Taylor et al., 1984).

Atmospheric sulphate

The sulphur geochemistry of any groundwater is initially controlled by the recharge environment, i.e. atmospheric inputs, although this contribution is usually minor compared with sulphate additions in the subsurface. Nonetheless, atmospheric sulphur plays a role, particularly in the sulphur cycle in soils and shallow groundwaters.

The sulphur isotopic composition of atmospheric sulphate is controlled by emissions from fossil fuel combustion, biological release of sulphur-bearing compounds (e.g. DMS or dimethylsulphide which will oxidize to sulphate) and, at least in coastal areas, by sulphate from sea spray. Sulphur from combustion of both petroleum and coal typically has $\delta^{34}S$ values that range from slightly negative to about +10‰ CDT (Fig. 6-1). Exceptions exist and especially in some Paleozoic crude oils from North America values as high as +16‰ are known.

For central Canada, Nriagu and Coker (1978) report a $\delta^{34}S$ range for sulphur fallout between about +2 and +9‰, with a clear seasonality and distinction between rural and urban areas (Fig. 6-5). The lower values appear to be dominated by biological sulphur whereas the higher ones reflect fossil fuel sulphur.

The oxygen isotopic composition of the fallout will depend on the oxidation process, described in detail by Holt et al. (1972). It is evident that atmospheric oxygen with its high ^{18}O contents ($\delta^{18}O = +23.5‰$) plays an important role and it is not surprising that positive δ-values in the range of +10‰ are most common. Atmospheric sulphate is thus isotopically quite distinct from secondary and evaporitic sulphate (Fig. 6-3) and as such can be recognized in groundwaters.

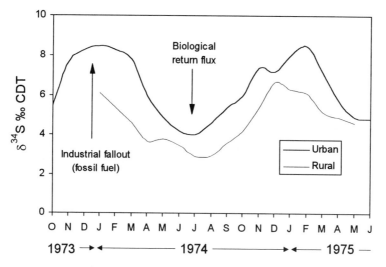

Fig. 6-5 Seasonal variations in the $\delta^{34}S$ composition of atmospheric sulphur fallout in the Great Lakes basin (after Nriagu and Coker, 1978)

Sulphate reduction

Hydrogen sulphide, H_2S, brings to groundwater that most unpleasant "rotten egg" odour at concentrations less than 1 mg/L. More importantly, it is an explosive gas that, like methane, can collect in water distribution systems at low pH. It is corrosive, attacking iron piping in reactions like:

$$H_2S + Fe° \rightarrow FeS + H_2$$

H_2S becomes highly toxic at higher atmospheric concentrations. Maximum acceptable concentration in drinking and aquatic waters is 0.05 mg/L as H_2S (Environment Canada, 1995).

In water, hydrogen sulphide dissociates into HS^- and S^{2-} according to pH (Fig. 6-6). Its solubility depends on pH and the availability of other metals for the precipitation of sulphide minerals. At high pH, HS^- and S^- dominate, and concentration is limited only by the solubility of sulphide minerals.

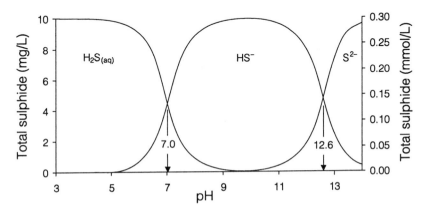

Fig. 6-6 The dissociation of H_2S in water as a function of pH, for example of 10 mg/L total sulphide.

The bacterial reduction of sulphate to sulphur (S°) or sulphide is at the root of the strong variations in the isotopic composition of sulphate. Given an accessible organic carbon substrate, bacteria such as *Desulfovibrio desulfuricans* will use sulphate as an electron acceptor and produce dissolved sulphide and mineralized carbon:

$$2CH_2O + SO_4^{2-} \rightarrow 2HCO_3^- + H_2S \quad \text{(oxidation of fixed carbon)}$$

or $$CH_4 + SO_4^{2-} \rightarrow HCO_3^- + HS^- + H_2O \quad \text{(oxidation of reduced carbon)}$$

Reaction of ^{32}S during sulphate reduction is faster than ^{34}S, due to the lower energy needed to break ^{32}S–O bonds. As reaction proceeds, ^{34}S gradually accumulates in the residual sulphate reservoir. The ^{34}S difference between coexisting dissolved sulphate and sulphide ($\Delta^{34}S_{SO_4-H_2S}$) has three controls: the $\delta^{34}S$ of the sulphate source (most often gypsum in the host rock), the fractionation factor, and Rayleigh distillation effects.

Fractionation is variable and depends on the environmental conditions under which the bacteria function and the bacterial communities involved. While *Desulfovibrio desulfuricans* has been used in most laboratory investigations, Krouse et al. (1970) found that other sulphur reducers dominate in natural springs, presumably with different affinities to mass selection. More importantly, they identified species that converted sulphate to intermediary forms such as sulphite (SO_3^{2-}), which was converted by others to sulphide. In this case, two fractionation steps may compound to produce differences up to 40‰. More typically, bacterial enrichment factors

($\epsilon^{34}S_{SO_4-H_2S}$) are closer to 25‰. This is much less than the thermodynamic enrichment for abiotic exchange. The following equations are approximations of the calculations by Sakai (1968), developed for geothermal systems.

$$10^3 \ln\alpha^{34}S_{SO_4-H_2S} = 3.0\,(10^6\,T^{-2}) + 14.009\,(10^3\,T^{-1}) - 11.197$$

and

$$10^3 \ln\alpha^{34}S_{SO_4-HS} = 3.0\,(10^6\,T^{-2}) + 15.670\,(10^3\,T^{-1}) - 13.592$$

At 25°C, the enrichment factors are 70‰ for SO_4^{2-}-H_2S exchange, and 73‰ for SO_4^{2-}-HS^-. Inorganic sulphate reduction may be found in some geological settings (mantle, geothermal, etc), but these contributions to most groundwaters are not significant. The $\delta^{34}S$ data for SO_4^{2-} and H_2S in a variety of groundwaters show that bacterial sulphate reduction must dominate in most systems (Fig. 6-7).

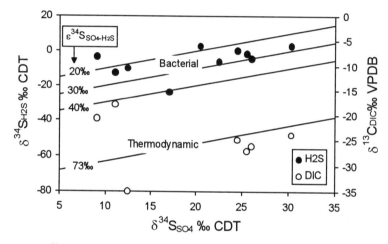

Fig. 6-7 Fractionation of ^{34}S between dissolved sulphate and sulphide in groundwaters, with the $\delta^{13}C$ for associated DIC. Values for $\delta^{13}C_{DIC}$ are generally depleted below the –15 to –10‰ range for normal bicarbonate waters, reflecting the role of organic carbon in sulphate reduction, (data from Krouse et al., 1970; Phillips, 1994).

In Chapter 2 it was shown that isotope effects in kinetic systems under closed system conditions can be described by a Rayleigh distillation process. Since the reduction of sulphate in most groundwaters does not occur in contact with sulphate minerals (which could replenish sulphate lost) the Rayleigh model normally does apply, i.e. an asymptotic enrichment in $^{34}S_{SO_4}$ does occur (Fig. 6-8). In the case where sulphide remains with the groundwater (closed system), the difference between the accumulating H_2S and the residual SO_4^{2-} increases dramatically as reaction proceeds. Continual replenishment of dissolved sulphate would lower $\Delta^{34}S_{SO_4-H_2S}$ towards the value of $\epsilon^{34}S_{SO_4-H_2S}$ for the particular system.

An enrichment in ^{18}O of the residual sulphate also accompanies reduction. Experimental observations suggest that at least during early reaction $\epsilon^{34}S$ is about 2.5 to 4 times greater than $\epsilon^{18}O$ (Pierre, 1989). However, Fritz et al. (1989) found that this ratio increases in the later stages of reaction, and that the $\delta^{18}O$ of the residual sulphate approaches a constant value. The authors attribute this to the involvement of sulphite during the biologically mediated reaction and isotope exchange with the water. The maximum enrichment appears to correspond to the thermodynamic enrichment factor in the system sulphate-water (Fig. 6-9).

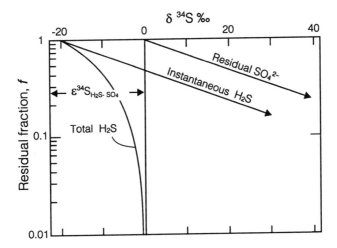

Fig. 6-8 Isotope fractionation during bacterial sulphate reduction. The diagram is constructed using a kinetic fractionation factor of –20‰. Total H₂S describes the isotopic changes in the product if accumulation occurs, i.e. the product remains in the system.

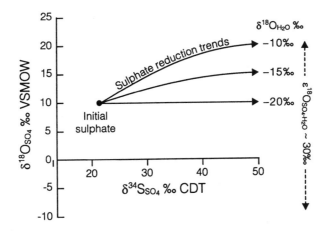

Fig. 6-9 Effect of sulphate reduction on the isotopic composition of sulphate. The $\delta^{34}S$ values of the sulphate follow a Rayleigh enrichment, often exceeding 30 to 40‰ as the sulphate is consumed. Conversely, $\delta^{18}O$ values approach a constant value that closely corresponds to the thermodynamic equilibrium between water and sulphate. This is about 30‰ at normal groundwater temperatures (Fritz et al., 1989).

In natural systems, the evolution of $\delta^{34}S$ and $\delta^{18}O$ during sulphate reduction can be complicated by continual additions to the residual SO_4^{2-} pool and variations in the source of sulphate. Fig. 6-10 presents such data for sulphate-reducing groundwaters throughout the Cretaceous Umm er Radhuma aquifer in Oman. In this case, the enrichment in $\delta^{34}S$ was used to calculate the amount of sulphate reduction that had taken place (discussed in Chapter 8 — Southern Oman).

Sulphate-water ¹⁸O exchange

Isotope exchange is slow in the aqueous sulphur system, especially for ¹⁸O-exchange between sulphate and water at normal groundwater temperatures. Chiba and Sakai (1985) show that the half-time of sulphate-water exchange at normal groundwater temperatures is on the order of 10

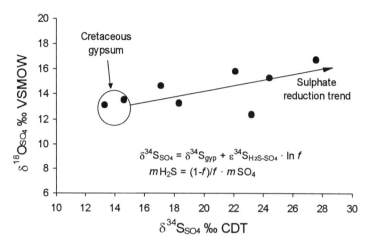

Fig. 6-10 Sulphate reduction in the Umm Er Radhuma carbonate aquifer, Oman (Clark et al., 1996). A Rayleigh equation was used to estimate the extent of reaction (mH$_2$S), using a value for ε^{34}S of 20‰.

million years. However, exchange is rapid in geothermal fluids. Low temperature exchange increases significantly at low pH, where bisulphate (HSO$_4^-$) and sulphuric acid have higher concentrations. Like carbonate, exchange of ^{18}O takes place through the dehydration/hydration reactions:

$$SO_4^{2-} + 2H^+ \leftrightarrow HSO_4^- + H^+ \leftrightarrow H_2SO_4 \leftrightarrow SO_{3(aq)} + H_2O$$

The second dissociation constant is $10^{-1.92}$, and so, at pH 1.92 bisulphate and sulphate have equal concentrations. Exchange is increasingly rapid below ~pH 3, a point of caution when acidifying sulphate samples to remove carbonate prior to analysis.

At neutral pH, the sulphate molecule is very stable and does not easily exchange its oxygen. Therefore, in most groundwaters both ^{18}O and ^{34}S reflect the source of sulphate and the processes that affect its concentration. Only in the case of extensive sulphate reduction in aquifers, as discussed above, will exchange of ^{18}O between water and the residual sulphate take place (Fritz et al., 1989).

Nitrogen cycling in rural watersheds

Nitrogen is a biologically active element and participates in a multitude of reactions that are important to life, and that affect water quality. Decay of biomass releases organic nitrogen, which oxidizes to nitrate (NO$_3^-$), a contaminant in drinking waters. At concentrations above 10 mg-N/L, NO$_3^-$ interferes with the O$_2$ carrying capacity of hemoglobin in infants (methemoglobinemia). Nitrate may also be implicated in the formation of carcinogenic nitrosamine compounds in humans. As ammonium (NH$_4^+$), it is toxic to aquatic life and contributes to oxygen demand. Most guidelines for aquatic life limit organic and inorganic nitrogen at 1 mg/L.

Within the nitrogen cycle, both kinetic and thermodynamic fractionation processes are potentially relevant to tracing N sources and sinks. Isotope fractionation between the various N-bearing compounds provides the basis for ^{15}N as a tool in isotope hydrogeology (e.g. Kreitler,

1975; Mariotti and Letolle, 1977 or Flipse and Bonner, 1985). In a related field, research in agriculture and soil science relies on the use of ^{15}N-enriched compounds to trace biological reactions.

The use of isotopes to trace nitrogen reactions in hydrology gained further attention when it became possible to routinely measure the ^{18}O-contents of nitrate (Amberger and Schmidt, 1987). The combination of ^{15}N and ^{18}O now provides a tool that enables us to distinguish between nitrates of different origin, to recognize denitrification processes and to discuss the N-budget in the soil-water system.

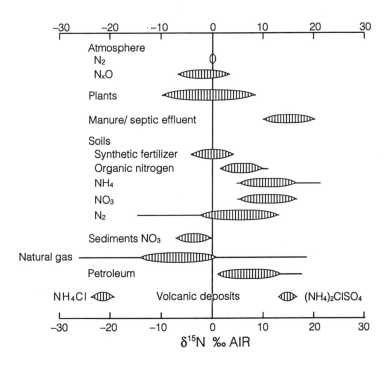

Fig. 6-11 The range for ^{15}N in natural materials (Amberger and Schmidt, 1987; Böttcher et al., 1990; Létolle, 1980).

The geochemistry of nitrate

Even the simplest forms of life incorporate N in their cell structure. Algae, for example, have a chemical composition close to:

$$C_{106}H_{263}O_{110}N_{16}P(S)$$

which is close to the average chemical composition of humic substances and DOC (Table 5-4). The reduction or "fixation" of nitrogen to bio-accessible amino groups ($N_2 \to NH_2$) requires energy to break the N–N bond of atmospheric N_2 (redox state = 0). N-fixation can be done by blue-green algae, as well as N-fixing bacteria in the root nodules of *Leguminosae*. Most plants derive their N from soil ammonium, NH_4^+ (–III redox state) to synthesize organic N.

As biomass decays, organic N can be transformed back to ammonium (NH_4^+) for recycling, which makes compost and manure such good fertilizers. Commercial urea fertilizers

(NH_2CONH_2) decompose in water to NH_4^+ which can in part be lost by volatilization. Under aerobic conditions, NH_4^+ is oxidized by nitrification to nitrate, NO_3^-. N changes redox state from –III to +V:

Nitrification $\quad NH_4^+ + 2O_2 \rightarrow NO_3^- + 2H^+ + H_2O$

Nitrate is the most stable form of nitrogen, after dissolved N_2 gas, in most groundwaters. Minor concentrations of intermediary nitrite, NO_2^- (+III redox state), can also be present.

Nitrate salts are highly soluble, and offer no practical limit to NO_3^- concentrations in water. Attenuation of NO_3^- contamination is limited to dilution and denitrification. The latter is a viable biological reaction requiring anoxic conditions and an accessible organic substrate, represented here as generic DOC:

Denitrification $\quad NO_3^- + {}^5/_4 CH_2O \rightarrow \frac{1}{2}N_2 + {}^5/_4 HCO_3^- + \frac{1}{4}H^+ + \frac{1}{2}H_2O$

Denitrification is generally carried out by the bacterium *Thiobacillus denitrificans,* although others can denitrify in the absence of organic carbon, using electron sources such as Mn^{2+}, Fe^{2+}, sulphide and CH_4.

Isotopic composition of nitrate

From these principal N-transforming reactions arises the distribution of isotopes that we observe in nitrate and ammonia in groundwaters. For simplicity, let's start with the atmospheric N_2 reservoir, which is defined as the ^{15}N standard ($\delta^{15}N_{N_2}$ = 0‰ AIR). Organic nitrogen fixation has only a minor fractionation effect on ^{15}N, which is depleted in the fresh organic matter by 1 to 5‰ (Létolle, 1980). The manufacture of urea fertilizers from atmospheric N_2 is also accompanied by only minor fractionation (Fig. 6-12).

The fractionation of ^{15}N through the food web is proportional to the trophic level of the organisms (Minagawa and Wada, 1984). Initially low ^{15}N contents in algae and other primary producers are magnified by over 10‰ by subsequent consumers through the food chain. The catabolic reaction (destructive metabolism) of amino acids (organic N) in soils as well as in animals produces NH_4^+, which is depleted in ^{15}N by several permil. In the case of animals, this imparts a reciprocal enrichment on the solid waste (manure). Nitrification of NH_4^+ adds a further fractionation, producing NO_3^- that is up to 10‰ more depleted. Complicating the story are the fractionation effects of volatilization, which favour $^{14}NH_3$ and enrich the residual ammonium solution.

Denitrification proceeds with a series of intermediary steps involving various nitrogen oxides. Net fractionations up to $\varepsilon^{15}N_{N_2-NO_3}$ = –20‰ were observed by Wada et al. (1975) and Létolle (1980). In groundwaters of the "Fuhrberger Feld" catchment (case study below), Böttcher et al. (1990) determined a value of –15.9‰.

The $\delta^{18}O$ composition of nitrate adds another perspective on the origin of NO_3^-. Experimental work has shown that in biologically formed nitrate, only one oxygen atom comes from atmospheric O_2. The other two come from the water (e.g. Hollocher, 1984), which is considerably more depleted in ^{18}O. This contrasts with nitrate in synthetic fertilizers, which receives its oxygen primarily from atmospheric O_2. No significant ^{18}O-fractionation effects appear to accompany nitrification. Since the two oxygen sources differ significantly, "natural" and "synthetic" nitrate should show differing isotopic compositions (Fig. 6-12). This is not always the case (Amberger and Schmidt, 1987), although $\delta^{18}O_{NO_3}$ has proven to be an excellent tracer of denitrification. Böttcher et al. (1990) found a remarkably consistent Rayleigh enrichment, from which they calculated $\epsilon^{18}O_{product-NO_3} = -8.0‰$. Together with $\delta^{15}N_{NO_3}$, these are excellent tools to demonstrate denitrification in groundwaters (Fig. 6-12). Where denitrification proceeds via the oxidation of organic carbon, a reciprocal $\delta^{13}C$ depletion in the DIC should be observed.

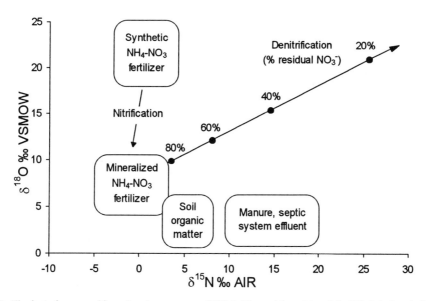

Fig. 6-12 The isotopic composition of various sources of NO_3^-. The enrichment trend for NO_3^- during denitrification is shown with % of original concentration remaining. $\delta^{18}O$ values for nitrate vary according to the $\delta^{18}O$ of local groundwaters. Ranges are for groundwaters with $\delta^{18}O_{H_2O} \sim -10‰$ VSMOW.

Nitrate contamination in shallow groundwaters

In agricultural regions, rural groundwater quality can suffer the combined impact of fertilizers and septic tank effluents. Nitrate levels in water wells are generally elevated, often exceeding water quality limits by three- to fourfold. Aravena et al. (1993a) document that it is possible to distinguish septic tank nitrogen from agricultural nitrogen and to define, on the basis of such analyses, a contaminant plume issuing from a septic tank weeping tile system. Regional agricultural inputs include NH_4NO_3 fertilizer, manure and soil organic N.

Oxidation to nitrate in the shallow groundwaters of this region provide a background contamination of 160 mg-NO_3^-/L, with $\delta^{15}N = 4.6 \pm 0.8‰$. Septic tank NH_4^+, characterized by the enriched $\delta^{15}N$ of catabolic organisms (humans), becomes nitrified within the shallow groundwaters. Concentrations vary from 100 to over 200 mg-NO_3/L. The $\delta^{15}N$ data along a section through the contaminated groundwaters (Fig. 6-13) clearly delineate this plume. The authors show that in this case, the oxidation of reduced nitrogen to nitrate within the groundwater system yields similar $\delta^{18}O$ values for both inputs. The difference between septic tank and non-plume nitrate was only about 1‰.

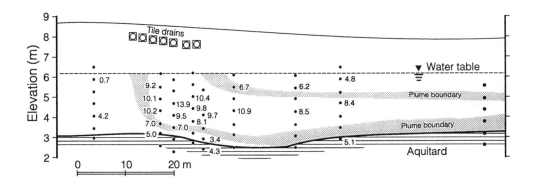

Fig. 6-13 The distribution of $\delta^{15}N_{NO_3}$ in groundwaters beneath a septic tank tile field (tiles shown as circles in squares) in Cambridge, southern Ontario. Nitrate is elevated in both the effluent and in regional groundwaters contaminated by agricultural NO_3^- (~160 mg-NO_3^-/L), and so the plume was define by Na^+ in the effluent. Nitrate in the septic effluent is distinguished from regional NO_3^- by $\delta^{15}N$ values that are close to 10‰. This contrasts with agricultural NO_3^- which is more depleted ($\delta^{15}N$ = 4.6‰) due to chemical fertilizer inputs. Section length is 90 m, with 4× vertical exaggeration (from Aravena et al. 1993a).

The "Fuhrberger Feld" Study

Denitrification is perhaps the only practical mechanism, beyond dilution, for remediation of NO_3^--contaminated groundwaters. The process requires a reduced substrate to provide the energy needed by NO_3^--reducing bacteria. Organic carbon is the most common, and can be supplied by DOC in septic tank plumes, organic-rich soils, or from the aquifer itself. Other viable substrates include sulphide minerals within the aquifer. The case study of the Fuhrberger Feld aquifer in Germany provides a textbook example.

The drinking water supplies for the city of Hannover in northern Germany rely heavily on groundwater. Most is extracted from an unconsolidated Quaternary aquifer that yields about $18 \cdot 10^6$ m³/yr of good quality groundwater. However, fear exists that the quality cannot be sustained. The region supports intensive agriculture, with an associated input of up to 200 mg-NO_3/L to the groundwater. Interestingly, increases in agricultural activity over the past decades are manifested in the Hannover water supply not by an increase in nitrate, but by sulphate. The nitrate in the water remains below 2 mg/L, whereas less offensive sulphate has increased from about 80 to over 250 ppm (Fig. 6-14).

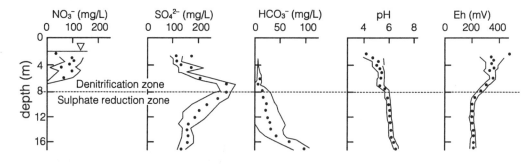

Fig. 6-14 Geochemical evolution in the Fuhrberger Feld aquifer system, showing zones of denitrification and sulphate reduction (from Böttcher et al., 1992).

Responsible for this twist of geochemistry is a process of denitrification based on oxidation of pyrite within the aquifer (Böttcher et al., 1992). At present, a minor decrease in sulphate is seen, although this may not be a long-term trend. When the oxidizable sulphide has been used up, denitrification will be limited and nitrate will increase in the main pumping wells. To asses the potential impact on water quality and establish possible remedial strategies, it was necessary to understand the geochemical evolution of the groundwater in the system.

Denitrification and ^{15}N

The groundwater system is characterized by recharge environments under forest and agricultural land and well-defined redox zones within the system: nitrate inputs of up to 200 ppm are typical for the agricultural recharge but are very low under forest. At less than 5 m depth in the aquifer (the water table in the recharge area is 6 to 10 m below ground surface), denitrification commences. Virtually all nitrate is transformed by the bacterium *Thiobacillus denitrificans* within the first 10 m below the water table (Fig. 6-14). Isotope analyses on the (residual) nitrate document that indeed denitrification takes place (Fig. 6-15) and that the attenuation of nitrate with depth is not simply mixing with groundwater which originates under the forest. So clear is this process, that a Rayleigh function (recall from Chapter 2) could be applied to both the $\delta^{15}N$ and $\delta^{18}O$ data to determine the enrichment factors that dominate in this environment (Fig. 6-16).

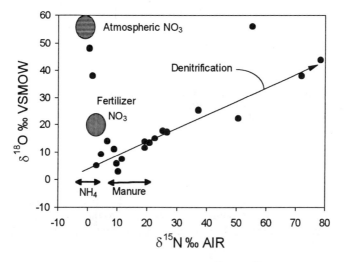

Fig. 6-15 Denitrification in the Fuhrberg aquifer shown by increases in $\delta^{15}N$ and $\delta^{18}O$ values (Böttcher et al., 1990).

The following reaction appears to dominate the denitrification in this zone:

$$5FeS_2 + 14NO_3^- + 4H^+ \rightarrow 7N_2 + 10SO_4^{2-} + 5Fe^{2+} + 2H_2O$$

and, less important in terms of denitrification

$$5Fe^{2+} + 7H_2O + NO_3^- \rightarrow 5FeOOH + \tfrac{1}{2}N_2 + 9H^+$$

This strong denitrification is dependent on several prerequisites: the presence of bacteria, a sufficient stock of oxidizable sulphur and appropriate redox conditions. The latter was believed to depend on the presence of lignite pebbles in the unconsolidated sediments. Since the lignite contains measurable amounts of reduced sulphur (Strebel et al., 1990), it was thought that both the organic matter and its sulphide were important for the denitrification process (note that these groundwaters contain on average 23 mg/L DOC). However, sulphur mass-balance calculations and the $\delta^{34}S$ data indicate that this sulphur originates primarily in the oxidation of finely disseminated pyrite and not the organic sulphur of lignite.

The limited role of lignite is also documented by isotope analyses on the DOC. The latter consists primarily of fulvic acids and has close to modern ^{14}C activities. This proves that the DOC substrate originates in the soil zone and not by bacterial decomposition of lignite, which is a highly refractory carbon source.

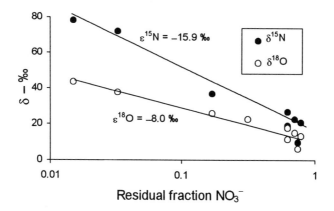

Fig. 6-16 Determining ^{15}N and ^{18}O fractionation during denitrification according to a Rayleigh function. Fraction NO_3^- remaining (f) determined as $f = NO_3^-{}_{measured}/NO_3^-{}_{original}$ (Böttcher et al., 1990).

Sulphate reduction at depth

Recharge under forest has very low nitrate concentrations and, therefore, no in-growth of sulphate is seen. However, forest recharge has close to 100 ppm sulphate. It originates most likely in fallout that has gone through redox cycling in the forest soils. Recharge under agricultural land has less sulphate but receives significant amounts by in-growth during denitrification.

In the deeper portion of this aquifer where nitrate concentrations are below detection, the reaction sequence is followed by sulphate reduction. The existing redox regimes are shown in Fig. 6-14 with a clear distinction between the zones of denitrification and sulphate reduction. Here, bacteria use sulphate to oxidize soil-derived DOC, which is reflected by the increase in HCO_3^-.

It is interesting to note that the existing wells and piezometers did not encounter a zone of methanogenesis, which often follows the sulphate reduction. The reduction of sulphate in the deeper portions of this aquifer is detected by the smell of H_2S and the decrease of sulphate concentrations (Fig. 6-14) as well as by enrichments in $\delta^{34}S$ and $\delta^{18}O_{SO_4}$ (Böttcher et al., 1990).

Source of chloride salinity

Saline groundwater is most often dominated by chloride salts, with sulphate and other anions as minor components. Sodium chloride salinity is characteristic of coastal groundwaters affected by seawater intrusion, or aquifers with remnant seawater. Na-Cl facies dominate in sedimentary basin brines, where Cl⁻ concentrations can exceed 100 g/L. Leaching of marine sediments is also a source of Na-Cl salinity. Less common is Ca-Cl salinity, although this facies dominates in brines from crystalline basement rocks where feldspar alteration is the source for Ca^{2+}.

Ionic ratio indicators

In geochemical studies, chloride is generally considered to be a conservative element. As with all halides, Cl is most stable in its –I valence state, and does not participate in any important redox reactions. It does not easily adsorb onto clays, nor do chloride minerals precipitate until Cl⁻ concentrations reach over about 200 g/L. Chloride is rather inert biologically, and so does not participate in the cycles that C, N and S do.

Variations in Cl⁻ concentration in natural waters are largely due to mixing within the hydrological cycle, as well as to evaporation, salt dissolution, ion filtration and diffusion. Tracing the source of salinity with only Cl⁻ concentration as a diagnostic tool is somewhat limiting. However, Na:Cl, Ca:Cl, $\delta^{18}O$:Cl and Br:Cl ratios, to name a few, are useful to diagnose mixing trends and geochemical processes in the subsurface. The Na:Cl ratio, for example, is generally preserved during seawater dilution due to the conservative nature of both ions in most groundwaters. Similarly, Br is conservative and so a good indicator of seawater salinity is its Br:Cl mass ratio of 0.0034. The stable isotopes ($^{37}Cl/^{35}Cl$) of chlorine can provide a different perspective.

Chlorine isotopes — $\delta^{37}Cl$

As chloride seldom participates in biological reactions, there is less fractionation than with ^{34}S, ^{15}N, ^{18}O or ^{13}C. Its conservative, inorganic nature leaves only physical processes for isotope partitioning. Even over a wide range of crustal temperatures and pressures, ionic chloride is relatively stable in aqueous solutions. However, Cl⁻ movement through these environments can be fractionating (Kaufmann, 1989). Eastoe and Guilbert (1992) found that Cl isotopes fractionate in hydrothermal systems. The most important fractionating processes seem to be physical mechanisms such as ion filtration, diffusion, geothermal boiling, brine evaporation and salt deposit formation. However, little work has been done to constrain the relative importance of these processes in fractionating ^{37}Cl. Until recently, studies of $\delta^{37}Cl$ in nature have focused on delineating ranges of values (Fig. 6-17). Eggenkamp (1994) summarizes recent work and develops fractionation and diffusion models to account for observed distributions.

Significant fractionation of ^{37}Cl occurs during seawater evaporation and chloride salt precipitation. The sequence of salts that precipitate begins with halite after loss of 91% of the original seawater volume. Potassium and magnesium salts do not precipitate from brines until 99% concentration. The ^{37}Cl enrichment between NaCl, KCl and $MgCl_2$ salts and their respective brine were measured experimentally (at 20°C) by Eggenkamp et al. (1994a):

$$\varepsilon^{37}Cl_{NaCl-solution} = 0.235 ‰$$

$\epsilon^{37}Cl_{KCl-solution} = -0.050 ‰$

$\epsilon^{37}Cl_{MgCl_2-solution} = -0.067 ‰$

It is interesting to note that halite is enriched over the residual brine, while K- and Mg- salts are marginally depleted. The authors point out that this reversal in fractionation does not allow strong Rayleigh depletions for $\delta^{37}Cl$ in late-stage evaporites. This implies that low $\delta^{37}Cl$ values in groundwaters must be due to diffusional fractionation.

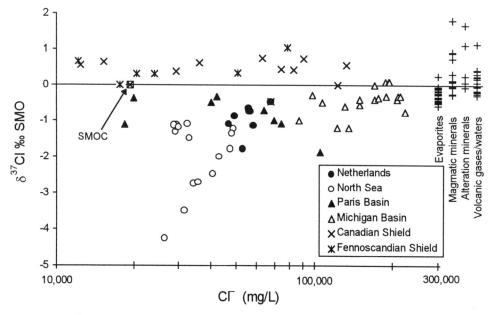

Fig. 6-17 The $\delta^{37}Cl$ composition of chloride brines from sedimentary basins and crystalline shield settings. Data from: Eggenkamp et al. (1994b) — Netherlands, Westland and Paris Basin; Coleman et al. (1993) — North Sea (depletion trend in these data identified as two end-member mixing); Kaufman et al. (1987) — Canadian Shield; Kaufman et al. (1993) — Michigan Basin; Eggenkamp (1994a) — evaporites, magmatic and alteration minerals; Magenheim et al. (1994) — magmatic rocks. Frape et al. (1992) show Canadian Shield brines with negative $\delta^{37}Cl$ values, suggesting that distinctions between basin and Shield brines are not so clear.

Much effort has been expended in trying to identify the origin of Cl⁻ in sedimentary basin brines (formation brines) and brines from crystalline shield environments. In these efforts, definition of the range in $\delta^{37}Cl$ of brines continues to improve (Fig. 6-17). General consensus holds that for the sedimentary basins, Cl⁻ originates in evaporites and remnant ancient seawater. Shield brines may have gained their salinity from leaching of fluid inclusions and halide-bearing minerals such as amphiboles (Kamineni, 1987) or leaching of highly modified Proterozoic seawater from shear zones (Guha and Kanwar, 1987).

Bottomley et al. (1997) present new δ^6Li data that support an origin of brines in the northern Canadian Shield as evaporated Paleozoic seawater. For mine waters with salinities exceeding 200 g/L, the δ^6Li values vary over a narrow range from −32.1‰ to −40.1‰, which is close to the value of modern seawater (−32.3 ± 0.5‰, Table 6-1). By contrast, the few data available for crystalline rocks are all heavier than −15‰.

Other sources of chloride salinity in groundwater include volcanogenic Cl⁻ (e.g. Giggenbach, 1992), alteration or dissolution of Cl-bearing minerals (Fig. 6-17), and of course, seawater infiltration. Unless fractionated by diffusion during infiltration, remnant seawater should preserve its characteristic $\delta^{37}Cl$ value of 0‰ SMOC.

The number of $\delta^{37}Cl$ measurements made for low-salinity groundwaters does not rival that of brines, although new data are available. Methods now even allow $\delta^{37}Cl$ of the low Cl⁻ concentrations in precipitation to be measured, which provide insights to the origin of atmospheric aerosols and atmospheric circulation (Tanaka and Rye, 1991).

Landfill Leachates

Municipal landfills in most regions must meet design and operational criteria to minimize the risk of groundwater contamination by leachate. Monitoring networks are also established to detect leakage from failed or old, unlined landfills. High salinity and elevated concentrations of certain compounds are used as evidence for leachate contamination. However, some rather extreme and characteristic values for environmental isotopes in leachates can also be measured and used as evidence for contamination.

The solutes present in landfill leachate are very much dependent on the landfill refuse. Municipal landfills contain a preponderance of organic material from household waste, vegetation, wood and paper. Decay and dissolution of organics establishes redox conditions that are highly reducing, favouring the generation of methane, hydrogen sulphide and even hydrogen gas (see Fig. 5-8 and reactions on page 130). Leachates with high DOC will generate large quantities of these gases. The isotopic fractionations associated with these bacterially mediated reactions are huge. Values for the $\delta^{13}C$ and δ^2H in the reaction products fall outside of the ranges associated with natural groundwaters.

While the presence of CH_4 in groundwaters cannot be taken alone as evidence of landfill leachate contamination, the $\delta^{13}C$ in the associated DIC provides support. Most landfills where strong methane production occurs have coexisting CO_2 or DIC that is enriched in ^{13}C (Fig. 6-18). According to the work of Lansdown et al. (1992) this suggests production via CO_2 reduction. Noteworthy is the trend at some sites to depleted $\delta^{13}C$ values in the CO_2. Hackley et al. (1996) attribute this to the subsequent oxidation of methane. Sources of oxygen and aerobic bacteria are be available in the zone of mixing between leachate and local groundwaters.

In groundwater systems with very high DOC loadings, production of methane can lead to a positive shift in the δ^2H of the water itself. The relevant equations involved in methanogenesis are given in Chapter 5. The δ^2H composition of methane is highly depleted from the water in which it is generated, due to the large fractionation factor ($10^3 \ln\alpha_{H_2O-CH_4}$ = 816‰ @ 25°C; Table 1 — front cover). However, a measurable positive shift requires that a large quantity of methane be generated. Siegel et al. (1990) shows a slight positive δ^2H-shift in groundwaters beneath a landfill in New York (Fig. 6-19).

A much stronger enrichment is observed in the leachates from landfills in Illinois (Hackley et al., 1996). With the strong enrichment in $\delta^{13}C$, it can be expected that the carbonate alkalinity (HCO_3^-) will increase. Although CO_2 is consumed by reduction to CH_4, there is also considerable DIC produced by the precursor reactions that involve the fermentation of high molecular weight organic compounds and the production of H_2.

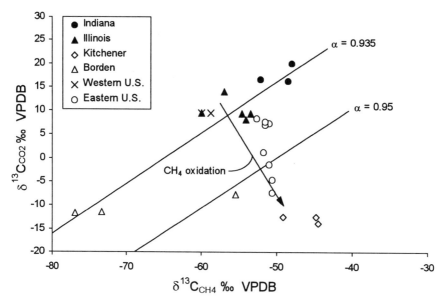

Fig. 6-18 The $\delta^{13}C$ composition of coexisting CH_4 and CO_2 in landfill gases and leachate. The strong enrichment in $\delta^{13}C_{CO_2}$ suggests CH_4 production via CO_2 reduction. General trend shown for oxidation of CH_4 by aerobic bacteria. Data from the following sources: Indiana, Western U.S. and Eastern U.S. — Games and Hayes, 1974; Illinois — Hackley et al., 1996; Kitchener and Borden — Barker and Fritz, 1981b.

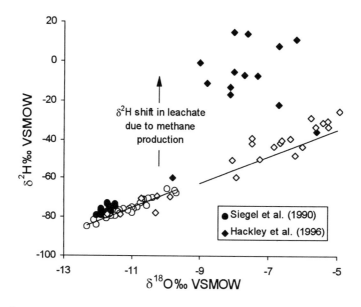

Fig. 6-19 Positive δ^2H shift in landfill leachates and groundwaters due to methane production (Siegel et al., 1990 — New York, leachate contaminated (●) and uncontaminated groundwaters (o); Hackley et al., 1996 — Illinois, landfill leachate (♦) with local precipitation data (◊) for Chicago from the IAEA GNIP).

The ^{14}C activity in dissolved carbon compounds and CH_4 from landfills provides another tracer. Hackley et al. (1996) measured ^{14}C activities up to 170 pmC in Illinois landfills. They attribute these activities, which greatly exceed modern atmospheric levels, to carbon released from organic refuse that dates to the era of atmospheric thermonuclear bomb tests. These authors also compile tritium data for landfill leachate from various sites, with levels up to 8000 TU. Such high levels exceed the ^{3}H activities measured in precipitation at the peak of the bomb testing era. Leaching of ^{3}H from luminescent paints that contain tritiated hydrocarbons is considered the most probable source.

Degradation of Chloro-organics and hydrocarbon

Leaking underground storage tanks are estimated to be the largest source of soil and groundwater contamination in North America by hydrocarbon compounds. Some of the more insidious groundwater contaminants are the chlorinated organics manufactured as solvents. The most common solvents include perchloroethylene (PCE, $Cl_2C=CCl_2$), trichloroethylene (TCE, $ClCH=Cl_2C$) and 1,1,1-Trichloroethane (TCA, CH_3CCl_3), used in a variety of industries. TCE, used in dry cleaning, is perhaps the most ubiquitous. With densities >1, these DNAPLs (dense, non-aqueous phase liquids) sink below the water table, leaving a residual pure phase that contributes to chronic contamination for decades.

PCE and TCA have solubilities of 200 and 480 mg/L, respectively. The solubility for TCE is 1100 mg/L, compared to the maximum acceptable concentration (MAC) for drinking water of 0.05 mg/L (Environment Canada, 1995). A developing field of environmental isotope geochemistry is that of fingerprinting contaminants in groundwater. Such forensic geochemistry can identify perpetrators of pollution, and may offer insights into transport processes and fate. The isotopes used are ^{13}C and ^{37}Cl.

Bartholomew et al. (1954) pioneered chlorine isotope research on organic compounds, showing that Cl–C bonding favoured the ^{37}Cl isotope. Fractionation of ^{37}Cl during manufacturing of chlorocarbons was demonstrated by Tanaka and Rye (1991). Variations in manufacturing processes and source materials then affect their ^{37}Cl and ^{13}C contents (Warmerdam et al., 1996). This serves to distinguish products from different manufacturers (Fig. 6-20). Noteworthy is the strong variation in both δ^{13}C and δ^{37}Cl for a single compound, reflecting differences in manufacturing processes. Measurement of these isotopes on solvents extracted from groundwaters at spill sites offers a tool to assign responsibility, and the cost of cleanup.

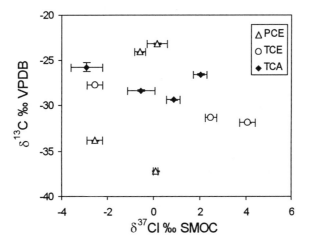

Fig. 6-20 The isotopic composition of chloro-organic solvents from four different manufacturers. One σ standard deviations shown for each sample, based on 2 to 10 analyses (modified from Van Warmerdam et al., 1996).

It is yet to be established whether any isotope effects accompany dissolution of the pure phase. A more ambitious area of research is to use isotopes to observe degradation of these contaminants (either by biodegradation or reductive dehalogenation) in groundwaters.

Remediation of leaked and spilled products by extractive technologies (pump and treat or by soil removal) are expensive and often ineffective. Biodegradation, on the other hand, is an in situ technology that can operate passively or by addition of nutrients and air to accelerate the bacterial oxidation of the contaminant. Loss of the contaminant can also occur by simple volatilization, but this can contravene air quality regulations and signals an inefficient biodegradation system. The efficiency of biodegradation can be tested by measuring CO_2 and $\delta^{13}C_{CO_2}$ in the soil gas.

Controlled biodegradation of gas-condensate (C_5 to C_{30} hydrocarbons) in laboratory experiments showed a steady increase in CO_2 from less than 3% to over 12% of the gas phase, accompanied by a decrease in O_2 from 18% to 11% (Aravena et al., 1996). Throughout, the $\delta^{13}C$ of the CO_2 remained within 1‰ of the bulk value for the condensate. Two field studies (Fig. 6-21) show an increase in the CO_2 content of the soil atmosphere as the contaminants are degraded. The $\delta^{13}C_{CO_2}$ decreases with the exponential increase in P_{CO_2}, approaching asymptotically the $\delta^{13}C$ value of the contaminant. The isotope effects during degradation may include selective oxidation of certain components of the hydrocarbon by bacteria, rather than the quantitative degradation of the compounds (Stahl, 1980).

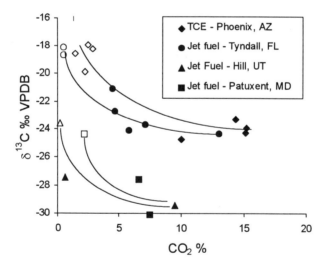

Fig. 6-21 Biodegradation of solvent and jet fuel in contaminated soils. Open symbols represent uncontaminated soils associated with contaminated soils (filled symbols). Data sources: TCE — Suchomel et al. (1990); Jet fuel — Aggerwal and Hinchee (1991).

Sensitivity of Groundwater to Contamination

Federal and regional jurisdictions are becoming increasingly aware of the vulnerability of certain groundwater resources to contamination from land use activities. Consideration is now given to the sensitivity of groundwater resources to contamination from land use, particularly in recharge areas. Groundwaters with short response times following meteoric events are highly sensitive to contamination from the surface. Rapid recharge minimizes retention time in the soil, where sorption and biodegradation attenuate contaminants. Further, strong response to meteoric events indicate phreatic conditions, with little to no protection from aquitards. Assessing the response to meteoric events, and the protection offered by aquitards can be aided by the monitoring of stable isotopes, both within the aquifers of concern, and within adjacent aquitards.

Temporal monitoring with stable isotopes

Time series monitoring provides a qualitative assessment of the geologic buffer zone that may exist between the surface and an aquifer, and supports other sources of data such as water level fluctuations and geochemistry. Seasonal variations in the ^{18}O and ^{2}H content of precipitation provide the most useful meteoric signal to monitor in groundwater. They are established through monthly or even weekly sampling of groundwaters. A sampling program for local precipitation is recommended, and is particularly important for detailed studies. Where sampling of local precipitation is not possible, data from regional monitoring stations (the IAEA monitoring network) can be used.

The preservation of a seasonal or event-scale meteoric signal in groundwater was discussed in Chapters 3 and 4. Figs. 4-1 and 4-2 show the increased attenuation of seasonal signals with depth, indicating an increase in the buffer zone for groundwater protection. In Fig. 4-4, the fully attenuated seasonal signals in the deep aquifer suggest that it is less vulnerable to activities at the surface than the unconfined aquifer.

Insights to transport pathways from surface can also be gained through temporal monitoring of other isotopes. An example is the seasonal variation of $\delta^{13}C_{DIC}$ in groundwaters from an agricultural watershed in eastern Ontario (Cane, 1996). Over 7500 water wells tap a regional carbonate aquifer at a depth of 25 m, which is overlain by sandy till. The watershed is heavily cultivated with corn for cattle feed and so there is concern that nitrate, organics and pesticides could be transported to the aquifer if there is significant local infiltration.

Corn is a C_4 plant and is enriched in ^{13}C relative to the natural C_3 type vegetation of the region (see Fig. 5-1 and discussion on page 119). Monitoring of $\delta^{13}C_{DIC}$, it was hoped, would help distinguish between regional groundwater and local contributions infiltrating through the fields. Biweekly to monthly sampling (Fig. 6-22) shows a remarkably consistent pattern of seasonal shifts in the $\delta^{13}C$ of DIC. When the water table was low during summer and mid-winter, values are enriched, in the range that would be expected following carbonate dissolution by CO_2 from C_4 vegetation. The low water level at these times widens the unsaturated zone and increases the aerobic oxidation of crop residues tilled into the subsurface. The enriched $\delta^{13}C$ values indicate that the DIC must be derived from corn, and so local infiltration through the fields is important, signalling a high potential for contamination from pesticides, nitrate and other agrochemicals. During periods of high water table and saturated soils, the DIC is dominated by the natural C_3 vegetation, although the increases in DOC indicate that infiltration from the agricultural areas still persists (Fig. 6-22).

Aquitards — impermeable or leaky barriers?

The movement of groundwater and solutes in aquitards has implications for contaminant migration to adjacent aquifers. In low permeability materials such as unfractured clays, glacial tills, shales, and intact crystalline rocks, advective flow is generally less than about 1 cm/yr, and molecular diffusion becomes the dominant transport mechanism. Although they are not great aquifers, groundwater flow and transport in aquitards are important aspects of contaminant hydrogeology. Aquitards offer protection to underlying aquifers from contaminant sources on the surface such as landfill sites or development. Aquitards are also important in the siting of repositories for hazardous and radioactive waste.

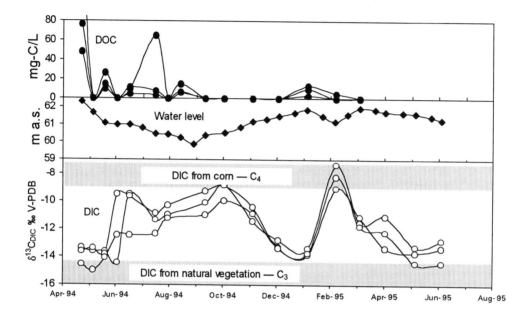

Fig. 6-22 Variations in $\delta^{13}C$ for DIC monitored in three domestic water wells in a shallow carbonate aquifer in the Raisin River agricultural watershed, Cornwall, Ontario. Shaded areas represent approximate ranges for DIC from C_3 and C_4 vegetation after dissolution of limestone. Infiltration through corn fields (C_4 vegetation) is suggested by seasonal shifts to enriched values. Saturated soils during spring and late fall preclude aerobic activity in the fields, as seen by increases in DOC levels at these times (from Cane, 1996).

A curious example using time-series data demonstrates that aquitards do not always provide the protection that they appear to. Stable isotopes were monitored over an 18-month period for groundwaters in the Jock River agricultural watershed of eastern Ontario. The regional limestone aquifer, confined throughout most of the basin by marine clay, is heavily exploited by local communities and farmers. Data from the unconfined and confined portions of the aquifer were presented in Fig. 4-4 to show the principle of "critical depth" for recharge. Although seasonal variations in the confined groundwaters have essentially been attenuated, sampling in June 1992 shows that leakage of surface waters into this confined aquifer can occur.

During this particularly hot and dry spring, evaporation from the wetlands discharge area produced an enrichment in ^{18}O and ^2H (inset in Fig. 6-23). Oddly, this evaporated water is observed in the confined groundwater too. Anomalously enriched values were measured for all confined bedrock samples for June 1, 1992 (Fig. 6-23). Heavy exploitation by local groundwater users apparently caused a temporary reversal of hydraulic gradients, and the wetland discharge area began recharging this aquifer — a potentially disastrous scenario depending on the quality of water in the wetland.

Diffusion across aquitards

Determining the rate of solute movement through aquitards is critical to long term safety assessment. The distribution of isotopes across aquitards provides information on solute diffusion rates. Oxygen-18 can be used if a sufficiently strong gradient exists. Chlorine isotopes are proving to be useful because of the measurable fractionation and the conservative nature of Cl⁻. Geochemically reactive isotopes such as ^{34}S and ^{13}C are unlikely to show the effects of diffusion.

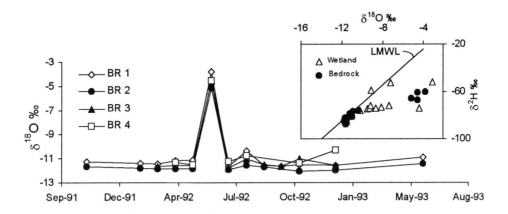

Fig. 6-23 Monitoring groundwaters in a carbonate aquifer confined by marine clay, Jock River Basin, eastern Ontario. These groundwaters have flowed well beyond the "critical depth" and seasonal variations are fully attenuated. The overlying clay barrier offers protection from contamination in this agricultural watershed, although infiltration from waters in the wetlands discharge zone can occur under conditions of over-pumping (modified from Velderman, 1993).

When advective flow is low (low permeability or low hydraulic gradient), solute movement is controlled by molecular diffusion D across a concentration gradient $\partial C/\partial x$. Fick's first law defines the flux, j, over a given concentration gradient:

$$J = -D\frac{\partial C}{\partial x}$$

Isotope fractionation during diffusion arises from the difference in the diffusion coefficient, D, of the two isotopic species, i.e. D_{16O} is marginally greater than D_{18O} and so has greater kinetic energy.

During gaseous self-diffusion (recall from Chapter 2), the ratio of diffusivities (or velocities, v) equals the square of the mass ratio, giving the fractionation factor for the heavy (m*) and light (m) isotopes:

$$\alpha_{m^*-m}\frac{D^*}{D} = \frac{v^*}{v} = \sqrt{\frac{m}{m^*}}$$

For aqueous-phase diffusion, the fractionation factor is simply the ratio of the heavy to light isotope diffusion coefficients. However, in the case of diffusion through a medium, Graham's law must take the medium's mass into account. The hydrated form must also be considered. Using the example of $\delta^{37}Cl$ for Cl⁻ in water, the species $Cl(H_2O)_4^-$ with masses 107 and 109 is likely the dominant form of chloride (Desaulniers et al., 1986). This gives a maximum fractionation of:

$$\alpha^{37}Cl = \frac{D_{^{37}Cl}}{D_{^{35}Cl}} = \sqrt{\frac{107(109+18)}{109(107+18)}} = 0.9974$$

In this formulation, 18 is the mass of H_2O through which the Cl^- is diffusing. Such fractionation is consistent with observed diffusion profiles, from which diffusion ratios can be calculated. A value for $\Delta^{37}Cl = -2.6$ and $-2.2‰$ was measured by Desaulniers et al. (1986) who calculated $\alpha^{37}Cl$ values of 0.9974 and 0.9978 in marine clays near Montreal. Eggenkamp et al. (1994b) determined a similar value of $\alpha^{37}Cl = 0.9977$ for Cl^- diffusion through freshwater sediments of Kau Bay, Indonesia.

The element of time is found in the amount of isotope fractionation (Fig. 6-24), and shape of the diffusion profile (inset). However, the boundary conditions are also important. In the simple case here, diffusion is from a constant concentration and constant $\delta^{37}Cl$ boundary into a sediment with no initial Cl^- (infinite initial-concentration gradient). Where diffusion takes place from high concentration to low concentration (finite gradient) the depletion in the $\delta^{37}Cl$ profile is limited by the isotope ratio in the down-gradient Cl^-. A minor advective flow component will also affect the profiles.

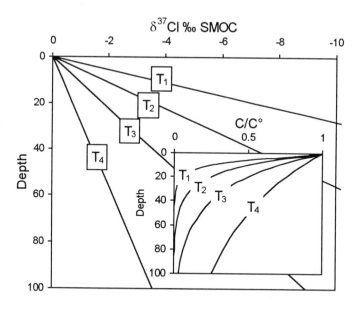

Fig. 6-24 Diffusion profiles using the example of $\delta^{37}Cl$ and Cl^- (inset). Depth and time units are arbitrary. A fractionation factor of 1.0026 was used to determine isotope depletion along the diffusion profile for increasing time. Conditions assume downward diffusion with constant concentration on the upper boundary.

Heat flow theory has provided the analytical solutions to Fick's laws (Carslaw and Jaeger, 1959) to model solute diffusion (Darcy's law for advective flow is also derived from heat flow). Theoretical diffusion profiles for a variety of solute boundary conditions are developed by Eggenkamp (1994) and Desaulniers et al. (1986).

Desaulniers et al. (1981) examined the downward movement of modern meteoric waters through low permeability clays and clay tills in southwestern Ontario (Fig. 6-25). Hydraulic conductivities are low (<10^{-7} cm/s) over the 30-m thickness of the clay. The hydraulic head shows a slight downward gradient, although advective groundwater velocities were determined to be less than about 0.03 cm per year. Modern tritium-bearing groundwater is found only in the upper weathered zone (see Chapter 7 for discussion on ^{3}H in modern waters). The δ^{18}O profile shows diffusive mixing between recent meteoric waters in the upper zone (–10‰) and a late glacial groundwater (–17‰) at the base of the profile (Fig. 6-25).

The age of the sediments themselves (~12–15 ka) limits the age of these paleogroundwaters. Modelling diffusion shows the slow downward diffusion of ^{18}O (and ^{2}H) from the shallow groundwaters towards the ^{18}O-depleted late-glacial groundwaters. The observed mixing curve was established with a 10 to 15 ka period of molecular diffusion. The *upward* diffusion of ^{37}Cl from the deeper paleogroundwaters towards the modern low-Cl$^-$ waters produced a similar profile after 10 ka (Desaulniers, 1986).

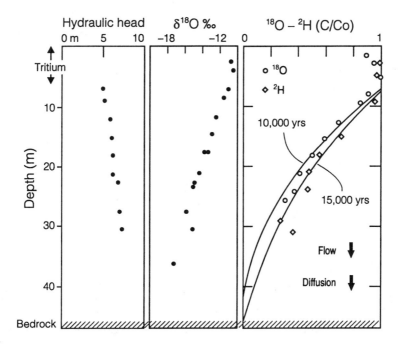

Fig. 6-25 The vertical distribution of hydraulic head and δ^{18}O in clay till at Sarnia, Ontario. The third panel shows normalized ^{18}O and ^{2}H profiles with results of diffusion modelling for both isotopes (modified from Desaulniers, 1981). Upper zone of modern tritium-bearing pore waters indicated on left.

Summary of Isotopes in Contaminant Hydrogeology

As a summary to water quality problems in hydrogeology, this section presents the roles that environmental isotopes can play in monitoring and understanding the processes involved. Environmental isotopes can play a role in most contaminant hydrogeology studies by responding to questions of groundwater recharge, mixing, residence time and response to meteorological events. Here, we summarize some of these and other "contaminant-specific" isotope methods in the more common domains of contaminant hydrogeology research. This is a growing field, and this section will no doubt be outdated shortly after publication.

166   Chapter 6   Groundwater Quality

Contamination in agricultural watersheds

The application of both fertilizers and pesticides/herbicides threatens the quality of groundwater and surface water in agricultural watersheds. The reliance by farming families and rural communities on potable groundwater supplies compounds the risk. The predominance of septic systems and municipal waste dumps in such watersheds adds a further complication to understanding sources and pathways of groundwater contaminants.

Recharge origin and mixing in groundwater
- Monitoring ^{18}O and ^2H is a fundamental tool to determine the recharge origin and mixing of groundwater in water-supply aquifers.
- The ^{13}C in DIC and DOC distinguishes between sources of carbon, monitors solute pathways, and identifies recharge through cultivated fields. This approach benefits from the difference between C_3 and C_4 type vegetation, and differences in recharge conditions.
- ^{14}C is a less routinely employed isotope in contaminant studies, but offers potential in distinguishing carbon sources in groundwaters.
- In this area of investigation, the usefulness of time-series data spanning at least one to several years provides a wealth of information on the dynamics of recharge.

Nitrate and ammonium from fertilizers, feedlots and manure stockpiles
- ^{15}N and ^{18}O in NO_3^- and nitrogen compounds to determine sources (fertilizers vs. septic effluent), and to document whether denitrification may be occurring.
- DIC and DOC participate in the denitrification process. Monitoring these species is of primary interest.

Chlorocarbon pesticides
- The use of ^{13}C and ^{37}Cl has yet to be proven, but may help determine the fate of pesticides after application. Distinguishing biodegradation, volatilization and export with foliage are future possibilities.

Septic effluent
- Septic systems are potential sources of DOC, NO_3^-, and fecal coliform bacteria in well water. Distinguishing between domestic and agricultural sources relies on the conjunctive use of ^{15}N and ^{18}O in NO_3^-. They also are used to show denitrification.
- ^{13}C in DOC and DIC can be important to trace carbon sources.
- Cl^- and potentially ^{37}Cl can signal septic effluent in agricultural watersheds.

Soil deterioration
- Soil salinization and water logging are problems in irrigated fields. Soil acidification by oxidation of high S soils is also a problem for arable lands recovered from coastal areas.
- ^{18}O and ^2H in shallow groundwaters and soil moisture columns can be used to quantify evaporative losses and water budgets.
- ^{34}S and ^{18}O in SO_4^{2-} bring insights into sources of salinity and oxidation processes.
- Carbon cycling and distinguishing active (labile) vs. refractory C in soils is traced with ^{13}C and ^{14}C in DIC and DOC.

Groundwater resources
- In arid regions, as well as temperate zones, groundwater resources are important for irrigation and potable water supply. Over exploitation and supply are at issue.

- ^{18}O and ^2H are important in groundwater provenance studies.
- ^3H, ^{14}C and other dating methods are necessary to assure the long-term viability of exploitation schemes. This is particularly so in arid regions where fossil groundwaters are generally used.

Sanitary landfills

In the past, municipal garbage dumps (sanitary landfills are only a recent technology) were unlined and sited with little regard to the local hydrogeology. The consequence is that leachate and gas migration is now threatening groundwater resources. Licensing modern sanitary landfills requires monitoring programs in the event of a failure in the liner or leachate collection systems. Environmental isotopes in conjunction with geochemistry play a role in identifying leachate migration from older, non-engineered sites or modern landfills.

Identification of leachate contamination
- Methane production from DOC is an indication of the strongly reducing conditions in landfill leachate. Both ^{13}C and ^2H in CH_4 are indicators of biogenic methane and can be used to identify the source.
- The strong enrichment in ^{13}C of DIC generated through methanogenesis in landfills is characteristic of leachates. This would contrast with the low values for uncontaminated groundwaters.
- The δ^2H content of leachate water can be enriched by extensive methanogenesis, and offers additional evidence of that process.
- $\delta^{15}N$, $\delta^{34}S$ and $\delta^{18}O$ can be used to identify the origin of N- and S- species in groundwaters where landfill leachate or other sources may occur.
- $\delta^{37}Cl$ and other isotopes in conjunction with chemistry may prove useful in identifying leachate contamination of groundwaters.

Siting of new facilities
- Site evaluations and baseline monitoring are now universally required for licensing new sanitary and other landfill facilities, to demonstrate the security of groundwaters from contamination. The application of $\delta^{18}O$ and δ^2H is necessary to determine groundwater origin.
- Where aquitards are present, ^{18}O, ^2H, ^3H and possibly ^{37}Cl can be used to determine rates of diffusive transport and whether rapid movement from surface through fractures and discontinuities may take place.

Fuel and solvent contaminated sites

Remediation of soils and aquifers contaminated with organic liquids can be achieved through methods such as biodegradation and reductive dehalogenation. Volatilization is discouraged because of restrictions on emissions to the air. Attenuation through sorption is also less desirable since the contaminant is not removed from the aquifer. Isotopes can be useful in monitoring the mechanism, pathway and effectiveness of contaminant loss. They may also help fingerprint the contaminant and establish "ownership" for clean-up.

- Low ^{13}C contents of fuels and solvents produced from fossil fuel contrast with that of DIC in most natural groundwaters. The $\delta^{13}C$ of DIC and DOC can then help identify products of degradation. An approach using carbon mass-balance equations is available to quantify contaminant degradation.

- The build-up of soil CO_2 can often provide evidence for biodegradation. The addition of $\delta^{13}C$ measurements can help determine the substrate and efficiency of biodegradation.
- Cl^- and ^{37}Cl in degradation products down gradient of treatment walls may prove to be a useful monitoring tool.

Siting hazardous waste facilities

Site characterization for toxic chemical and radioactive waste storage and disposal facilities requires a rigorous evaluation of transport in the geosphere/hydrosphere. Environmental isotopes have been used extensively to characterize groundwater movement and reactions with the far-field geological environment (e.g. IAEA, 1983; Pearson et. al. 1991). The range of applications embraces the entire subject of environmental isotope hydrogeology, including the tracing of groundwater origin and movement, diffusion, age, and solute-specific transport and reaction.

Problems

1. Given that the $\delta^{34}S$ of sulphur in the Earth at the time of its formation was a primeval value close to that found in meteorites (i.e. 0‰ CDT), account for the wide spectrum of $\delta^{34}S$ we now observe in crustal rocks.

2. Mine engineers are concerned about the significant inflows of saline groundwaters to the deeper adits in a mine situated on the Pacific coast in western Canada. Extension of current mining activities further under the sea in this Permian, sedimentary-volcanics hosted massive-sulphide deposit, depends on your assurances that seawater is not the source of these inflows. These analyses (in mg/L) are for samples collected from various levels within the mine, as well as local shallow groundwaters.

Sample	pH	HCO_3^-	SO_4^{2-}	Cl^-	Ca^{2+}	Na^+	$\delta^{34}S$	$\delta^{18}O_{SO4}$	$\delta^{18}O_{H2O}$	δ^2H_{H2O}
Shallow groundwaters										
	6.20	16	16	194	25	108	21.1	9.5	-10.7	-81.46
Mine waters										
MF1	6.52	17.9	90	975	52	552	5.3	2.1	-10.3	-78
MF2	6.89	20.5	222	1950	79	1210	0.5	-0.3	-9.7	-76
MF3	7.12	22.0	272	2370	136	1523	0.1	0.8	-9.5	-74
MF4	7.03	22.4	207	2560	104	1612	1.9	1.5	-9.3	-76
MF5	6.81	22.2	223	2700	96	1568	1.1	-0.6	-9.5	-73
MF6	6.95	23.9	260	3110	113	1961	2.2	2.5	-9.2	-73
MF7	7.03	23.3	184	4700	98	2500	3.0	0.2	-8.2	-69
MF8	7.23	25.4	317	3860	141	2310	3.6	1.4	-8.7	-70
MF9	7.18	26.2	367	4580	160	2494	4.2	3.3	-8.4	-69
MF10	7.30	27.5	420	4580	180	2440	3.3	3.0	-8.2	-69
MF11	7.23	30.3	393	5340	160	2900	5.2	4.9	-7.8	-67
MF12	7.48	38.5	506	6900	204	4100	7.1	6.4	-7.1	-64
MF13	7.52	37.4	939	16100	350	9110	11.8	12.6	-2.5	-39
MF14	7.54	31.8	1057	16500	432	9000	12.5	12.9	-2.4	-42
MF15	7.61	28.8	992	17910	382	9930	13.1	13.0	-2.0	-46
MF16	7.65	30.2	864	15700	366	8300	11.6	13.0	-2.9	-41
MF17	7.56	30.3	1057	18000	420	9260	12.1	12.9	-1.8	-42
MF18	7.55	30.0	1093	17200	452	9700	11.7	12.8	-2.5	-46

- Identify the geochemical and isotopic characteristics of the saline end member and determine its origin. Is modern seawater infiltrating into the mine?

- What are the mixing ratios for the low salinity mine waters.

- What are the sources of sulphur in this system?

- Is sulphate reduction taking place in these waters?

- What is the origin of the high salinity mine waters?

3. The following data for groundwaters collected from monitoring wells situated along the flow path from a landfill site. The high sulphur contents indicate suggest leakage from the landfill, where both high S-organics and gyp-roc (gypsum wall board) from construction waste have been put. What is the origin of sulphur in the leachate, and account for its evolution along the flow path. The groundwater temperature here is close to 25°C.

Sample	pH	HCO_3^-	SO_4^{2-}	HS^-	Ca^{2+}	$\delta^{34}S$	$\delta^{18}O_{SO4}$	$\delta^{18}O_{H2O}$
TR-0m	7.55	32	1020	0	398	11.8	9.6	-12.20
TR-20m	7.69	89	927	33	127	14.0	10.9	-11.70
TR-35m	7.86	166	800	77	43	16.7	15.8	-11.88
TR-85m	7.75	224	705	110	51	15.0	13.3	-11.65
TR-150m	8.18	337	520	175	16	21.8	16.8	-12.20
TR-260m	8.44	345	507	180	12	26.0	18.5	-11.80
TR-350m	8.67	459	320	245	8	29.7	19.0	-11.90
TR-500m	8.57	466	309	249	5	28.1	17.9	-12.10
TR-750m	9.10	508	240	273	4	36.6	17.9	-11.70

Explain the decrease in Ca^{2+} and increases in pH and HCO_3^- along the flow system. Calculate the gypsum and calcite saturation indices to support your interpretations.

4. Groundwaters were sampled from an industrial landfill containing gyp-roc (gypsum) which has led to extensive production of H_2S. The oxidation of organics has raised the temperature of the landfill to 18°C. Leachate from the landfill has an average δ^2H value of –68‰, which is enriched over that of local precipitation by 3‰. Can you determine the amount of H_2S production in moles per litre?

Chapter 7

Identifying and Dating Modern Groundwaters

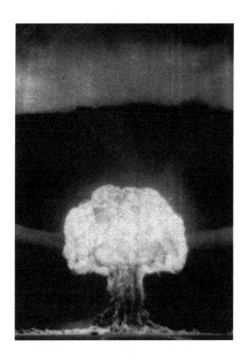

The "Age" of Groundwater

Groundwater may not necessarily age as well as some wines, but some very good aquifers in the world host groundwaters many thousands of years old. The "age" of a groundwater has important implications for water resource management. Using groundwaters that are not actively recharged is mining. On the other hand, groundwater that is part of the modern hydrological cycle is continuously renewed. Its exploitation is potentially sustainable. Deep and/or regional groundwaters can be mixtures of modern and older inputs. However, the presence of even a minor component of modern recharge is important because it indicates a hydraulic connection with an active flow system. Increased pumping may then increase the proportion of the modern contribution.

The age of a groundwater can be a misleading term for two reasons. (1) Only tritium is part of the water molecule and can thus actually "date" the water. All other dating methods rely on dissolved constituents whose abundance in water is controlled by physicochemical and biological processes. (2) Hydrodynamic mixing and convergence of groundwater flow paths integrate a variety of recharge origins and ages. Only in well-defined and usually regional artesian aquifers will age gradients along the flow path be preserved. As a consequence, one should not talk about the age of a water but about *"groundwater mean residence times"* (MRT). Yet, the term "age" is simpler and has slipped into common usage, as well as into this book.

"Modern" groundwater

In practical terms, "modern" groundwaters are those recharged within the past few decades and so are part of an active hydrological cycle. Classical methods are often the best indication of whether groundwaters are actively recharged. Evidence from hydrogeological mapping, seasonal fluctuations in water level, temporal variations in geochemistry or stable isotopes, and anthropogenic pollution (e.g. nitrate) are indications of active recharge. However, isotopes are used when such hydrogeological information is ambiguous, and more importantly, to constrain the age of recently recharged water.

Tritium has become a standard for the definition of modern groundwater. The era of thermonuclear bomb testing in the atmosphere, from May 1951 to 1976, provided the tritium input signal that defines modern water. Its decay from natural, pre-bomb levels is such that it cannot normally be detected in groundwaters recharged before about 1950. Modern groundwaters are then younger than about 45 years relative to the mid-1990s. Tritium-free groundwaters are considered "submodern" or older.

The tools for dating groundwater

A number of isotope methods can be used to assess mean residence times. Seasonal variations in stable isotopes provide a measure of age. The more routinely applied techniques are based on the decay of radionuclides (Table 1-3). Those with a long half-life (^{14}C, ^{36}Cl, ^{39}Ar and ^{81}Kr) can be used to date paleogroundwaters, and are discussed in Chapter 8. Short-lived radioisotopes (^{3}H, ^{32}Si, ^{37}Ar, ^{85}Kr and ^{222}Rn) and those produced by man's nuclear activities over the past four decades (^{3}H, ^{14}C, ^{36}Cl and ^{85}Kr) indicate modern recharge.

While the methods for analysis of ^{3}H and ^{14}C are now routine, other radioisotopes require complicated sampling and/or analytical techniques. The "submodern" period in the dating range between modern waters and paleogroundwaters is problematic. While this >45 to ~1000 year range can potentially be filled by ^{39}Ar dating ($t_{1/2}$ = 256 yr), this method requires rather ideal aquifers, very large samples, complicated sample preparation techniques and special counting facilities. Very few laboratories can afford this and, therefore, ^{39}Ar dating has not developed into a routine tool.

Stable Isotopes

The strong correlation between temperature and stable isotopes in meteoric waters (Chapter 3) provides a seasonal signal that can be used to date groundwaters. The amplitude of seasonal variations in $\delta^{18}O$ and $\delta^{2}H$ is attenuated during groundwater recharge (Fig. 4-3). The preservation of seasonal variations signifies short mean residence times.

The attenuation of seasonal variations during infiltration and groundwater flow is a function of the physical/hydraulic parameters of the unsaturated zone and/or aquifer medium. The rate of attenuation is proportional to the length of the flow path travelled, and, by consequence, to time. Time can be determined by monitoring the seasonal variation in $\delta^{18}O$ or $\delta^{2}H$ in precipitation and in the groundwater over at least 1 year.

The precipitation input is approximated by a sinusoidal function. Its decreasing amplitude in groundwater can be expressed as a function of time by the following equations:

for precipitation $\delta_{prec}(t) = A \sin 2\pi t$

and

for groundwater $\delta_{gw}(t) = B \sin (2\pi t + \alpha)$

where $\delta_{prec}(t)$ and $\delta_{gw}(t)$ are the input and output isotopic composition at time t, and A and B represent the amplitude (in ‰) of the precipitation and groundwater variations, respectively, and α is the phase shift of the output function as compared to the input function. Damping of the amplitude is defined as C = B/A, and the mean residence time (MRT) of the water is:

$$MRT = 1/2\pi (1-C)^{1/2} / C \qquad \text{Burgman et al. (1987)}$$

or can be determined on the basis of the phase displacement α:

$$MRT = \tan \alpha / 2\pi \qquad \text{Stichler (1980)}$$

Application of this interpretation to field data shows that measurable, seasonal variations generally disappear in the reservoir within 2 to 3 years, depending on the magnitude of the input variations.

This approach can also be applied to baseflow in streams if the residence time of water in a catchment basin is to be determined. Comparison of precipitation input with discharge for several rivers in Sweden over a 10-year period yielded mean residence times of 2 to 27 months (Burgman et al., 1987). Taylor et al. (1989) studied in a similar manner the catchment response to precipitation inputs in the Weimakariri River in New Zealand. Their data (Fig. 7-1) show mean residence times of water in the catchment of about 1 year. Note that the isotope variations for the 5-year model output are within the analytical uncertainty.

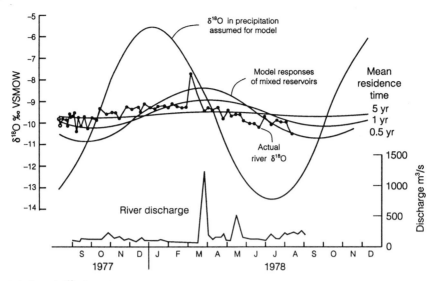

Fig. 7-1 Variations of $\delta^{18}O$ in Weimakiriri River compared to the sinusoidal input from precipitation. Discharge values for mean residence times in the well mixed catchment of 0.5, 1 and 5 years are shown (from Burgman et al., 1987).

Tritium in Precipitation

Tritium (^3H or T) is probably the most commonly employed radioisotope used to identify the presence of modern recharge. It is a short-lived isotope of hydrogen with a half-life of 12.43 years (Unterweger et al., 1980). It is directly incorporated into the water molecule ($^1H^3HO$ or ^1HTO) and so is the only radioisotope that actually dates groundwater. Tritium is produced naturally by cosmic radiation, although much greater production accompanied the atmospheric testing of thermonuclear bombs between 1951 and 1980. By ~1990 most of this "bomb" tritium had been washed from the atmosphere, and tritium levels in global precipitation are now close to natural levels. Small releases from nuclear power plants and weapons plants may preclude reaching the natural levels measured in pre-1951 water. Over the past three decades, the presence of thermonuclear tritium in groundwater was used as clear evidence for active recharge. While thermonuclear tritium can still be found in some slowly moving groundwaters, the largely natural ^3H signal is now relied upon for dating modern groundwaters.

Cosmogenic tritium

Natural tritium is formed in the upper atmosphere from the bombardment of nitrogen by the flux of neutrons in cosmic radiation, following the reaction:

$$^{14}_7N + ^1_0n \rightarrow\ ^{12}_6C + ^3_1H$$

The tritium thus formed combines with stratospheric oxygen to form water:

$$^3H + O_2 \rightarrow\ ^3HO_2 \rightarrow\ ^1H^3HO$$

Tritium decays to ^3He by beta release:

$$^3H \rightarrow\ ^3He + \beta^-$$

Tritium concentrations are expressed as tritium units (TU) where:

$1 \text{ TU} = 1 \ ^3\text{H per } 10^{18}$ hydrogen atoms

and $\quad 1 \text{ TU} = 0.118 \text{ Bq·kg}^{-1} \quad (3.19 \text{ pCi·kg}^{-1})$ in water $\quad$ (IAEA 1983)

Natural tritium levels in precipitation are very low and represent a secular equilibrium between natural production and the combination of decay in the atmosphere plus loss to the hydrosphere and oceans. Production rates and concentrations in precipitation are a function of geomagnetic latitude, with greater production at higher latitudes. Very few measurements exist of natural, pre-bomb tritium in precipitation. Precipitation near Ottawa had about 15 TU, as measured in laboratory reagents mixed with Ottawa River water before 1951 (Brown, 1961). Kaufman and Libby (1954) used vintage wines to determined pre-bomb ^{3}H concentrations of 3.4 to 6.6 TU for precipitation at the lower latitudes of the Naples NY, Bordeaux and Rhône regions. In Fig. 7-2 we see the concentrations of cosmogenic tritium measured in meteoric water prior to 1951, and the increases observed during the early tests of hydrogen-fusion devices.

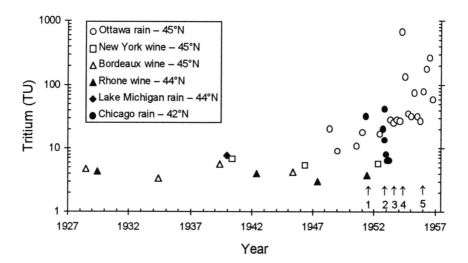

Fig. 7-2 Tritium levels in precipitation prior to and during the earliest atmospheric tests of thermonuclear devices. Ottawa rain — Brown, 1961; wines — Kaufman and Libby, 1954; Chicago rain, Lake Michigan — Kaufman and Libby, 1954). Atmospheric tests: 1 — George, 75 kilotons, U.S.A.; 2 — Ivy-Mike, 10 megatons, U.S.A.; 3 — RDS-6s, 400 kilotons, U.S.S.R.; 4 — Castle series, total 47 megatons, U.S.A.; 5 — second Soviet thermonuclear test, 2 megatons.

Thermonuclear (bomb) tritium

On May 9, 1951, the world's first thermonuclear flame was ignited by a ^{2}H-^{3}H-^{235}U device code-named George[1]. The Hagiwara-Fermi-Teller concept of hydrogen fusion had been successfully tested. On the first day of November, 1952, The Ivy-Mike shot demonstrated that megaton energy releases could be achieved by hydrogen fusion and the thermonuclear superbomb was born. The atmospheric detonation of these devices began a period of anthropogenic ^{3}H production which raised concentrations in the stratosphere by several orders of magnitude.

---

[1] Physics Today, November 1996 <http://www.aip.org/pt/>.

Early configurations of the hydrogen bomb used the radiation of ^{235}U-fission to compress the deuterium-tritium fuel to initiate fusion. Subsequent designs use the high neutron production of ^{235}U-fission to split lithium deuteride, and the heat of fission to then ignite the deuterium and tritium products:

$$^{235}U \xrightarrow{fission} Kr + Ba + 2n$$

$$^{6}Li^{2}H \xrightarrow{fission\ of\ ^{6}Li} {}^{3}H + {}^{2}H + {}^{4}He + heat\ of\ U\text{-fission}$$

$$\downarrow \quad \downarrow$$

Fusion

$$\downarrow$$

$$^{5}He + energy$$

$$\downarrow$$

$$^{4}He + n$$

$$^{14}N \rightarrow {}^{12}C + {}^{3}\mathbf{H}$$

^{5}He is not very stable ($t_{½} = 6.7 \cdot 10^{-22}$ s) and decays by neutron emission to stable ^{4}He. The tremendous neutron flux at the end of this reaction chain activates atmospheric nitrogen, producing ^{3}H. Although fission from detonation of uranium and plutonium bombs began on July 16, 1945, the neutron flux from these comparatively small and low-level fission tests, had no impact on meteoric ^{3}H.

Table 7-1 Tritium released in the atmosphere by major thermonuclear tests (Gonfiantini, 1996; Rath, 1988)

Year	Country	Fusion Energy Megatons of TNT	Tritium released TBq*
1951	USA	0.075	0.1
1952	USA	6.0	4.4
1953	USSR	~0.2	~0.1
1954	USA	17.5	13
1955	USSR	1.5	1.1
1956	USA, USSR	15.3	11.3
1957	UK, USSR	10.0	7.4
1958	UK, USA, USSR	31.1	23
1961	USSR	96.9	72
1962	USA, USSR	140.8	104
1963	USA, USSR	0.21	0.15
1963	*Limited Test Ban Treaty*		
1967	China	1.3	1
1968	China, France	2.6	1.9
1969	China	1.0	0.7
1970	China, France	1.2	0.9
1973	China	0.9	0.7
1974	China	0.15	0.1
1976	China	1.75	1.3

* Tera-becquerels (10^{12} Bq)

Atmospheric testing of nuclear devices between 1952 and 1962 (Table 7-1) generated a tremendous quantity of atmospheric tritium (Fig. 7-3). This substantial input created a tritium reservoir in the stratosphere which contaminated global precipitation systems for over four decades. A 1963 Soviet-American treaty banned the atmospheric testing of thermonuclear devices, although minor Chinese and French tests continued until 1980. The final year of megaton tests (1962) generated a huge peak, which appeared in precipitation in the spring of 1963. This 1963 peak became a marker used in many hydrological studies. Concentrations of ^3H in precipitation are now largely back to natural, cosmogenic levels.

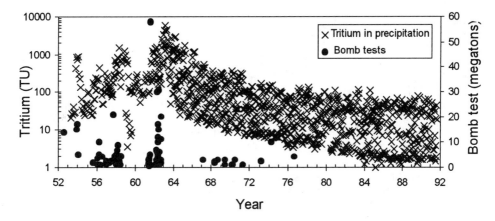

Fig. 7-3 Tritium in precipitation from thermonuclear bomb tests since 1952. Tritium data for selected stations in North America and Europe, from the IAEA GNIP data base. Bomb test data summarized from various sources by Rath (1988) and Gonfiantini (1996).

Even at the peak, anthropogenic tritium fallout was not a radiological threat to health. Canadian drinking water standards impose a limit on ^3H of 7,000 Bq·kg^{-1}, or about 60,000 TU. Minor amounts of ^3H are presently released to the atmosphere by nuclear power plants, nuclear fuel reprocessing facilities, preparation of weapons-grade nuclear material and by the manufacturers of tritiated paints and liquids used in illuminating dials and signals.

The longest record of atmospheric tritium concentrations was initiated by R.M. Brown for precipitation at Ottawa, Canada, and begins in 1953 (Fig. 7-4). The IAEA has established several long-term records from its monitoring stations throughout the world. Numerous short-term records at American stations were generated during the peak input years between 1960 and 1980. All data are available at <www.iaea.or.at:80/programs/ri/gnip/gnipmain.htm>.

Two features should be observed in this figure; the consistent annual fluctuation in ^3H, and the rapid decline from the 1963 peak. The greatest transfer of tritium from the stratosphere to the troposphere occurs during the spring in mid-latitude zones. This is due to seasonal disturbances in this boundary by displacement of the jet stream as upper level air circulation reorganizes in the spring. This "spring leak" annually recharges the hydrosphere (Fig. 7-4). The decline of tritium in precipitation is not due only to decay, which decreases tritium levels by only 5.5% per year (shown below in Fig. 7-8). A major factor is attenuation by the oceans and groundwaters, which have since become major reservoirs of thermonuclear ^3H.

There are also strong variations in the global distribution of tritium. Circulation in the stratosphere is latitudinally constrained, resulting in a latitudinal banding of tritium in rainfall

(Eriksson, 1966). Fig. 7-5 shows the strong variation in ^{3}H monitored at Ottawa, Cape Hatteras in North Carolina, and in the tropics at Barbados. Note the highest levels found at the most northerly station. As virtually all atmospheric testing was conducted in the northern hemisphere (with the exception of the French tests in the South Pacific), there is also a several year lag during mixing across the equator.

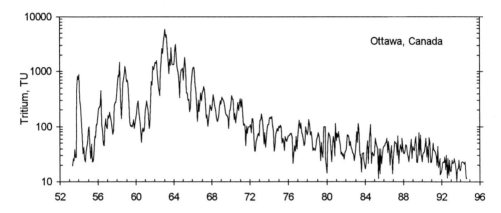

Fig. 7-4 Tritium in precipitation at Ottawa as measured in composite monthly samples (monitoring record established by R.M. Brown, AECL, Canada). Decreases from the peak in 1963 is due to attenuation in the oceans. The flattening of the decline after about 1980 and the sharp decline in 1990 likely reflect local activities in southern Ontario.

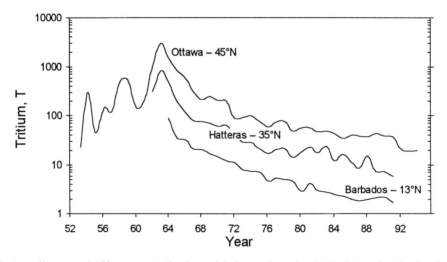

Fig. 7-5 Annually averaged tritium concentration in precipitation at three sites in North America showing the strong latitudinal variation.

Nuclear reactor tritium

Considering the high neutron fluxes in nuclear reactor cores, it is not surprising that tritium is also produced in such settings by neutron activation of nitrogen. Pathways into the environment occur through steam and water releases from power plants. Although these are only point sources from a global perspective, emissions can be locally high, and result in high-^{3}H halos in surface waters, groundwaters and plant cellulose. For example, 1000-MW boiling water reactors discharge on the order of 10^{10} Bq/year (20 to 30 Ci/yr), which is in the range of some of the smaller nuclear bomb tests. Fast breeder reactors (reactors that create more fissionable daughter nuclides than parents consumed) can leak up to five times as much.

Geogenic production of ^3H

A neutron flux is present not only from cosmic radiation, but also in the subsurface, due to spontaneous fission of U and Th. In rocks with appreciable amounts of lithium, this results in ^{3}H production through the fission of ^{6}Li according to:

$$^6Li + n \rightarrow {}^3H + \alpha$$

Geogenic tritium is then incorporated directly into groundwater, where concentrations depend on the Li content of the host rock, and porosity (Gascoyne and Kotzer, 1995). The short half-life and low production rate generally preclude significant accumulation of geogenic ^{3}H, although some rocks can have measurable quantities. Andrews et al. (1989) calculate that levels up to 0.7 TU should exist for groundwaters in the Stripa granite, which may account for the low levels measured in the deeper samples. In most aquifers, geogenic ^{3}H is close to or less than the low-level analytical detection limit of 0.1 TU.

In contrast, the high neutron fluxes found in zones of uranium mineralization can generate levels higher than cosmogenic ^{3}H. In the ore zone at Cigar Lake, a UO_2 deposit at the base of Proterozoic sandstones in Saskatchewan, groundwaters have geogenic ^{3}H concentrations exceeding 250 TU (Fabryka-Martin et al., 1994). The Oklo natural reactors of Gabon also have elevated tritium in groundwaters due to subsurface production.

Dating Groundwaters with Tritium

Kaufman and Libby (1954) first recognized the potential for dating groundwater with cosmogenic tritium. As geogenic ^{3}H in most groundwaters is negligible, measurable ^{3}H in groundwaters virtually always signifies modern recharge. When levels are high (>~30 TU), thermonuclear bomb ^{3}H is implicated, indicating recharge during the 1960s. Groundwaters containing levels of tritium that are close to detection (~1 TU) are most often submodern or paleogroundwaters that have mixed with shallow modern groundwaters in the discharge zone (e.g. springs) or within boreholes. Both qualitative and quantitative approaches to dating groundwaters with ^{3}H are possible:

1. *Velocity of the 1963 ^{3}H peak* — identification of the bomb-spike preserved in groundwaters clearly identifies their age.

2. *Radioactive decay* — calculating the time for decay from a known input level to the measured level.

3. *Exponential model for input function* — determining the "recharge attenuated" tritium levels for a given groundwater flow system, and applying the decay equation.

4. *Time series analysis* — repeat sampling from specific points over several years allows monitoring of the bomb spike as it passes through the aquifer, giving an indication of mean residence time.

5. *Qualitative interpretation* — measurable ^{3}H ≡ component of modern recharge.

Velocity of the 1963 "bomb peak"

During the era of thermonuclear bomb testing, the 1963 bomb peak became a classic "marker horizon" in aquifers. As this era recedes further into the past, the ^{3}H peak has either moved through most actively circulating aquifers or has been attenuated by dispersion and mixing. Nonetheless, it may be preserved in some less active hydrogeological settings.

The tritium fallout pattern in precipitation (Fig. 7-4) can be preserved only in hydrogeological systems where advective mixing is minimal. This occurs for recharge waters that transit thick non-saturated zones. Here, new water moving into storage in the upper layers displaces older water downward along the profile. The result is the reverse of the classic curve for ^{3}H in precipitation (Fig. 7-6).

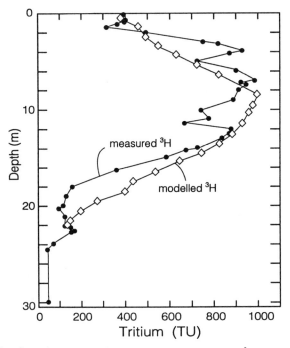

Fig. 7-6 Tritium in soil moisture in a sand aquifer, Denmark. The measured ^{3}H profile is shown with the profile generated with a piston flow model incorporating dispersion (modified from Andersen and Sevel, 1974).

If a specific year can be identified at a given depth in a recharge profile, then the soil moisture content above this level compared to total precipitation since that year represents the percent infiltration. Dinçer et al. (1974) use this approach and show that net recharge through dune sands in Saudi Arabia was as high or higher than 30%. Unfortunately, the thermonuclear bomb peak has long since moved through the non-saturated zone in most aquifers. Only examples with either low infiltration rates or thick non-saturated zones may still preserve the ^{3}H profile.

Hydrodynamic dispersion and mixing below the water table tend to attenuate variations in the ^{3}H input function. Nonetheless, some groundwaters have preserved the 1963 tritium bomb peak sufficiently well that its position can be used to determine groundwater flow rates below the water table. This method provides estimates of flow velocities and mean circulation times, which can be compared to hydrodynamic velocities. Michel et al. (1984) used the 1963 tritium peak to determine the velocity of groundwater moving into a confined sandy aquifer from a special waste disposal site in Gloucester, Ontario. Tritium values in 1982 (Fig. 7-7) show that three tritium peaks are preserved in the aquifer: immediately below the special waste compound, at 250 m along the flow path through the aquifer, and further down-gradient in the sand. The 250-m peak is 1963 recharge, which gives a flow velocity of 4.4 cm/day. The close peak is due to tritium-labeled organic compounds in the waste sorbing on the sand. The most distal peak represents rapid flow through coarse basal sediments overlying bedrock.

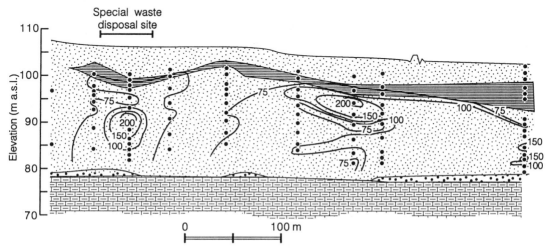

Fig. 7-7 Distribution of tritium in groundwaters in a confined sandy aquifer down-gradient of a special waste compound (Michel et al., 1984). Aquifer is confined by clay silts at about 5 m depth.

A much better definition of the tritium fallout peak in recharge environments can be obtained if it is possible to analyze for ^{3}He — the decay product of tritium, as well as ^{3}H. Several studies have been undertaken to document the usefulness of this approach especially since only 40 cm^3 of water collected in absolutely airtight Cu-tubes is required for these measurements (Ekwurzel et al., 1994). The ^{3}H/^{3}He dating technique is discussed in more detail below.

Radioactive decay

Dating of groundwater by decay of tritium is based on the assumption that the tritium input into a groundwater is known and that the "residual" tritium measured in a groundwater is the result of decay alone, according to the decay equation:

$$a_t{}^3H = a_o{}^3H \, e^{-\lambda t}$$

Here, $a_o{}^3H$ is the initial tritium activity or concentration (expressed in TU) and $a_t{}^3H$ is the residual activity (measured in a sample) remaining after decay over time t. The decay term λ is equal to ln2 divided by the half-life $t_{1/2}$. Using tritium's half-life, $t_{1/2}$ = 12.43 years, this equation can be rewritten:

$$t = -17.93 \ln \frac{a_t {}^3 H}{a_o {}^3 H}$$

For the example of $a_o{}^3H$ = 10 TU, this gives the decay series:

years:	0	10	20	30	40	50	60
TU:	10	5.7	3.3	1.9	1.1	0.6	0.4

Thus, the useful range for dating with ^{3}H is less than about 50 years when analyses are performed by the enriched method (limit of detection = 0.8 TU). Pre-bomb tritium could be measured if low-level measurement techniques such as propane synthesis or ^{3}He ingrowth (± 0.1 TU) are used. Rarely do we find groundwaters that are uncontaminated by mixing for precise dating by ^{3}H decay. High-precision analyses are important where minor mixing with modern waters, or *in situ* production of ^{3}H are to be measured. Tritium levels in modern precipitation are now close to stable, natural levels, and the possibility of using the decay equation for young groundwaters becomes more straightforward.

The strong seasonal and annual variations seen in precipitation (Fig. 7-4) represent a complicated input function and make it difficult to determine the initial ^{3}H at the time of recharge. For piston flow (like a train moving people on a single track compared with cars on a multi-lane freeway), with little mixing or dispersion along the flow path, a recharge date can be estimated from decay lines (Fig. 7-8). Back-extrapolation along a decay line from a measured ^{3}H value for a given sampling date would intersect the precipitation curve at the year(s) when infiltration took place. The problem with this approach is its simplicity (not a characteristic of natural systems) which assumes that only one year's precipitation has contributed to the water sample. More commonly, the tritium input is a multi-year average. We need to calculate this more realistic input function with a model that accounts for mixing and decay in the recharge environment.

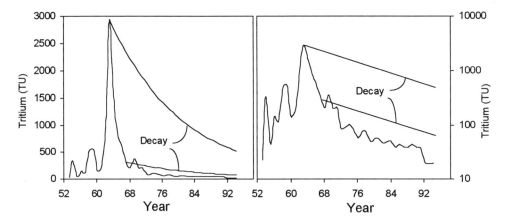

Fig. 7-8 Decay of tritium from given inputs from precipitation (on normal and semi-log plots). The input function is more complicated than this simple decay function would imply, due to mixing and decay in the recharge area. The decline of tritium is far greater than accounted for by decay alone, due to mixing with oceans and groundwater.

Input function for ^3H in groundwater

Consider a parcel of groundwater in a regional aquifer. In most cases it is a composite of several years' precipitation which have contributed to this water through mixing within the unsaturated zone, and by flow through the recharge area where additions to the water table continually contribute to the groundwater. Once this groundwater has moved into a confined zone, or to depths where it no longer receives additions from the surface, its ^3H content will decrease by decay. The ^3H concentration in this groundwater parcel will be a function of its residence time in the recharge environment.

The multi-year ^3H input function can be determined from: (1) the weighted contribution of ^3H from each year, with (2) a correction for the decay of each year's precipitation contribution during storage in the recharge area. For example, let's assume that our parcel of groundwater accumulated over three years in the recharge area, with 25% from year 1, 50% from year 2 and 25% from year 3. At the end of year 3, when our groundwater moves past the recharge zone, its ^3H concentration will be equal to 25% of the ^3H level in year 1 precipitation times 25%, with 3 years of decay (5.5% per year) plus 50% of the ^3H in year 2, decayed over 2 years, plus 25% of year 3 ^3H decayed over 1 year.

The input function for tritium in precipitation can be approximated by this approach, using a decay model of recharge that is essentially a smoothing calculation. The model assumes that each groundwater component along the flow system has received the same weighting of multi-year contributions from precipitation in the recharge environment. The weighting of each year is approximated by a normal distribution that puts the greatest weight on the precipitation from the central years, with lowest contributions from the first and last years.

Fig. 7-9 shows three recharge models for the annually averaged tritium curve for Ottawa. The recharge curves are plotted for the year when the groundwater becomes shut off from additional recharge. Increasing the number of recharge years in the model attenuates the peak fallout level, and increases inputs in post 1962 recharge relative to the input curve (1 year MRT).

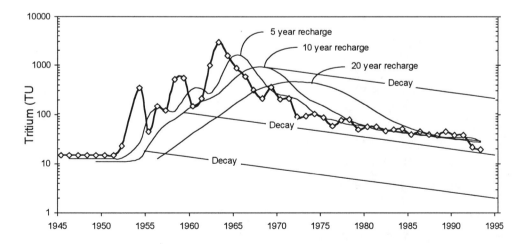

Fig. 7-9 Tritium fallout (weighted annual average) in precipitation collected at Ottawa (heavy line), with decay models for initial tritium in groundwater. Models assume that recharge to a given parcel of groundwater takes place over 5, 10 and 20 year periods and are plotted for the year that the groundwater becomes closed to further contributions from precipitation. The decay lines are valid for "closed system conditions" following the movement of the groundwater beyond the recharge area.

Extrapolation from a measured value for groundwater sampled at a given date to the most appropriate recharge input curve provides an estimate of groundwater residence time beyond the recharge environment. For example, lets assume that in 1995 we measured 2 TU in groundwater from an aquifer with an estimated 10 year period of accumulation in the recharge area. Back extrapolation along the bottom decay curve in Fig. 7-9 intersects the 10 year recharge curve at the end of 1954. Total mean subsurface residence time would then be about 40 years plus the mean residence time in the recharge area (about 5 years).

The MRT selected for the model depends on the hydrogeological setting in the recharge environment. Systems with thick unsaturated zones and large, unconfined recharge areas may mix recharge over 5 to 10 years or more before a given groundwater parcel becomes closed to further inputs. In contrast, systems with recharge regions of limited areal extent may have little mixing in the recharge area. In this case, the recharge period may be less than 5 years.

Time series analysis

Sequential sampling of groundwater for tritium analysis can be used to identify the bomb-spike and provide a good indication of mean residence time. Using a smoothed atmospheric ^{3}H input function, the ratio of two samples from the same piezometer or well (corrected for decay) will indicate the relative location of the 1963 peak. This can be expressed as a ratio of the tritium level at the early sampling, corrected for losses due to radioactive decay between sampling times, to the level in the later sampling:

$$\frac{^3H_{early} \cdot e^{\lambda t}}{^3H_{later}}$$

where t is the time between samplings in years (t = t_{later} − t_{early}), and λ is the decay constant (λ = ln $2/t_{½}$) which for tritium is 0.05576. A ratio greater than 1 would suggest that the peak has passed the observation point, whereas if this ratio is less than 1, then the observation point is sampling the leading edge of the spike. Sequential sampling with a frequency as low as every 5 to 10 years would then be useful.

Fritz et al. (1991) carried out such a study for groundwater in the Waterloo aquifer, a complex of glaciofluvial sands, clays and tills that supply water to the Kitchener-Waterloo region. While a sampling of piezometers in 1976 proved inconclusive in determining the rate of groundwater flow, subsequent sampling in 1979, 1981 and 1988 showed that most wells registered a drop in ^{3}H levels, after correction for decay. Thus, most of the 1976 groundwaters had sampled the tail of the bomb spike. This study was able then to refine estimates for groundwater mean residence time from younger than 30 years to within a decade.

Qualitative interpretation of ^{3}H data

In the three decades since the last major tests, thermonuclear bomb tritium has been greatly attenuated by the oceans, and levels are now approaching that of natural atmospheric production. This evolution of the input function plus mixing in some aquifers with pre-bomb groundwater conspire against refined interpretations of groundwater ages. A quantitative interpretation of groundwater mean residence times may not be possible, and only qualitative interpretations can be made:

For continental regions:

<0.8 TU	Submodern — recharged prior to 1952
0.8 to ~4 TU	Mixture between submodern and recent recharge
5 to 15 TU	Modern (<5 to 10 yr)
15 to 30 TU	Some "bomb" ^{3}H present
>30 TU	Considerable component of recharge from 1960s or 1970s
>50 TU	Dominantly the 1960s recharge

For coastal and low latitude regions:

<0.8 TU	Submodern — recharged prior to 1952
0.8 to ~2 TU	Mixture between submodern and recent recharge
2 to 8	Modern (<5 to 10 yr)
10 to 20	Residual "bomb" ^{3}H present
>20 TU	Considerable component of recharge from 1960s or 1970s

Tritium in alluvial groundwaters — an example from Oman

Shallow alluvial groundwaters in Oman are recharged primarily by infiltration of flash flood events, whereas interchannel areas receive direct infiltration and subsurface flow. Recharge is sporadic, and shallow groundwater can have a wide range of ages. Tritium becomes an excellent indicator of the occurrence of modern inputs.

Fig. 7-10 shows the tritium concentrations in groundwater in the alluvial fans flanking the Northern Oman Mountains. In this case, recharge is dominantly from infrequent convective rainfall occurring in the mountains. Tritium levels in these groundwaters decline with increased distance along the principal wadi courses. In the most up-gradient areas, ^{3}H contents are in the range found for modern precipitation at nearby Bahrain. However, they are below detection in the thick alluvial sediments found towards the *serir* (interior plain).

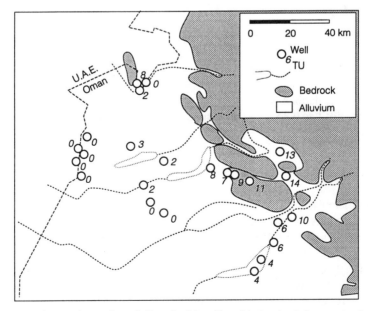

Fig. 7-10 Tritium contents in groundwaters from shallow alluvial aquifers of the interior drainage system in northern Oman.

If groundwater ages increased rather uniformly down-gradient, one would expect an *increase* in ^{3}H down-gradient, despite decay, because the thermonuclear ^{3}H input function is decreasing faster than natural decay (Fig. 7-8). If this were the case, plug flow would dominate. So what do these tritium levels tell us? Groundwater in the upper reaches of the alluvial aquifers comprises mixtures of recharge from the past few years only (<~5 year mean residence time). Groundwater in the interior plain with ^{3}H levels below detection must have been recharged prior to 1952, and can be considered sub-modern or older. Groundwater in the interfluvial regions (fossil alluvial channels) are sub-modern in age, which demonstrates that direct infiltration of rainfall is not an efficient recharge mechanism in such regions.

Deep groundwaters - mixing in fractured rock

Groundwater found at depth in crystalline or sedimentary basin settings have high salinities developed through water-rock interaction over long (perhaps geologic) time scales. When measurable ^{3}H above 1 TU is found in these brines this signifies a hydraulic connection with the surface. Although leakage through shafts, adits and boreholes can be implicated, connected fractures and faults are often responsible. This has important implications for contaminant transport, and in particular for the security of potential radioactive waste repositories. Monitoring tritium in deep mine settings demonstrates that such short-circuits can be established over short time periods.

Tritium levels (and ^{14}C) increase over time in some mine inflows (Moser et al., 1989; Douglas, 1997), indicating that near-surface groundwaters are mixing at increasingly greater depths. Frape and Fritz (1987) show ^{3}H increasing from less than detection to near modern levels (37 TU in the Con mine in Yellowknife, and 70 TU in Sudbury) over several years of active discharge. Douglas (1997) shows a decrease at the Con mine by 1996, signifying the passage of the 1964 peak.

However, measurements at the Stripa site, a closed iron mine in central Sweden which was under study as an analogue for a radioactive waste repository, show small amounts of tritium in some groundwaters up to 1000 m deep. For this situation, Andrews et al. (1989) show that a minor component of ^{3}H may be due to subsurface production. This can occur in rocks with high U, Th and K levels (granites, shales) due to neutron activation of N. The short half-life will maintain concentrations generally very close to detection and at Stripa below 0.7 TU, the maximum level of production. Such occurrences of tritium in the deep groundwater has important implications with respect to the degree of isolation provided at this depth. Similar observations and calculations were made for the saline waters of the KTB deep drillhole in Germany where fluids had 0.35 TU. This can be attributed to *in situ* production.

Groundwater Dating with ^{3}H - ^{3}He

The earlier discussion leaves us with the impression that our ability to produce quantitative tritium ages for groundwaters is fading as fast as the thermonuclear bomb peak. However, by measuring ^{3}H together with its daughter ^{3}He, true ages can be determined through calculations that do not rely on the complicated tritium input function. The drawback to this approach is that ^{3}He is not a routinely sampled nor measured isotope (Clarke et al., 1976). The method was first introduced to the field of hydrogeology in 1979 (Torgersen et al., 1979). Schlosser et al. (1988) provide an excellent review and application and Ekwurzel et al. (1994) compare this technique with dating based on chlorofluorocarbons (CFCs) and ^{85}Kr.

Helium–tritium systematics

The decay of tritium from an initial concentration ^3H_o after some time t is predicted by:

$$^3H_t = {^3H_o}\, e^{-\lambda t}$$

However, determining t requires that we know ^3H_o. The decay of tritium leads to an ingrowth of ^3He, which would then be:

$$^3He_t = {^3H_o}(1- e^{-\lambda t})$$

By combining both equations, we can cancel the dependency on the input concentration of tritium ^3H_o:

$$^3He_t = {^3H_t}(e^{-\lambda t} -1)$$

The helium concentration at time t, ^3He_t, is expressed in TU (1 ^3He per 10^{18} H). The measured ^3He must be corrected for atmospheric ^3He that is dissolved at the time of recharge. Note that atmospheric He is dominantly ^4He. This input is assumed to be at equilibrium with the atmosphere, and considers the following points:

- Atmospheric ^4He concentration is 5.24 ppmv (Glueckauf, 1946)
- Atmospheric $^3He/^4He$ ratio is $1.3 \cdot 10^{-6}$ (Coon, 1949, cited in Andrews, 1987)
- The solubility of atmospheric helium is temperature dependent, and for 10°C is $4.75 \cdot 10^{-8}$ cm^3 STP/cm^3 H$_2$O (Fig. 7-11).
- ^4He is slightly more soluble in water, with a fractionation factor, $\alpha_{w\text{-air}} \sim 0.983$ (Benson and Krause, 1976)

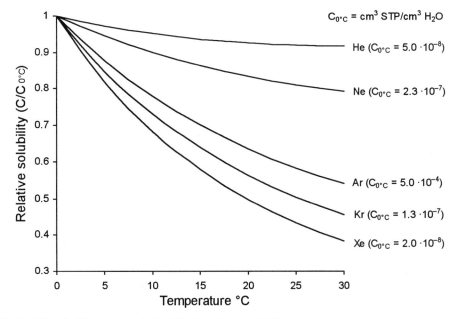

Fig. 7-11 Solubility of noble gases in water (modified from Andrews, 1992).

The measured value of ^{3}He, corrected for atmospheric ^{3}He, represents ^{3}He ingrown from ^{3}H decay, and is then used in the dating equation:

$$t = 12.43/\ln 2 \cdot \ln(1 + [^3He_t]/[^3H_t])$$

where $[^3He_t]/[^3H_t]$ is the concentration ratio of these two isotopes expressed in tritium units.

Applications of the ^{3}H-^{3}He method

Schlosser et al. (1988) apply this method to groundwater in shallow alluvial aquifers in Germany, using the bomb peak for verification (Fig. 7-12). In this example, the ingrown ^{3}He signal clearly follows the ^{3}H peak. They find that 77 to 85% of the tritogenic ^{3}He remained in the groundwater, the rest presumably lost by diffusion. Comparing their calculated ^{3}He/^{3}H age for the 1963 bomb peak in the groundwaters, they find only a 3 year (15%) discrepancy, which is not bad precision in groundwater dating.

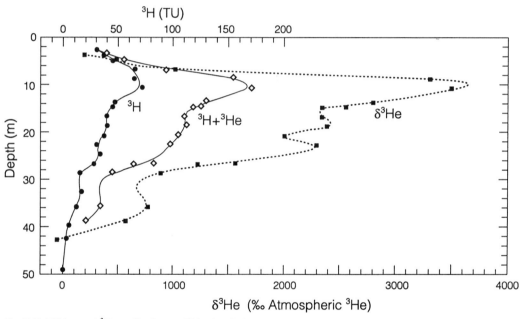

Fig. 7-12 Tritium and ^{3}He profiles from multi-level wells in Germany (modified from Schlosser et al., 1988).

Very similar observations on the retention of ^{3}He in the groundwater are made by Eckwurzel et al., (1994) and by Szabo et al. (1996) investigating shallow groundwaters in New Jersey. Because of rapid vertical flow (~1 m/yr), confinement of ^{3}He was very high — a prerequisite for the technique. The study demonstrates the value of the ^{3}H/^{3}He dating technique and compared it with a complementary dating tool — chlorofluorocarbons or CFCs.

Chlorofluorocarbons (CFCs)

Like tritium, hydrogeologists have managed to find a use for CFCs, an otherwise unwanted contaminant in our atmosphere. These compounds are resistant to degradation, making them a useful marker for modern groundwater. Atmospheric CFC concentrations have been increasing since the 1940s (Fig. 7-13), providing a characteristic input function. CFCs have been extensively used to trace oceanic circulation patterns over the past decade. Thompson and Hayes (1979) and subsequent studies have documented their usefulness for dating young groundwaters.

Szabo et al. (1996) note that apparent groundwater ages based on the ^{3}H-^{3}He dating technique were identical to CFC ages, within the margin of error of the methods. However, the CFC technique may require local adjustments to the input function inasmuch as local urban emissions of CFCs cause variations in its atmospheric concentration. The solubility of CFC compounds are also highly sensitive to temperature, and must be calculated for groundwater temperatures in the recharge environment according to the Henry's law constants (Warner and Weiss, 1985).

Measurements are carried out by analytical methods for measuring trace organic compounds. Compounds are stripped from solution with He, separated by gas chromatography and detected by electron capture. The analysis of CFC compounds is less complicated than for ^{3}H, although potential contamination by exposure to the atmosphere during sampling is a risk. Furthermore, retention in aquifers with a high organic fraction, and subsurface degradation can affect transport. CFC-12 appears to exhibit the most conservative behaviour (Cook et al., 1995).

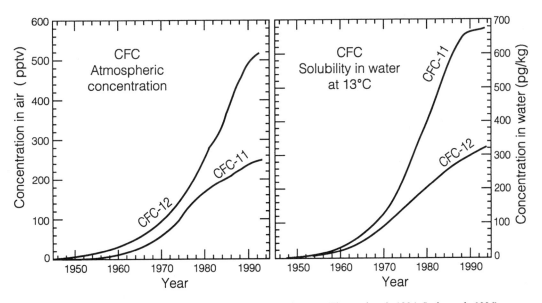

Fig. 7-13 Concentrations of CFCs in the atmosphere and dissolved in water (Ekwurzel et al., 1994; Szabo et al., 1996).

Thermonuclear ^{36}Cl

Thermonuclear bomb testing generated not only a whopping peak in tritium, but also in the radioactive isotope of Cl — ^{36}Cl. The high neutron flux of hydrogen bomb tests in marine settings during the early years of testing elevated concentrations by over two orders of magnitude above its natural atmospheric abundance. By consequence, ^{36}Cl can be used — like tritium — to identify modern recharge. However, unlike ^{3}H, its long half-life ($t_{½}$ ~ 300,000 years) precludes the use of ^{36}Cl decay to date modern waters.

Natural atmospheric production of ^{36}Cl occurs due to activation of atmospheric argon by cosmic radiation and arrives at the earth's surface as a dry fallout or in precipitation. Natural production and fallout varies with latitude from close to 20 atoms m^{-2} s^{-1} at mid latitudes to less than 5 near the equator and poles. It is also produced epigenetically through activation of Cl, K and Ca on the land surface by cosmic radiation (see Chapter 8). Thermonuclear ^{36}Cl was produced by activation of atmospheric Ar and marine Cl$^-$ by the high neutron fluxes accompanying low altitude tests in the 1950s. A record of this fallout is preserved in glacier ice (Fig. 7-14), reconstructed by Bentley et al. (1986).

High levels of ^{36}Cl in groundwater indicate, like tritium, that recharge has occurred since this time. Fig. 7-14 shows that since the period of maximum ^{36}Cl contamination, concentrations have been steadily declining. This washing out of atmospheric Cl has brought concentrations, at least by 1980, back closer to natural levels. However, storage and recycling of Cl in the biosphere seems to maintain a background ^{36}Cl activity chronically elevated above natural levels (Gwen Milton, Atomic Energy of Canada Ltd., pers. comm.).

Concentrations of ^{36}C in groundwater requires conversion from the fallout rate (atoms m^{-2} s^{-1}) to concentration in water (atoms per litre). For example, with a natural ^{36}Cl fallout rate of 15 atoms m^{-2} s^{-1}, annual precipitation of 750 mm, and 75 mm loss to evapotranspiration, the concentration in runoff and/or groundwater can be calculated. The annual fallout rate would be:

$$15 \text{ atoms m}^{-2} \text{ s}^{-1} = 473 \cdot 10^6 \text{ atoms m}^{-2} \text{ yr}^{-1}$$

Dividing by the corrected annual precipitation gives l:

$$473 \cdot 10^6 / (0.75 - 0.0075) = 700 \cdot 10^6 \text{ atoms m}^{-3} \text{ or } 7 \cdot 10^5 \text{ atoms L}^{-1}$$

Measurements of ^{36}Cl concentrations in groundwaters exceeding about 10^7 atoms L^{-1} (depending on rainfall and infiltration rates) is a good indication that there is a component of thermonuclear ^{36}Cl and hence, modern recharge. Remember that, unlike ^{3}H, ^{36}Cl is a solute only and so how and when the chloride was dissolved is important. Conjunctive use of ^{36}Cl and ^{3}H is of use in recharge studies (Chapter 4). Applications of ^{36}Cl as a dating tool for old groundwaters are found in Chapter 8.

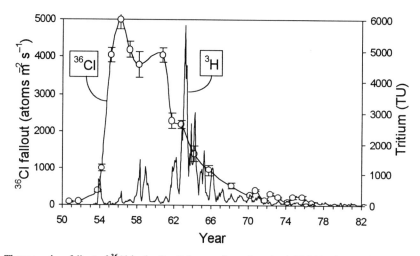

Fig. 7-14 Thermonuclear fallout of ^{36}Cl in the Dye-3 ice core from Greenland (70°N)), shown with tritium measured in precipitation at Ottawa, Canada (45°N) for comparison. The ^{36}Cl peak preceded the tritium peak due to the low altitude marine setting of the early tests, which activated marine Cl$^-$ (^{36}Cl data from Bentley et al., 1986). Sharp drop of the peak is due to washing of Cl$^-$ from troposphere by precipitation, and not decay (t$_{½}$ = 300 ka).

Detecting Modern Groundwaters with ^{85}Kr

Krypton has six stable isotopes (^{78}Kr, ^{80}Kr, ^{82}Kr, ^{83}Kr, ^{84}Kr and ^{86}Kr) and two radioisotopes; ^{81}Kr has a half-life of 210,000 years and ^{85}Kr has a half-life of 10.76 years. Like ^{3}He, sampling and measurement of ^{85}Kr preclude its routine use as an environmental isotope, although it is gaining interest in view of declining tritium levels in the atmosphere. It is naturally produced in very minor amounts from neutron activation of ^{84}Kr:

$$^{84}\text{Kr} (n,\gamma) \,^{85}\text{Kr}$$

However, nuclear activities and especially nuclear fuel reprocessing have released significant amounts of ^{85}Kr into the atmosphere so that since about 1950 a constant increase can be observed (Fig. 7-15). As a matter of fact, it was possible to monitor some Soviet nuclear activities with this isotope.

Krypton-85 is collected by vacuum extraction from water in the field (~100 L) and is measured by counting techniques to levels approaching 0.1dpm/cm^3 Kr STP (Ekwurzel et al., 1994). This is now a fairly routine technique and a number of laboratories in North America, Europe and elsewhere can count the extracted gas. This is important since its short half-life make its use in view of the decreasing tritium levels increasingly relevant for the identification of young groundwater (Ekwurzel et al., 1994).

The presence of ^{85}Kr is a clear indication that the groundwaters are young; its growing input function limits makes identification of very young waters possible (1 to 5 years). New approaches using the different input functions for ^{3}H and ^{85}Kr extend its applicability over a longer time range (Loosli et al., 1991). It should also be mentioned that ^{81}Kr with its long half-life of 210,000 years is also a potential dating tool, yet preparation and measurement procedures are so complex that its routine application will not be possible in the foreseeable future.

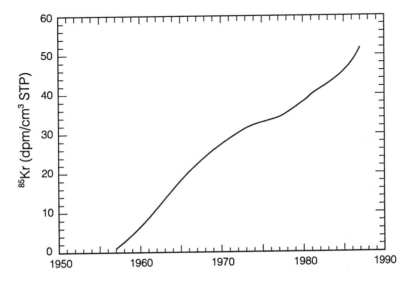

Fig. 7-15 The measured activity of ^{85}Kr in the atmosphere, originating from the nuclear fuel cycle (after Loosli et al., 1991).

Submodern groundwater

Tritium, helium and ^{85}Kr are useful dating tools for modern groundwaters. For old groundwaters (paleogroundwaters, fossil groundwaters), radioisotopes with longer half-lives are important (^{14}C and ^{36}Cl are discussed later). However, the submodern range between about 40 and ~1000 years B.P. is difficult to date. From a hydrogeological perspective, dating groundwaters within this range may indeed be important. Unfortunately, few radiogenic isotopes have a half-life in the range for submodern groundwaters. Argon-39 does ($t_{1/2}$ = 269 years), but its sampling and analysis is far from routine. Silica-32 also has an appropriate half-life ($t_{1/2}$ = 140 ± 20 years), but is complicated by mixing with the abundant Si sources in the subsurface.

Argon-39

Due to the low content of argon in the atmosphere and, hence, in groundwaters, degassing of a large sample volume is required in the field. Normally, a 2-L argon sample (at STP) is required that involves extraction by vacuum-degassing or boiling from up to a 15 m^3 volume of water (see solubility in Fig. 7-11). Analysis is by high pressure gas proportional counting over a period of 1-month.

Argon-39 is produced in the atmosphere by cosmic radiation on ^{40}Ar:

$$^{40}Ar + n \rightarrow\ ^{39}Ar + 2n$$

It has a half-life of 269 years. The natural atmospheric activity of ^{39}Ar is about 0.112 dpm L^{-1} Ar (Loosli and Oeschger, 1979) and is believed to be constant within 7% over at least the past 1000 years (Loosli, 1983). Effects of thermonuclear testing are not seen in modern atmospheric ^{39}Ar. Activities are expressed as a percent of modern atmospheric concentration. The dating range lies between about 50 and 1500 years and thus closes the gap between tritium and ^{14}C.

The principal advantage of ^{39}Ar is that, being a noble gas, it is inert and corrections for geochemical processes are not necessary. Furthermore, as the extraction is done with ^{40}Ar, the activity ratio can be measured, and quantitative extraction is not required. Subsurface production of stable ^{40}Ar by ^{40}K decay must be taken into consideration, as it dilutes the specific concentration of ^{39}Ar. Corrected ^{39}Ar activities in groundwater can then be translated into groundwater age estimates by the decay equation:

$$^{39}Ar =\ ^{39}Ar_o \cdot e^{-\lambda t}$$

The ^{39}Ar contents in ice cores have been measured in glacier ice from the Greenland ice sheet for comparison with dates established by the δ^{18}O record (Fig. 7-16). The excellent correlation shows the reliability of the input function, at least in uncomplicated systems. In groundwaters, complications arise due to subsurface production and/or mixing.

Subsurface production is problematic for groundwater dating. Andrews et al. (1989) summarize such production in crystalline terrain and other rocks, where substantial amounts of U, Th and K produce ^{39}Ar *in situ* through reactions such as:

$$^{39}K + n \rightarrow\ ^{39}Ar + p$$

$$^{38}Ar + n \rightarrow\ ^{39}Ar + \gamma$$

and $\quad ^{42}Ca + n \rightarrow \,^{39}Ar + \alpha$

However, should the neutron flux be known, then the theoretical "secular equilibrium" concentrations can be calculated and compared to measured values. Since decay and ingrowth follow exponential laws it will take about five half-lives to establish "equilibrium." Where this is found a minimum age of about 1400 years is determined.

One of the first, detailed investigations of ^{39}Ar was carried out for groundwater in Germany, as a tool to complement dating with ^{14}C and ^{3}H (Loosli and Oeschger, 1979). Since then Loosli and colleagues have published a number of studies in which they employ ^{39}Ar as a dating tool in comparison with other analyses. The most important studies were most likely done during hydrogeological investigations within the Swiss nuclear waste program (Pearson et al. 1991, Scholtis et al., 1996).

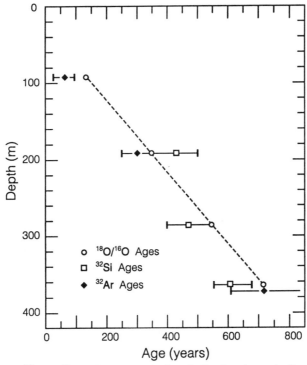

Fig. 7-16 Comparison of ^{39}Ar and ^{32}Si ages for ice samples from the Dye-3 station on the Greenland ice sheet. ^{32}Si ages were calculated using a half-life of 650 years. This is now believed to be closer to ~140 years (modified from Oeschger and Loosli, 1977).

Forster et al. (1984) compared ^{39}Ar data from a sandstone aquifer in the Saar region of Germany. The aquifer was chosen because it could be assumed that subsurface production was negligible and therefore a direct comparison with other dating techniques would be possible. Some of the results are shown in Table 7-2.

Last but not least, it should be mentioned that the extraction of ^{39}Ar from groundwater also permits the determination of the $^{40}Ar/^{36}Ar$ ratio. This ratio has a constant value of 295.5 in the atmosphere. However, most aquifer rocks contain K-bearing minerals and therefore the β^--decay of the long-lived ^{40}K ($t_{1/2} = 1.25 \cdot 10^9$ yr) to ^{40}Ar can cause accumulation of this argon isotope. Therefore, the $^{40}Ar/^{36}Ar$ ratio in groundwater is 295.5 or higher. This applies especially to very old groundwater which thus can be identified (Pearson et al., 1991) although absolute dating is not possible with this method unless the ^{40}Ar production rate is known. The ubiquitous yet variable presence of K-minerals as well as the possibility of subsurface gas migration and accumulation in deep aquifers makes this an impossible task. However, if considered with other isotopic evidence such as He concentrations, ^{3}H/^{4}H ratios and ^{36}Cl determinations, the ^{40}Ar/^{36}Ar ratios contain valuable information on the age and "functioning" of an aquifer system and its groundwaters.

Table 7-2 Comparison of ages by ^{14}C, ^{39}Ar, ^{85}Kr and ^{3}H for groundwaters in the Saar sandstone, Germany (from Forster et al., 1984)

Locality	Dating tool	Mean residence time (years)	
		Exponential model	Piston flow
Kirkel	^{14}C	< 2000	< 2000
	^{39}Ar	125 ± 35	110 ± 25
	^{85}Kr	> 60	> 25 ± 5
	^{3}H	> 60	> 25
Altheim 2	^{14}C	< 2000	< 2000
	^{39}Ar	770 ± 80	430 ± 30
	^{85}Kr	> 60	> 25
	^{3}H	> 60	> 25
Blickweiler	^{14}C	2400–4200	2000–3600
	^{39}Ar	> 2000	> 2000

Silica-32

This environmental isotope has only recently received serious attention for dating groundwaters, mainly because of the difficulties in establishing its half-life. The ubiquitous geogenic sources of silica have also played more than a minor role in discouraging its use. Its half-life is now believed to be 140 ± 20 yr (Kutschera et al., 1980; Alburger et al., 1986). The difficulties in measurement of ^{32}Si are its low concentration in water, which is on the order of µBq L^{-1}. These difficulties precluded a good determination of its atmospheric production for earlier studies. Measurements are now made by counting decays of its daughter, ^{32}P, following extraction of dissolved Si and 3-month storage to allow ^{32}P ingrowth (Morgenstern et al., 1995).

Few studies have demonstrated the dating potential of ^{32}Si. Oeschger and Loosli (1977) used it to compare with ^{39}Ar ages in ice cores from Greenland (Fig. 7-16), although the half-life at this time was greatly overestimated. Andrews et al. (1983) found measurable ^{32}Si in only the youngest (^{3}H-bearing) groundwater of the Bunter Sandstone of the English East Midlands (see discussion of this aquifer in Chapter 8). Morgenstern et al. (1995) apply the method to groundwaters in limestone aquifers in Estonia. They find cosmogenic ^{32}Si activity to be ~5 mBq m^{-3} in rainwater. However, this decreases to about 2.5 ± 0.5 mBq m^{-3} in shallow groundwaters due to sorption in soils. Further, although groundwaters at 50 m depth in the limestone aquifer contain tritium, their ^{32}Si activity is less than detection. This study concludes that geochemical reaction of Si during recharge and groundwater flow overwhelms any losses due to decay. This leaves little hope that ^{32}Si will in the near future be useful in resolving groundwater ages in the critical submodern period of 50 to 1000 years.

Problems

1. A survey of environmental isotopes in groundwater undertaken in 1990 has just been approved for a second phase. A series of archived samples collected in June 1990 for ^{3}H are now to be analysed. The measurements obtained in December 1996 for the three samples were OM1 = <0.8 TU, OM2 = 46.1 TU and OM3 = 10.5 TU.
   - What were their ^{3}H concentrations at the time of sampling?
   - On the basis of these data alone, what can you say about the age of these waters.

2. A bottle of 1948 vintage Hungarian wine was opened in 1992 for ^{3}H analysis by the ^{3}H → ^{3}He ingrowth method. An aliquot were taken and distilled to isolate the H$_2$O, degassed to remove all He in solution, and sealed in a glass ampoule for a 36 month period. The sample was then opened in 1995 and the ingrown ^{3}He content measured. The ^{3}He content was $2.8 \cdot 10^{-16}$ cm^3 STP/cm^3 water. What would the ^{3}H activity have been for the precipitation that nourished the vineyards in 1948?

3. A series of piezometers (TK series) have been installed to various depths in a sandy glacial till aquifer near Ottawa to test for rates of groundwater flow. Piezometric levels show that flow is vertically downward. The following tritium contents were measured in samples taken in July, 1996.

Site	Depth (m)	^{3}H (± 0.8) TU
precipitation	surface	19 (n=6)
TK-0.5	0.5	20.2
TK-1	1.0	21.8
TK-2.5	2.5	21.3
TK-4	4.0	19.9
TK-6	6.0	29.9
TK-8	8.0	47.1
TK-10	10.0	113.8
TK-12	12.0	40.8
TK-14	14.0	1.9
TK-16	16.0	<0.8

   - Account for the distribution of tritium through this aquifer.
   - Calculate the groundwater velocity along this vertical profile through the till.
   - What are the initial ^{3}H contents at the time of recharge for each sample, and how do they compare with the Ottawa record for ^{3}H in precipitation (Fig. 7-4). Account for any discrepancies.

4. A series of shallow groundwaters (10°C) at 10 m depth were sampled from an unconsolidated, sandy aquifer around a nuclear waste repository. Analysis of ^{3}H and ^{3}He concentrations (^{3}He values not corrected for atmospheric ^{3}He) gave the following results:

Sample	Distance from recharge (m)	^{3}H (TU)	^{3}He (10^{-14} cm^3 He/cm^3 water)
CR1	56	705	11.39
CR2	91	696	7.96
CR3	348	762	50.81
CR4	444	681	84.80
CR5	548	737	87.79

   Conversion: 1 cm^3 ^{3}He/cm^3 water = $4.0177 \cdot 10^{14}$ TU (^{3}He equivalent)

- Plot the distribution of ^{3}H and ^{3}He with increasing distance from the recharge area.
- Calculate groundwater residence times for each of the samples, and determine the groundwater velocity for each water mass.
- Is there any evidence for diffusive loss of tritigenic ^{3}He from the groundwaters? Explain.
- Can residence times for these waters be calculated using only the ^{3}H data? Explain.

5. What ^{36}Cl concentration would you expect to find in the groundwaters sampled at 10 m depth in the profile given in problem 4 above, assuming that there are 950 mm of precipitation annually, and evapotranspiration is 10%?

6. Groundwaters sampled from a confined sandstone aquifer dominated by piston-flow yielded tritium concentrations of <0.1 TU in 1996.

   - What is the minimum age for this groundwater?
   - What do you propose to constrain the age of this groundwater to better than older than ~60 years?

7. The δ^{18}O composition of precipitation and groundwater in an alluvial aquifer was monitored over a three year period, giving the following average monthly values. Plot these data and determine the mean residence time (MRT) of the groundwater according to the approaches of Burgman (1987) and Stichler (1980) discussed at the beginning of this chapter.

Month	precipitation	groundwater
Jan	-14.8	-10.3
Feb	-14.0	-10.4
Mar	-12.4	-10.4
Apr	-11.4	-10.9
May	-8.4	-11.0
Jun	-8.1	-11.5
Jul	-7.4	-11.6
Aug	-7.9	-11.8
Sept	-9.7	-11.6
Oct	-11.6	-11.5
Nov	-12.7	-10.8
Dec	-14.2	-10.6

Chapter 8
Age Dating Old Groundwaters

Constraining the age of groundwaters that are clearly sub-modern or older can be important in establishing the long-term potential for aquifer recharge. For groundwater development and management policy, the question of renewability is most important. However, the exploitation of a nonrenewable resource can have serious political and sociological implications, as documented for the Mexico City (where the City centre has sunk by several meters), Northern Chile (where the fossil water resources of the Pampa del Tamuragal have been critically depleted by the City of Iquique) and in Libya where the exploitation of the Nubian sandstone aquifer to grow wheat in the northern Sahara desert have drawn water levels down several hundreds of meters. Determining whether a groundwater is 2,000 or 20,000 or even 200,000 years old is thus relevant to more than paleohydrologic and paleoclimatic studies.

Terminology for groundwaters that are older than "modern" as identified by the methods discussed above is important because of the implications for recharge. *Fossil* groundwaters have been recharged in the past by meteorological processes that no longer prevail (pluvial paleoclimates). However, paleogroundwaters in many regional flow systems may be old due simply to their low velocities and long flow paths, and may actually receive modern inputs in the recharge environment. None of these terms addresses the aspect of induced recharge under the stress of exploitation. While most paleogroundwaters are not part of active flow systems, heavy pumping can potentially induce flow from adjacent aquitards, poorly connected aquifers or from surface water sources.

Age dating old groundwaters begins with a determination that they are tritium-free, and so have no modern component (Chapter 7). Tritium-free groundwater can be considered sub-modern (recharged >50 years ago) or older, and have not incorporated any significant amount of modern water during discharge. The only absolute, albeit indirect, dating techniques for groundwater involve the decay (or in-growth) of long-lived radionuclides. By far the most routinely applied is ^{14}C, which is transported as dissolved inorganic carbon (DIC) or dissolved organic carbon (DOC).

Geochemical techniques for dating, such as the extent of water-rock interaction, or the degree of salinization are also important, but not quantitative. The $\delta^{18}O$-δ^2H signature of a paleogroundwater may also indicate age, by showing a change in climate controls on precipitation. The hydrodynamic characteristics of the groundwater flow system itself are important in estimating subsurface mean residence times. It is important to note that one technique alone can be misleading if applied without an understanding of the geochemical and hydrodynamic processes in the aquifer. A collaboration of methodologies is the best approach.

Stable Isotopes and Paleogroundwaters

Changes in temperature and precipitation patterns are the basis of climate change. Given the good correlation we see between isotopes in precipitation and groundwater, climate change should be recorded in fossil or paleogroundwaters. Temperate latitude climates have experienced significant changes in temperature since late Pleistocene time, whereas precipitation variability has dominated low latitude paleoclimate change. Such climate changes are manifested by a shift in the stable isotope content of precipitation, and in deuterium excess. This "paleoclimatic effect" is one of the most important tools in identifying paleogroundwaters.

In temperate regions, the dominant effect of climate change is in the position of precipitation on the local meteoric water line. Late Pleistocene paleogroundwaters from temperate regions (e.g. North America or Europe) will be isotopically depleted with respect to modern waters and shifted along the GMWL towards negative values. Fig. 8-1 shows the depletion in ^{2}H for paleogroundwaters as compared with modern groundwaters for various regions of Europe. The recharge of such groundwaters can be affected by the distribution of ice sheets and permafrost. Canada was ice covered until about 10 to 15 ka B.P., and so paleogroundwaters generally postdate glaciation. Often they contain a component of isotopically-depleted glacial meltwater as remnant pore water (e.g. glacio-lacustrine clays studied by Desaulniers et al., 1981, and Remenda et al., 1994) or glacial meltwaters that have infiltrated Shield terrain during deglaciation (Douglas, 1997). Less commonly, they may have been recharged at an earlier interstadial or interglacial when the land was ice-free and could then be isotopically enriched (e.g. the Milk River aquifer; Hendry and Schwartz, 1988).

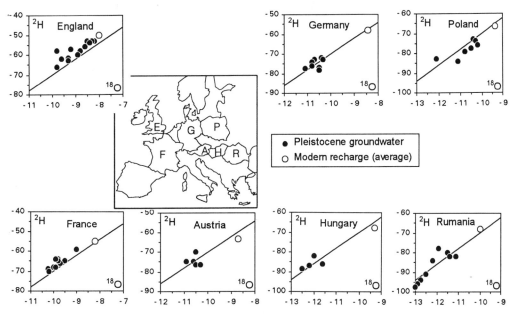

Fig. 8-1 The δ^{18}O and δ^2H composition of Pleistocene groundwaters in Europe compared with the average of modern local precipitation and the GMWL. From regional data compiled by Rozanski (1985).

Unlike a depletion in ^{18}O and ^{2}H as observed for temperate region paleogroundwaters, the paleoclimatic effect in arid regions is manifested by a displacement of the meteoric water line. Recall that variations in humidity during primary evaporation affect the deuterium excess value, d. In arid regions like the Eastern Mediterranean and North Africa, the modern MWL is characterized by a deuterium excess value of 15 to 30‰. However, pluvial (humid) climates have characterized these regions in the past, e.g. the early Holocene pluvial which has been observed by high lake stands in North and East Africa (Street and Grove, 1979) and by lacustrine sediments found in the sand dunes of the Empty Quarter of Oman and Saudi Arabia (McClure, 1976).

Under such conditions, the meteoric water line is closer to the global line. So, pluvial-climate groundwaters tend to plot on or even below the GMWL. This has been observed in many deep artesian groundwaters from such regions that have been identified by other means (^{14}C) to be paleogroundwaters (Fig. 8-2). Such paleogroundwaters are also characterized by a depletion in ^{18}O and ^{2}H with respect to modern waters. The significance of an observed paleoclimatic shift in groundwaters is that it shows them to be fossil, and not part of actively recharged flow systems. These resources are finite and their exploitation is mining.

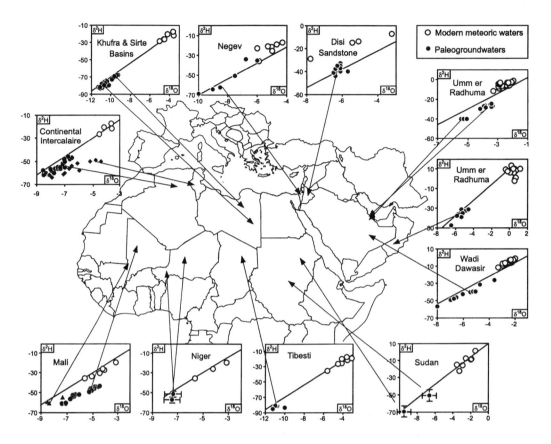

Fig. 8-2 Stable isotope signature of various paleogroundwaters from the Middle East and North Africa. Continental Intercalaire — Gonfiantini et al., 1974; Kufra and Sirte Basins — Edmunds and Wright, 1979; Sonntag et al., 1979; Negev — Gat and Dansgaard, 1972; Disi Sandstone — Lloyd, 1980; Umm er Radhuma, Saudi Arabia — Moser et al., 1978; Wadi Dawasir — Hötzl et al., 1980, Umm er Radhuma, Oman — Clark et al., 1987; Tibesti — Sonntag et al., 1979; Southern Sahara — Dray et al., 1983.

Groundwater Dating with Radiocarbon

Libby discovered radiocarbon in atmospheric CO_2 in 1946. He determined its half-life to be 5568 years (the Libby half-life) and recognized then its potential as a dating tool. Godwin subsequently refined the determination of its half-life to 5730 years.

Atmospheric $^{14}CO_2$ mixes with all living biomass through photosynthesis, as well as with meteoric waters and oceans (and carbonates formed in such waters) through CO_2 exchange reactions. By consequence, any carbon compound derived from atmospheric CO_2 since the late Pleistocene is potentially eligible for radiocarbon dating. Radiocarbon provides the chronology on which archeologists have reconstructed our history in the Holocene. It has also provided the basis of climatic reconstructions in the late Pleistocene and Holocene.

As we will see, ^{14}C is also the leading tool in estimating the age of paleo- and fossil groundwaters. The method is based upon the incorporation of atmospherically derived ^{14}C from the decay of photosynthetically-fixed carbon in soil. Radiocarbon in the soil can be taken into solution as dissolved inorganic carbon (DIC = $CO_{2(aq)} + HCO_3^- + CO_3^{2-}$) or as dissolved organic carbon (DOC). Neither approach is without complications. Let's first look at the basics of ^{14}C.

Decay of ^{14}C as a measure of time

Radiocarbon dating is based on measuring the loss of the parent radionuclide (^{14}C) in a given sample. This assumes two key features of the system. The first is that the initial concentration of the parent is known and has remained constant in the past. The second is that the system is closed to subsequent gains or losses of the parent, except through radioactive decay. If we can be sure of these two conditions, then time is precisely measured by the exponential loss of the parent according to its half-life (which is simply a statistical determination of the relative stability of a given nuclide). This is represented by the decay equation:

$$a_t = a_o \cdot e^{-\lambda t}$$

where a_o is the initial activity of the parent nuclide, and a_t is its activity after some time, t. The decay constant, λ is equal to $\ln 2/t_{½}$. For ^{14}C, $t_{½}$ is 5730 years, and this equation simplifies to:

$$t = -8267 \cdot \ln\left(\frac{a_t{}^{14}C}{a_o{}^{14}C}\right)$$

From this relationship, $a_t{}^{14}C$ is half of $a_o{}^{14}C$ after one half-life, and one quarter of $a_o{}^{14}C$ after two (Fig. 8-3). The usual expression of ^{14}C activity is as a percent of the initial ^{14}C activity ($a_o{}^{14}C$), i.e. percent modern carbon (pmC; see Chapter 1).

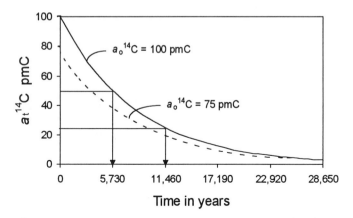

Fig. 8-3 The decay of ^{14}C and illustration of half-life. Decay shown also for case where $a^{14}C < 100$ pmC (75 pmC here).

Assuming our two conditions are met ($a_o{}^{14}C$ is known and closed system decay), dating with ^{14}C is a simple calculation. Since the time of Libby, two independently dated records have helped establish $a_o{}^{14}C$ over the past 30,000 years. These are tree rings and corals. These records of $a_o{}^{14}C$ during the Holocene and late Pleistocene are discussed below. The potential ^{14}C dating range is then limited only by analytical precision, which now allows measurements to about 10 half-lives or almost 60,000 years. However, the poor preservation and subsequent contamination of old material — our second condition for ^{14}C dating — makes the effective dating range much shorter. For organic materials such as vegetation and collagen, the effective range is generally less than about 50,000 years. For groundwaters, the range is limited to about 30,000 years, and in most cases to a much shorter time. Reaction and evolution of the carbonate system strongly dilute the initial ^{14}C activity in DIC and DOC. The result is an artificial "aging" of groundwaters by dilution of ^{14}C. Unraveling the relevant processes and distinguishing ^{14}C decay from ^{14}C dilution is an engaging geochemical problem.

Production of ^{14}C in the atmosphere

Before tackling ^{14}C dilution, let's start with a look at the generation of atmospheric ^{14}C and the variation of $a_o^{14}C$ over the past 30,000 years. Cosmic radiation produces many nuclides in the upper atmosphere, including ^{14}C. Primary cosmic rays are high-energy particles, electrons and photons from the Sun and beyond, that continually shower Earth (Gregory and Clay, 1988). The lower energy cosmic rays are mainly protons, with other light nuclei present as well. Over 90% of cosmic radiation is attenuated in the atmosphere. Collisions of these high-speed particles with atmospheric gases generate through *spallation* secondary particles of which neutrons are a large component.

Radiocarbon is produced in the upper atmosphere through one of many nuclear reactions, the bombardment of nitrogen atoms by this secondary neutron flux:

$$^{14}_{7}N + ^{1}_{0}n \rightarrow ^{14}_{6}C + ^{1}_{1}p$$

where n = neutron and p = proton.

This ^{14}C then oxidizes to carbon dioxide and mixes with other atmospheric gases, resulting in a constant flux of $^{14}CO_2$ to the troposphere where it dissolves in the oceans or is consumed by vegetation during photosynthesis. Decay of vegetation and root respiration return much of the ^{14}C to the atmosphere. From Table 8-1, the largest storehouse of ^{14}C is by far the oceans, in the form of HCO_3^-. Accumulation in the troposphere and the hydrosphere/biosphere is balanced by radioactive decay and burial. This balance, or "secular equilibrium," is relatively robust over short periods (decades to 100s of years) and amounts to an atmospheric concentration of $^{14}CO_2$ on the order of 10^{-12} of the total atmospheric CO_2. This ^{14}C concentration (activity) is defined as "modern" ^{14}C and is the basis of the radiocarbon standard (see Chapter 1). Thus, not all atmospheric CO_2 is ^{14}C-active, only a very small fraction. This secular equilibrium can be altered by both natural processes and man's activities.

Table 8-1 Terrestrial carbon reservoirs

Form	Mass (10^{18} g)	Living biomass equivalent
^{14}C-free		
Marine Carbonates	60,000	107,000
Sedimentary Hydrocarbon	15,000	26,800
Recoverable Coal, Oil and Gas	4	7.1
^{14}C-active		
Oceanic DIC	42	75
Dead Vegetation	3	5.4
Atmospheric CO_2	0.72	1.3
Life on Earth	0.56	1

(From Berner and Lasaga, 1989.)

Neutron fluxes in the subsurface from the spontaneous fission of uranium and other elements can produce hypogenic ^{14}C by neutron activation of ^{14}N or neutron capture by ^{17}O and α decay.

In zones of exceptionally high neutron fluxes, hypogenic ^{14}C be important. Calculations of *in situ* neutron fluxes and ^{14}C production were made for the ore zone in the Cigar Lake uranium deposit in Saskatchewan. Secular equilibrium for groundwater in rocks with 40 weight percent U was determined to be 29 pmC. With over 50 weight percent U, this rose to over 200 pmC (Fabryka-Martin et al., 1994). However, the subsurface neutron flux in most rocks is very low. When considered with the large reservoir of carbon for dilution in the subsurface, hypogenic production is a negligible source of ^{14}C in groundwaters.

Natural variations in atmospheric ^{14}C

A constant production and stable concentration of atmospheric ^{14}C would require that the secondary neutron flux from cosmic radiation has been constant. In fact, it has not. Dendrochronology studies show strong variations in the ^{14}C activity of atmospheric CO_2 during the Holocene. Counting rings provides a firm chronology (t in the dating equation). Measurement of $a^{14}C$ in tree rings by AMS, then provides a measure of $a_o^{14}C$ through time. Fig. 8-4 shows that ^{14}C activity in the atmosphere has varied by over 10% during the Holocene. This record has been extended into the late Pleistocene by measuring the ^{14}C content of corals (Bard et al., 1990). A reliable chronology of the corals was provided by U/Th disequilibrium dating by TIMS (thermal ionization mass spectrometry). This work shows that atmospheric ^{14}C was a whopping 40% higher during the last glacial maximum. In addition to this systematic decrease since ca. 30,000 years ago are second-order excursions, the so-called "Suess wiggles" with ca. 200-year period (Suess, 1980). There is even an 11-year cycle to ^{14}C production that matches the sunspot cycles.

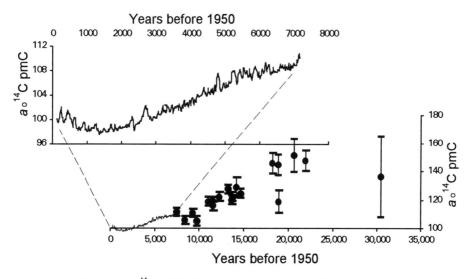

Fig. 8-4 Composite of atmospheric ^{14}C activity from tree rings, determined using their dendrochronological age (from Pearson et al., 1986), and from shallow marine corals, based on their U/Th age (from Bard et al., 1993). Holocene data show the ca 200-year period Seuss variations that are related to changes in solar output. The strong decrease from ca. 30,000 years BP to present are related to changes in the Earth's geomagnetic field.

Why the variations? The short-term cycles have been related to variations in solar output (Stuiver and Quay, 1980; Damon et al., 1989). Satellite measurements of solar output since 1979 document variations that follow the 11-year sunspot cycle (Hoyt et al., 1992). Historical records of sunspot activity also show strong correlation with atmospheric ^{14}C. However, they are weak compared to the long-term evolution in atmospheric ^{14}C. This is due to the changing structure of the Earth's geomagnetic field (Damon et al., 1989), which shields the Earth from much of the incoming flux of charged particles. This field is internally generated by the dynamo of the rotating/convecting Fe-Ni liquid outer core. Subtle variations in its dipole affect the flux of solar rays into the atmosphere and by consequence the production of ^{14}C. These huge variations in $a_o^{14}C$ will affect the calculated age of a radiocarbon-dated sample. If the standard 100 pmC is categorically used, one expresses the age in *radiocarbon years* rather than calendar years.

Anthropogenic impacts on atmospheric ^{14}C

While evidence now suggests that the variation of atmospheric ^{14}C has varied between at least 97 and 140 pmC, anthropogenic effects over the past century have been even greater. The combustion of fossil fuel has pumped over 70 ppmv of dead carbon into the atmosphere (now at ~360 ppmv), diluting ^{14}C by 25%. The record of this contribution in Vostok ice is clear from both the net increase in CO_2 and by the shifting isotopic signature from a high pre-industrial $\delta^{13}C$ value of the atmosphere (–6.4‰) towards that of fossil fuel (~–26‰) (Friedli et al., 1986).

While the industrial age has been diluting atmospheric ^{14}C, the nuclear age has been creating it. Since the 1950s, atmospheric weapons testing and nuclear power plants have been releasing additional radiocarbon to the atmosphere and biosphere (Fig. 8-5). The high neutron flux generated by thermonuclear bomb testing (see Chapter 7) activated ^{14}N to produce ^{14}C. The result was a considerable increase (a doubling at the peak in 1963) of the ^{14}C activity of atmospheric CO_2 that has subsequently been attenuation by exchange with the oceans. This peak is a very useful tracer to examine air sea transfers of CO_2 and ocean mixing. Like tritium, it can be a useful indicator of modern recharge to aquifers. Although thermonuclear bomb tests were global in impact, local increases in $a^{14}C$ are observed in the vicinity of nuclear power stations.

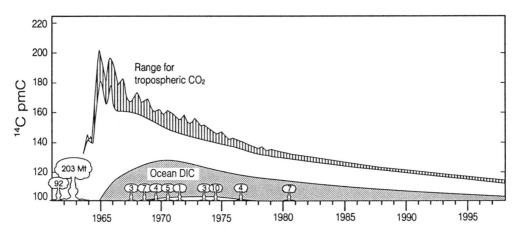

Fig. 8-5 Ranges for radiocarbon in tropospheric CO_2 and the DIC in near-surface seawater. The data show for the northern hemisphere an almost twofold increase in ^{14}C contents during the mid-1960s as a consequence of the atmospheric testing of thermonuclear bombs (annual total testing given in megatons). In the southern hemisphere the increase was lower, but since about 1970 the concentrations have been very similar around the globe. Oceanic ^{14}C increased through the uptake of atmospheric ^{14}C. Based on data to 1979 from IAEA (1983) and extrapolated to 1996. Bomb test data from Rath (1988).

Anthropogenic effects are of little relevance for paleogroundwater but such carbon is found in young systems. In modern groundwaters, radiocarbon activities above 100 pmC are often measured. This addition of radiocarbon assists in the recognition of very young groundwaters. It also is useful in establishing dilution by carbonate dissolution in groundwater recharge areas.

The ^{14}C pathway to groundwater in the recharge environment

Although rainwater contains some $^{14}CO_2$ from the atmosphere, it is the soil zone that gives recharging groundwater its radiocarbon signal. Atmospheric ^{14}C is incorporated into vegetation by photosynthesis and later released in the soil by decay and root respiration. The result is a huge reservoir of ^{14}C in the soil zone. Fig. 8-6 shows the pathway for atmospheric ^{14}C through the soil zone and into groundwater. Shown also is the fractionation experienced by ^{13}C along this pathway. Recall that the radiocarbon standard of 100 pmC is calibrated to wood grown in 1890 and has $\delta^{13}C = -25‰$. The $\delta^{13}C$ value of atmospheric CO_2 (prior to significant contributions from fossil fuel) was about −6.4‰. This is 18.6‰ enriched over the value for vegetation of about −25‰. The strong fractionation during photosynthesis that causes this depletion of ^{13}C will also affect the ^{14}C. Saliège and Fontes (1984) determined that there is a ~2.3 × mass effect for ^{14}C with respect to ^{13}C fractionation. This means that unpolluted atmospheric CO_2 was 42.5‰ or 4.3% enriched over the standard 100 pmC.

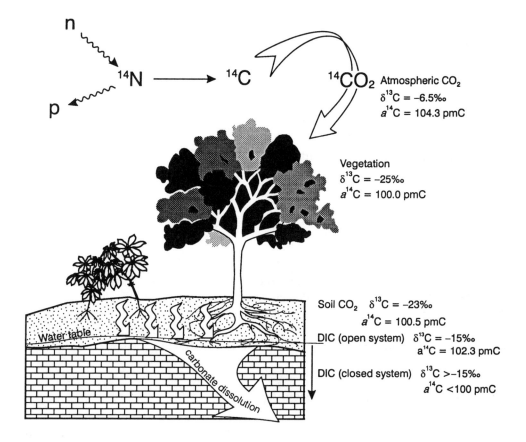

Fig. 8-6 The pathway and associated fractionation of ^{14}C and ^{13}C in CO_2 during photosynthesis, respiration in soils, and dissolution by groundwaters.

Bacterial degradation of organic litter, as well as organic substances released by living plants, pumps an enormous amount of CO_2 into soils. While the atmosphere today contains about 360 ppmv CO_2 ($P_{CO_2} \sim 10^{-3.44}$), soils contain between 3000 to 30,000 ppmv ($P_{CO_2} \sim 10^{-2.5}$ to $10^{-1.5}$). The amount of carbon dioxide that can dissolve will depend on the geochemistry of the recharge environment: the temperature, the pH of the water, the partial pressure of CO_2, and the weathering reactions that take place in the soil. Consequently, as groundwaters recharge through soils, they gain levels of ^{14}C-active DIC that are much higher than that provided by direct dissolution of atmospheric CO_2. The ^{13}C and ^{14}C in soil CO_2 are marginally enriched over the vegetation due to diffusion of lighter CO_2 from the soil to the atmosphere.

If the ^{14}C gained in the soils remains with the groundwater along the flow path, without subsequent dilution, its decay can be used as a measure of age. As this is rarely the case, dilution and loss by geochemical reactions both within the soil and along the flow path must first be addressed.

The most typical reactions include:

1. Calcite (limestone) dissolution, beginning in the recharge area
2. Dolomite dissolution
3. Exchange with the aquifer matrix
4. Oxidation of "old" organics found within the aquifer and other biochemical reactions
5. Diffusion of ^{14}C into the aquifer matrix

The dilution of ^{14}C through reaction is accounted for in the decay equation by the dilution factor or fraction, q. The ^{14}C-activity of DIC in the groundwater recharge environment following calcite dissolution ($a^{14}C_{rech}$) is equal to the modern ^{14}C in the soil ($a_o^{14}C$) times the dilution factor:

$$a^{14}C_{rech} = q \cdot a_o^{14}C$$

Thus, the decay equation corrected for dilution becomes:

$$a_t^{14}C = q \cdot a_o^{14}C \cdot e^{-\lambda t}$$

or $\quad t = -8267 \cdot \ln\left(\dfrac{a_t^{14}C}{q \cdot a_o^{14}C}\right)$

Unfortunately, the simplicity of this equation is deceiving. The dilution factor, q, is the elusive grail sought by isotope hydrogeologists in their efforts to date groundwater. The development of carbonate and ^{13}C evolution presented in Chapter 5 provides a basis for calculating q. Here we will explore approaches to tracing ^{14}C evolution and correcting radiocarbon ages through using a tool kit of aqueous geochemistry, ^{13}C, and a spreadsheet program for modelling.

Correction for Carbonate Dissolution

The approaches to correct apparent ^{14}C water ages have evolved over the past 30 years from "statistical models" and "mixing models," to "process-oriented models." The following discussion presents a few of the published models. These are followed by the development of algorithms useful for correcting the various dilution processes. Calculations require some familiarity with alkalinity and aqueous geochemistry. A brief review was given in Chapter 5. References for further reading include such texts as Garrels and Christ (1965), Freeze and Cherry (1979), Stumm and Morgan (1996) and Drever (1997).

In the approaches presented here, the diluting source of carbon is presumed to be ^{14}C-free. This is certainly the case with marine limestone, which is generally millions of years old. On the other hand, some soil carbonates may have measurable ^{14}C activities. Conversely, the oxidation of old carbon in soils can generate soil CO_2 with $a^{14}C$ that is slightly less than modern. In such cases, the correction factor must be modified accordingly.

Recall from Chapter 5, that the carbonate evolution of many groundwaters involves the dissolution of soil CO_2 with the subsequent dissolution of carbonate:

$$CO_{2(soil)} + H_2O + CaCO_3 \rightarrow Ca^{2+} + 2HCO_3^-$$

Two sources of DIC then dominate in recharge environments — the ^{14}C-*active* component from the soil used to date the water, and a (generally) ^{14}C-*free* carbonate which dilutes $^{14}C_{soil}$. This dilution is characteristic of non-carbonate aquifers as well. In sandy aquifers or tills, calcite is often present as a component of the matrix or as secondary cement and calcrete. In crystalline environments, fracture calcite is often available.

Under closed system conditions, the stoichiometry of calcite dissolution by carbonic acid imparts about a 50% dilution to the initial ^{14}C:

$$^{14}CO_2 + H_2O + CaCO_3 \rightarrow Ca^{2+} + HCO_3^- + H^{14}CO_3^-$$

Under open system conditions, the DIC is continuously exchanging with the infinite reservoir of ^{14}C-active soil CO_2. In this case, the initial ^{14}C activity of DIC ($a_o^{14}C_{DIC}$) remains unchanged at 100 pmC (or at the current initial ^{14}C activity). Unfortunately, the reality is somewhere in between. Most groundwaters reach calcite saturation during transition from open to closed conditions. The openness of the system and the contributions to the DIC pool can be reflected by the δ^{13}C of the DIC.

Fig. 5-6 shows the evolution of $\delta^{13}C_{DIC}$ towards enriched values as carbonate is dissolved. Under open system conditions, this is due to the increase in fractionation of ^{13}C between soil gas and DIC as the pH rises ($\epsilon^{13}C_{DIC-CO_2} \approx 9‰$ @ 25°C, Fig. 5-5). As the pH increases, the strong ^{13}C enrichment between $CO_{2(soil)}$ and HCO_3^- becomes increasingly important and the $\delta^{13}C_{DIC}$ goes up. However, it is still controlled by the $\delta^{13}C_{CO2(soil)}$ and $a^{14}C$ is still 100 pmC.

Under fully closed conditions, the increase in $\delta^{13}C_{DIC}$ is due solely to mixing between DIC from soil ($\delta^{13}C \approx -12$ to $-20‰$) and marine carbonate ($\delta^{13}C \approx 0‰$; Fig. 5-12). The closed system evolution of $\delta^{13}C_{DIC}$ is more dramatic because here the influence of calcite dissolution is felt. We can calculate ^{14}C dilution by the additional enrichment in $\delta^{13}C_{DIC}$ from this dilution.

Statistical correction (STAT model)

Statistical models assume that after the initial carbon uptake in the soil zone by infiltrating water, some ^{14}C dilution will occur through the addition of ^{14}C-free carbon. Statistical evaluations are possible if geochemical evolution can be averaged over the recharge area to estimate an "initial" value for the ^{14}C activity of the aqueous carbonate. This initial value represents the fraction of ^{14}C remaining after secondary carbon additions. An often quoted value is q = 0.85, i.e. 85% of the initial ^{14}C concentration remains after dilution. Better is an approach where a number of "characteristic" q values are reported (Vogel, 1970):

0.65–0.75	for karst systems
0.75–0.90	for sediments with fine-grained carbonate such as loess
0.90–1.00	for crystalline rocks

The lower q values emphasize the importance of carbonate reactions in aquifers with abundant carbonate whereas in crystalline rocks with little carbonate, q values close to 1 have been determined. The corrected age is then determined from the decay equation, using this estimate of q:

$$t = -8267 \cdot \ln \frac{a^{14}C_{DIC}}{q_{STAT} \cdot a_o^{14}C}$$

It is essential to carry out statistical evaluations for each particular study area, through measurement of $a^{14}C_{DIC}$ in tritium-bearing groundwaters in the recharge area. Assuming that the recharge conditions were similar in the past, and ignoring the slight fractionation in ^{14}C between soil CO_2 and DIC (about 0.2%), then the initial radiocarbon content can be estimated from the relationship:

$$q = \frac{a^{14}C_{DIC}}{a^{14}C_{soil}}$$

This approach was taken in Jordan (Bajjali et al., 1997), where $a^{14}C_{atm}$ was measured at 114 ± 6 pmC. Tritium-bearing groundwaters in the recharge area averaged 65 pmC. With a slight correction for fractionation during transfer to the soil (−16‰ × 2.3 or about −4%), the ^{14}C dilution by carbonate dissolution prior to decay is calculated as:

$$q = \frac{a^{14}C_{DIC}}{a^{14}C_{atm} \cdot 0.96} = \frac{65}{114 \cdot 0.96} = 0.59$$

The simple statistical approach can be problematic. Geochemical reactions that take place in groundwaters beyond the recharge area (above) must also be corrected for. The actual ^{14}C dilution is usually much larger (and thus the q-factors smaller) in older systems than in young ones. Geochemical models must then be considered.

Alkalinity correction (ALK model)

Where the recharge area cannot be studied, geochemical models can be used that are based on the parameters measured in the water being dated. An often quoted approach is the Tamers (1975) or "chemical" correction based on initial and final carbonate (DIC) concentrations. It was proposed for groundwater in which calcite is dissolved under closed system conditions (recall from Chapter 5), and is formulated as:

$$q_{ALK} = \frac{mH_2CO_3 + \frac{1}{2}mHCO_3^-}{mH_2CO_3 + mHCO_3^-}$$

For the vast majority of groundwaters, this leads automatically to a q-factor of about 0.5 because of the reaction:

$$H_2CO_2 + CaCO_3 \leftrightarrow 2HCO_3^- + Ca^{2+}$$

whereby most of the carbonic acid is consumed by limestone dissolution, and the original ^{14}C from soil carbon dioxide is diluted to about 50%. This model assumes fully closed system conditions, where no exchange with soil CO_2 during calcite dissolution occurs. This model is of limited interest due to its simplification of geochemical reaction.

Chemical mass-balance correction (CMB model)

The chemical mass balance model is a closed system model where carbonate dissolution takes place below the water table and the DIC does not exchange with the soil CO_2. The calculation compares the DIC gained from dissolving soil CO_2 to that measured in the groundwater sample. The correction factor "q" is calculated from:

$$q = \frac{mDIC_{rech}}{mDIC_{final}}$$

where $mDIC_{rech}$ is the ^{14}C-active DIC gained by dissolution of soil CO_2 during recharge, and $mDIC_{final}$ is the total carbonate content at the time of sampling (^{14}C-active + ^{14}C-dead). The model requires that $mDIC_{rech}$ be calculated from estimated P_{CO_2}-pH conditions for the recharge environment (Chapter 5). If the concentration of DIC has been measured in the recharge area groundwaters (where $a^{14}C_{DIC} = a^{14}C_{atm}$), then this can be used. However, if climates have changed since recharge, then estimates of $mDIC_{rech}$ are somewhat speculative. Further, this approach does not account for open-system carbonate dissolution during recharge. Nevertheless, it could be a worthwhile calculation to make and it may signal some geochemical complications to explore.

The recipe for applying a geochemical mixing model is:

(i) Calculate the initial carbonate content during recharge ($mDIC_{rech}$) by assuming pH and P_{CO_2} values and the application of carbonate equilibria presented in Chapter 5 (as in the example Table 8-2) to determine $[H_2CO_3]$ and $[HCO_3^-]$. For example, the initial concentration of DIC in groundwater with a pH of 6.3 recharging through a soil with $P_{CO_2} = 10^{-2.2}$ is determined according to :

$$K_{CO_2} = \frac{[H_2CO_3]}{P_{CO_2}} = 10^{-1.47}, \text{ and so } [H_2CO_3] = 10^{-1.47} \cdot 10^{-2.20} = 10^{-3.67}$$

$$K_1 = \frac{[HCO_3^-] \cdot [H^+]}{[H_2CO_3]} = 10^{-6.35}, \text{ and so } \frac{[HCO_3^-]}{[H_2CO_3]} = \frac{10^{-6.35}}{10^{-6.30}} = 0.89$$

$$[HCO_3^-] = [H_2CO_3] \cdot 0.89 = 10^{-3.72}$$

Thus, the initial DIC content is the sum of carbonic acid (dissolved CO_2) and HCO_3^- (assuming that activities [i] are equivalent to molalities m_i at low salinity):

$$mDIC_{rech} = 10^{-3.67} + 10^{-3.72} = 10^{-3.23} \text{ or 36 mg/L as bicarbonate, } HCO_3^-.$$

(ii) The final carbonate content (DIC$_{final}$) is best represented by the measured (titrated) alkalinity (designated as CMB-Alk model). When titrated alkalinities are unreliable the DIC$_{final}$ value can be calculated from chemical data (designated as the CMB-Chem model):

$$m\text{DIC}_{final} = m\text{DIC}_{rech} + [m\text{Ca}^{2+} + m\text{Mg}^{2+} - m\text{SO}_4^{2-} + \tfrac{1}{2}(m\text{Na}^+ + m\text{K}^+ - m\text{Cl}^-)]$$

where the chemical constituents are measured concentrations (in moles/L). This form of the CMB model was proposed by Fontes and Garnier (1979) to account for Ca^{2+} and Mg^{2+} added by carbonate dissolution, with a correction for Ca^{2+} added through the dissolution of gypsum ($m\text{SO}_4^{2-}$), and Ca^{2+} lost through ion exchange processes ($m\text{Na}^+ + m\text{K}^+ - m\text{Cl}^-$). Cl^- is included in this term to account for Na^+ gained through incorporation of salinity from seawater or halite dissolution.

The application of these ^{14}C dilution models is only justified in geochemically simple systems where no carbonate is lost from the groundwater. Further, both models assume that the system is closed with respect to soil CO_2 during carbonate dissolution. This is seldom the case, as weathering reactions generally begin during infiltration through the unsaturated zone under open-system conditions. A better approach is based on a ^{13}C and carbonate mass balance to quantify contributions to the DIC pool.

$\delta^{13}C$ mixing ($\delta^{13}C$ model)

Carbon-13 can be a good tracer of open and closed system evolution of DIC in groundwaters (Chapter 5). The large difference in $\delta^{13}C$ between the soil-derived DIC and carbonate minerals in the aquifer can provide a reliable measure of ^{14}C dilution by carbonate dissolution. The $\delta^{13}C$ mixing model allows for incorporation of ^{14}C-active DIC during carbonate dissolution under open system conditions, and subsequent ^{14}C dilution under closed system conditions.

Pearson (1965) and Pearson and Hanshaw (1970) first introduced a $\delta^{13}C$ correction based on variations in ^{13}C abundances. Any process that adds, removes or exchanges carbon from the DIC pool and which thereby alters the ^{14}C concentrations will also affect the ^{13}C concentrations. The q-factor was obtained from a carbon isotope-mass balance where:

$$q = \frac{\delta^{13}C_{DIC} - \delta^{13}C_{carb}}{\delta^{13}C_{soil} - \delta^{13}C_{carb}}$$

where: $\delta^{13}C_{DIC}$ = measured ^{13}C in groundwater
$\delta^{13}C_{soil} = \delta^{13}C$ of the soil CO_2 (usually close to –23‰)
$\delta^{13}C_{carb} = \delta^{13}C$ of the calcite being dissolved (usually close to 0‰)

However, this early version of the $\delta^{13}C$ model assumed that carbonate dissolution takes place under closed system conditions, precluding isotope exchange with the soil gas reservoir. The above equation is only correct if the $CO_{2(soil)}$ is taken up by the water without significant fractionation effects. This is the case in low pH environments where only about a 1‰ depletion accompanies dissolution of $CO_{2(soil)}$.

However, Fig.5-5 shows that at higher pH values (pH 7.5 to 10) the DIC in equilibrium with the $CO_{2(soil)}$ is enriched in ^{13}C. This enrichment with respect to the original $CO_{2(soil)}$ varies between

7 and 10‰ depending on temperature (Table 5-3). The $\delta^{13}C_{soil}$ is therefore replaced by a initial $\delta^{13}C$ value for DIC in the infiltrating groundwaters ($\delta^{13}C_{rech}$) defined as:

$$\delta^{13}C_{rech} = \delta^{13}C_{soil} + \varepsilon^{13}C_{DIC\text{-}CO2(soil)}$$

Here, $\varepsilon^{13}C_{DIC\text{-}CO2(soil)}$ is the pH-dependent enrichment between soil CO_2 and the aqueous carbon. An approximate value can be read from Fig. 5-5 or calculated from the distribution of DIC species and their respective enrichment factors (problem 3). The effect of this parameter is to control the amount of correction by the model according to the pH conditions during infiltration. The modified dilution model is then:

$$q_{\delta^{13}C} = \frac{\delta^{13}C_{DIC} - \delta^{13}C_{carb}}{\delta^{13}C_{rech} - \delta^{13}C_{carb}}$$

This is a neat algorithm because it allows for carbonate dissolution under open and closed system conditions. Any dissolution that takes place under truly open system conditions (i.e. exchanging carbon with the soil CO_2) will equilibrate both ^{14}C and ^{13}C with the soil CO_2. The dilution factor q is then a measure of the amount of carbonate dissolved under closed system conditions, when dilution of ^{14}C occurs.

The main problem with this approach is that the enrichment factor that you choose for $\varepsilon^{13}C_{DIC\text{-}CO2(soil)}$ can affect groundwater ages enormously. This enrichment factor is based on the pH of the groundwater during recharge, which a diligent hydrogeologist may measure *in situ*. However, this may not be an accurate representation of conditions in the past if the groundwaters are old and recharge conditions have changed. Let's look at an example by varying the pH of the recharge waters. Remember that:

$$\varepsilon^{13}C_{DIC\text{-}CO2(soil)} = mCO_{2(aq)} \cdot (\varepsilon^{13}C_{CO2(aq)\text{-}CO2(g)}) + mHCO_3 \cdot (\varepsilon^{13}C_{HCO3\text{-}CO2(g)})$$

where m is the mole fraction of the two carbonate species and the enrichment factors, $\varepsilon^{13}C$, are from the fractionation equations in Chapter 5 (Table 5-2). The mole fraction of these DIC species is a function of pH (see Fig. 5-2 and equations on page 116). From Table 8-2, we see that the variations in corrected ages will vary by almost 50%.

Table 8-2 Example of the influence of initial pH on ^{13}C and ^{14}C ages in the ^{13}C mixing model.

Input:	Case A	Case B	Case C
pH	6.0	6.4	7.0
$mCO_{2(aq)}/mHCO_3^-$	2.5	1	0.25
$\delta^{13}C_{DIC}$ (measured)	–12.5‰	–12.5‰	–12.5‰
$\delta^{13}C_{soil}$ (estimated)	–23‰	–23‰	–23‰
$\delta^{13}C_{carb}$ (estimated)	0‰	0‰	0‰
$a^{14}C$ pmC	35	35	35
Calculated:			
$\varepsilon^{13}C_{DIC\text{-}CO2(g)}$	1.5	3.4	5.7
$\delta^{13}C_{inf}$	–21.5	–19.6	–17.3
q	0.58	0.64	0.72
Ages (yr B.P.):			
Uncorrected	8700	8700	8700
Corrected	4175	4990	5960

Thus, by varying the pH assumed for the infiltrating groundwaters by 1 pH unit, the age will vary considerably. However, this is only over the range from pH 6 to 7. If we find that in the recharge area groundwater pH values are neutral or higher there is little variation because the soil CO_2 exchanges mainly with HCO_3^-.

The effect of dolomite dissolution

Calcite saturation is generally achieved in the recharge area for most groundwaters. This is particularly so for aquifers in carbonate rocks. Due to the slower kinetics, dolomite saturation is often achieved further along the flow path, assuming that there is at least a minor fraction of $MgCa(CO_3)_2$ available. As this takes place under closed system conditions, it imparts an additional dilution on the $^{14}C_{DIC}$ pool. The increase in mMg^{2+} beyond the recharge area reflects this additional dilution:

$$mMg_{excess} = mMg_{meas} - mMg_{rech}$$

The dilution factor for this process takes into account the fact that two moles of carbonate ions are released for each mole of Mg^{2+} in dolomite, and is approximated by:

$$q_{dol} = \frac{mDIC}{mDIC + 2mMg_{excess}}$$

The effect of this process on carbonate evolution and ^{13}C can be tested through the isotope mass balance equation, assuming that increases in DIC are buffered by calcite precipitation:

$$\delta^{13}C_{DIC} = \frac{\delta^{13}C_{rech} \cdot mDIC + \delta^{13}C_{dol} \cdot 2mMg_{excess}}{mDIC + 2mMg_{excess}}$$

However, the precipitation of calcite during dolomite dissolution is a complication to the model. As pH, DIC and Ca^{2+} concentrations increase, calcite is driven to supersaturated conditions. The secondary $CaCO_3$ is usually on the order of 2.5‰ enriched over the DIC. If the loss of calcite is not significant with respect to the total DIC pool, then this fractionation will not greatly alter its $\delta^{13}C$ value, and the $\delta^{13}C$ mixing model would then be appropriate.

Matrix exchange (Fontes-Garnier model)

The exchange of carbon isotopes between the DIC and carbonate minerals in the aquifer matrix is often considered as a cause for $\delta^{13}C_{DIC}$ enrichment and ^{14}C dilution. Groundwaters that are essentially at equilibrium with calcite will exchange carbonate across the mineral-solution interface where CO_3^{2-} and Ca^{2+} are in a continual process of recrystallization. Turner (1982) reviews this micro-scale interaction. As carbonate is cycled between the dissolved and solid phases, ^{13}C (and ^{18}O) are exchanged between the DIC and matrix. Constraints to such a process include the grain size of carbonate minerals in the aquifer, the volume and geometry of pores and the degree of "aging" of the mineral surfaces. The "reactivity" of the mineral surfaces decreases with increased crystallization and age.

Isotope exchange should be recorded by closed system enrichments to ^{13}C, and so radiocarbon ages can be corrected according to the ^{13}C mass-balance equation discussed above. However, if secondary calcite surfaces are involved, then the $\delta^{13}C$ will be depleted from that of the host limestone (or dolomite), and the effect may not be fully accounted for. A more complicated correction is proposed by Maloszewski and Zuber (1991) who model the retardation of ^{14}C as a sorption reaction. A limitation to their approach is the sensitivity of the model to distribution coefficient for ^{14}C and pore geometry, which vary between aquifers.

Fontes and Garnier (1979; 1981) developed a correction model (the F-G model) that uses cation concentrations to determine the contribution of ^{14}C-free matrix carbonate, and isotope mass balance to apportion $^{14}C_{DIC}$ into that exchanged with (i) CO_2 gas in the soil (open system exchange), and (ii) the carbonate matrix (matrix exchange). In their model, the total of matrix-derived carbonate is calculated as:

$$m\text{DIC}_{carb} = m\text{Ca}^{2+} + m\text{Mg}^{2+} - m\text{SO}_4^{2-} + \tfrac{1}{2}(m\text{Na}^+ + m\text{K}^+ - m\text{Cl}^-)$$

This accounts for carbonate dissolution based on Ca and Mg, with a correction for evaporite dissolution ($m\text{SO}_4^{2-}$) and cation exchange ($m\text{Na}^+ + m\text{K}^+$, with a correction for Na$^+$ from salt, $m\text{Cl}^-$). Their $m\text{NO}_3^-$ term is dropped here as it is negligible in old waters. This DIC is then apportioned into two components — that which has exchanged with the soil CO_2 (^{14}C active) in an open system and that which has exchanged with the carbonate matrix (considered to be ^{14}C-dead) under closed system conditions. The fraction of this DIC that has exchanged with soil CO_2 in an open system ($m\text{DIC}_{CO_2\text{-exch}}$) is calculated from the mass-balance relationship:

$$m\text{DIC}_{CO_2\text{-exch}} = \frac{\delta^{13}C_{meas} \cdot m\text{DIC}_{meas} - \delta^{13}C_{carb} \cdot m\text{DIC}_{carb} - \delta^{13}C_{soil} \cdot (m\text{DIC}_{meas} - m\text{DIC}_{carb})}{\delta^{13}C_{soil} - \varepsilon^{13}C_{CO_2-CaCO_3} - \delta^{13}C_{carb}}$$

In this equation, the value for $m\text{DIC}_{CO_2\text{-exch}}$ (moles of matrix-derived DIC that have exchanged with the soil CO_2) can be negative. This simply indicates that isotope exchange between DIC and the matrix is the dominant exchange process. The dilution factor then becomes:

$$q_{F-G} = \frac{m\text{DIC}_{meas} - m\text{DIC}_{carb} + m\text{DIC}_{CO_2\text{-exch}}}{m\text{DIC}_{meas}}$$

Which model do I use?

The most appropriate approach to correct $^{14}C_{DIC}$ ages depends on the geochemical system particular to the groundwaters, and the data available. The geochemical evolution may be (and likely is) more complicated than these simple relationships represent. At best, these models provide reasoned estimates of groundwater ages. More commonly, they can highly overestimate groundwater age, but are still an improvement over uncorrected ^{14}C ages.

The alkalinity mixing model (ALK) takes a very simplistic view of carbonate evolution, ignoring all reactions beyond simple dissolution of carbonate by soil CO_2. The chemical mass-balance models (CMB-Alk and CMB-Chem) are sensitive to the pH-P_{CO_2} conditions that the modeller specifies for the recharge area. The ^{13}C mixing model ($\delta^{13}C$) is also strongly dependent on recharge conditions, which affect the ^{13}C enrichment factor during dissolution of CO_2 and the evolution of $\delta^{13}C_{DIC}$. Without supporting data from the recharge area or other dating tools, such estimates are unreliable. The F-G model does not require specified input parameters, and so removes this uncertainty. However, it does not take into account additional sources of DIC beyond carbonate dissolution and exchange.

The best approach is to collect as much field data as possible, including samples from the recharge area, and compare results with various models. More complicated modelling, and other dating tools can then be considered. Using a spreadsheet facilitates these calculations, as the algorithms are easily entered. Sensitivity analysis for various input parameters can then be carried out by changing boundary conditions to extreme values and comparing with results for average values or best estimates. This is particularly important for the pH and P_{CO_2} specified for the recharge environment.

The following data (Table 8-3) for groundwaters in southern Oman provide a useful case study to test the calculations and compare model results (Clark et al., 1987; see Fig. 8-11 below). Note the presence of HS^- in these waters, which is a complication that will be discussed below using again these data. Major ion concentrations are given in milli-equivalence per litre (meq/L) where 1 meq/L = mmole/L × valence. The age estimates given by the major carbonate correction models are graphed for comparison in Fig. 8-7 and given in Table 8-4.

Table 8-3 Geochemical and isotopic data collected for groundwaters from the Umm er Radhuma carbonate aquifer, Oman

Site	^{14}C pmc	^{13}C ‰	T °C	pH	Ca^{2+}	Mg^{2+}	Na^+	K^+	HCO_3^-	Cl^-	SO_4^{2-}	HS^-
					←			Solute concentrations in meq/L				→
Recharge Area												
Average	69.8	−6.3	28.0	7.26	2.85	2.25	1.59	0.05	5.94	2.38	0.20	nd
Artesian Groundwaters												
55	3.3	−2.0	34.0	7.55	3.59	4.41	6.00	0.27	2.61	6.09	2.63	0.2
56	2.3	−2.1	35.0	7.45	5.39	4.29	5.43	0.27	2.00	6.46	4.65	0.8
57	4.2	−1.8	37.5	7.10	5.24	3.54	7.22	0.37	4.04	7.00	3.02	0.7
58	6.3	−4.8	37.0	7.76	3.99	5.04	6.96	0.32	3.12	7.23	5.83	1.6
62	3.0	0.0	38.0	7.55	3.65	6.29	10.83	0.22	1.87	13.80	3.81	0.6
63	7.3	−1.7	32.0	8.85	2.80	2.86	7.35	0.31	2.23	6.31	1.94	0.5
64	2.7	−0.6	34.0	7.55	2.99	4.24	10.70	0.51	0.75	11.23	5.46	1.8
66	7.9	−4.6	34.0	7.45	7.19	12.24	25.83	0.36	1.97	34.57	8.92	0.4

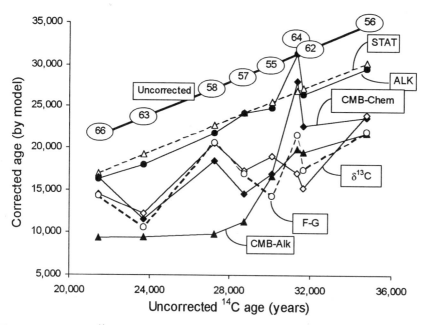

Fig. 8-7 Comparison of ^{14}C correction models for groundwaters from Oman (calculations from Tables 8-3 and 8-4)

Table 8-4 Application of ^{14}C age correction models — recharge parameters and results for groundwaters in Oman

	Recharge parameters					Calculated parameters		
T	pH	P_{CO2}	$\delta^{13}C_{soil}$	$a_o{}^{14}C$	$\delta^{13}C_{carb}$	mH_2CO_3	$mHCO_3^-$	$\delta^{13}C_{inf}$
25	6.5	0.007	–18‰	100pmC	2‰	0.00024	0.00034	–13.8‰

Site	Model results →						
	Uncorrected	STAT	AKL	CMB-Alk	CMB-Chem	$\delta^{13}C$	F-G
55	30,107	25,628	24,850	16,686	16,944	19,093	14,304
56	34,773	30,294	29,630	21,784	23,726	24,031	22,005
57	28,717	24,237	24,149	11,300	14,557	17,296	16,984
58	27,254	22,775	21,828	9838	18,463	20,533	20,688
62	31,665	27,186	26,412	19,503	22,643	15,213	17,443
63	23,696	19,217	17,992	9475	11,585	12,191	10,649
64	31,366	26,886	31,366	19,943	28,043	17,088	21,569
66	21,457	16,978	16,314	9405	16,459	14,497	14,290

Case study of the Triassic sandstone aquifer, U.K.

Groundwaters of the Triassic age Bunter sandstone in the English East Midlands were studied by Bath et al. (1979) and provide a good example to study radiocarbon dating with relatively few geochemical complications. The Bunter aquifer is a non-marine quartzose sandstone confined above and below by marls and mudstones, and dips eastward under the North Sea from outcrops in the East Midlands. The highlands of the outcrop region are situated just beyond the limit of glacier ice during the last glacial maximum, which is a consideration for the recharge of paleogroundwaters.

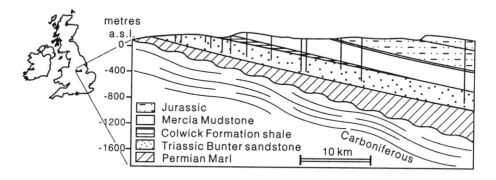

Fig. 8-8 Geological setting of the Triassic Bunter sandstone, eastern England (modified from Andrews et al., 1994).

The carbonate evolution of the groundwaters in a down-gradient direction is characterized by increases in $mHCO_3^-$ and $\delta^{13}C_{DIC}$ due to interaction with minor carbonate in the fluvial sediments. Radiocarbon activities decrease from 40 to 60 pmC in the unconfined aquifer region to less than 2 pmC in the deep groundwaters. Fig. 8-9 shows the inversely proportional evolution of ^{14}C and ^{13}C. Shifts in ^{14}C along the down-gradient trend are matched by an opposite shift in $\delta^{13}C$, demonstrating that much of the loss of ^{14}C is through geochemical reaction with carbonate in the sandstone matrix. The low $a^{14}C$ values in the recharge area are evidence that the carbonate system evolves under closed system conditions, and the ^{13}C mixing model presented above is appropriate for age corrections.

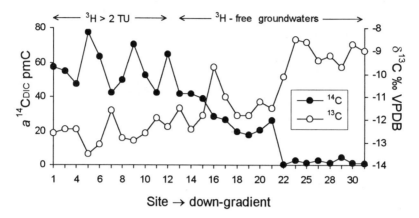

Fig. 8-9 Down-gradient evolution of a^{14}C and δ^{13}C in groundwaters from the Bunter sandstone. The inverse correlation of a^{14}C with δ^{13}C demonstrates the loss of ^{14}C by reaction with matrix rather than decay (Bath et al., 1979).

From these data, three groups of groundwaters can be identified: (i) tritium-bearing groundwaters with highly variable ^{14}C and ^{13}C contents (sites 1 to 12), (ii) tritium-free groundwaters with intermediary ^{14}C contents (18 to 42 pmC), and (iii) tritium-free groundwaters with very low a^{14}C (<5 pmC).

The ^{14}C modelling results match nicely with δ^{18}O data which show a paleo-recharge effect (Fig. 8-10). The tritium-bearing waters have 760-year-old to "future" ages, which reflect the degree of uncertainty in the correction model. The intermediate (ii) groundwaters fall in an age bracket of 1300 to 8000 years B.P., and the deeper (down-gradient) groundwaters of group (iii) have modelled ages in the 19- to 35-ka range.

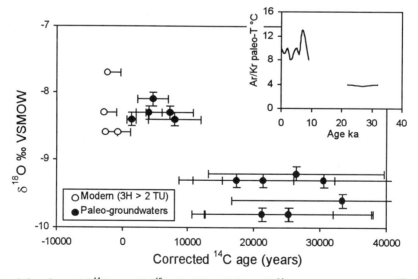

Fig. 8-10 Correlation of corrected ^{14}C ages with δ^{18}O. Confidence limits on the ^{14}C ages are ±50%, and for δ^{18}O are ±0.1‰. The strong shift in δ^{18}O to higher values signifies warmer climatic conditions at the beginning of the Holocene (after Bath et al., 1979). Paleotemperatures determined by Ar–Kr concentration relationships (recall Fig. 7-10) given in the inset diagram also document this paleoclimatic shift (after Andrews and Lee, 1979).

Comparison with $\delta^{18}O$ data substantiate the radiocarbon ages (Fig. 8-10). The Holocene groundwaters have $\delta^{18}O$ values similar to or enriched over modern groundwaters. Paleotemperatures established by Andrews and Lee (1979) from the relative concentrations of argon and krypton in these groundwaters (inset in Fig. 8-10) show that the Holocene was as warm as today or warmer (in the early Holocene hypsithermal). By contrast, the deeper groundwaters with very low ^{14}C activities were recharged under the cooler climatic conditions of the late Pleistocene.

The tremendous amount of geochemical reaction in the Pleistocene samples has unlikely been fully accounted for by the $\delta^{13}C$ correction model. In question is the value selected for $\delta^{13}C_{carb}$. In calculating the ages in Fig. 8-10 this parameter was assigned the typical marine value of 0‰, although the authors measured values averaging –7‰ in the carbonate cement of the sandstone aquifer. Using this value shifts the age of the group (iii) groundwaters to between 12 and 30 ka (although the ^3H groundwaters jump to future ages of a few thousand years). A more accurate assessment likely falls somewhere in between, and for this reason errors of ± 50% are indicated in Fig. 8-10. This would account for the apparent hiatus in groundwater recharge during the period of deglaciation between 10 and 20 ka.

Some Additional Complications to ^{14}C Dating

The $^{14}C_{DIC}$ acquired from the soil, now diluted by carbonate dissolution and exchange, may yet suffer additional dilution before following a simple radioactive decay over the ensuing millennia. Some additional reactions include:

- Matrix diffusion of ^{14}C (^{14}C loss)
- Sulphate reduction (^{14}C dilution)
- Incorporation of geogenic CO_2 (^{14}C dilution)
- Methanogenesis: $2CH_2O \rightarrow CO_2 + CH_4$ (^{14}C dilution)

All these processes conspire to reduce the initial ^{14}C content of DIC in groundwaters. Corrections to establish the pre-decay $a^{14}C$ require knowledge of aquifer characteristics (porosity and mineralogy) and geochemical evolution.

Matrix diffusion of ^{14}C

An often overlooked and problematic process affecting the ^{14}C activity in old groundwaters is the loss of ^{14}C by diffusion into dead-end pores (Sudicky and Frind, 1981; Neretnieks, 1981). Annoyingly, the impact of aquifer diffusion increases with residence time, and so compounds the error for older groundwaters.

Consider a dual porosity aquifer where transport in the connected pore or fracture network is advective, and by diffusion in the secondary pores and microfractures. Under steady-state conditions, a conservative solute diffusing into and out of the secondary porosity from the main channels will, over time, establish a uniform concentration. Not so for ^{14}C, which decays over time. Diffusion into secondary porosity, coupled with decay, will maintain a gradient of ^{14}C from the main channels towards the matrix. The influence of matrix diffusion on radiocarbon ages depends, among other factors, on the microstructure of aquifer porosity.

The principal factor affecting retardation by aquifer diffusion is the ratio of matrix porosity to fissure porosity, n_p/n_f. Maloszewski and Zuber (1984) show that the retardation factor, $(1+n_p/n_f)$, can be related to "real" and apparent (conventional) ^{14}C age through the simple relationship:

$$t_{real} = \frac{t_{apparent}}{1 + n_p/n_f}$$

It is difficult to characterize fracture and matrix porosity for an aquifer, although typically n_p is on the order of 0.05 to 0.1 while n_f is usually less than about 0.01, suggesting that reasonable diffusion retardation values $(1+n_p/n_f)$ would be >5 to 10.

For the case study of the fine-grained Calcaires Carbonifères aquifer in northern France and Belgium, uncorrected ^{14}C ages exceed hydrodynamic ages by over 2 orders of magnitude. Maloszewski and Zuber (1991) demonstrate that matrix diffusion imparts a retardation factor close to 10 and accounts for a portion of this discrepancy, the balance being due to normal carbonate dissolution reactions.

In the Cracow-Silesian region of Poland, groundwaters in a dual porosity carbonate aquifer (n_f = 0.016 to 0.035 and n_p = 0.02) were found to be $^{14}C_{DIC}$-free and so exceeded 30,000 years (Maloszewski and Zuber, 1991). These estimates were far greater than the hydrodynamic ages and contained no $\delta^{18}O$–$\delta^{2}H$ shift characteristic of Pleistocene recharge. Some wells with uncorrected ages of more than 5000 years ($a^{14}C$ = 40 pmC) also contained tritium, and were obviously modern. Taking into account a matrix diffusion retardation factor of 1.6 to 2.2, as well as a factor for matrix exchange, accounts for the loss of ^{14}C from these Holocene groundwaters.

Sulphate reduction

The oxidation of organics encountered by groundwaters within the aquifer complicates radiocarbon dating by adding ^{14}C-free carbon to the DIC pool (Clark et al., 1996). Sulphate gained from within sedimentary strata is perhaps the most common electron acceptor for oxidizing organic carbon. The reaction produces H_2S, which is characteristic of sulphate-reducing groundwaters. Other processes such as methanogenesis also contribute mineralized carbon to the DIC pool (see below).

It is not surprising that organics are often encountered in aquifers. Sedimentary rocks can retain minor sedimentary organic carbon following diagenesis. The Cretaceous limestone aquifers of northern Jordan, for example, contain up to several percent sedimentary organic carbon as kerogen (Bajjali et al., 1997). Alluvial aquifers may contain buried peat and other vegetation. Even crystalline aquifers can contain reduced carbon as methane or other gases.

Although organics encountered within aquifers are usually of geological age and ^{14}C-free, alluvial aquifers formed within the late Pleistocene or Holocene may contain organic carbon with measurable ^{14}C. The ^{14}C activity of DIC generated from such sources then has to be taken into account in any correction model (e.g. Aravena et al., 1995).

While the isotopic composition of organics is distinct from marine carbonates, their contribution to the DIC pool cannot be quantified by ^{13}C alone. With sulphate reduction, the DIC is a three-component mixture involving soil CO_2, aquifer carbonate and organic carbon. The concentration of H_2S can be used with $\delta^{13}C$ to quantify the DIC contributed from oxidation of organics. Redox evolution towards lower Eh values in groundwater was discussed in Chapter 5, where it was shown that dissolved O_2 is the first electron acceptor to be consumed as organics are oxidized. This is followed by NO_3^- and then SO_4^- in the chain of biologically mediated redox reactions. Generally, there is little available NO_3^- in groundwaters to buffer redox, and the next principal step in the evolution of a groundwater's redox potential is that of sulphate reduction.

Sulphate is not an uncommon anion to find in many aquifers. It may be supplied by marine aerosols that accumulate in the recharge environment. Oxidation of pyrite is another source of SO_4^{2-}. In sedimentary strata, evaporite beds or gypsum in vugs and fractures are common sources of sulphate. This is of particular importance in many aquifers where sulphate reduction occurs because such strata can be a source for both the organic carbon and the sulphate, thus providing a continuous input function to this geochemical reaction.

The production of DIC during sulphate reduction depends on the type of organic involved. Oxidation of humic substances (from soil, peat, brown coal etc.) in which the carbon has an overall redox state near 0 (fixed carbon) would follow a reaction such as:

$$2CH_2O + SO_4^{2-} \rightarrow H_2S + 2HCO_3^-$$

where 1 mole of HS^- would signify the contribution of 2 moles of DIC. If calcite precipitation is assumed, changes to mDIC will be minimal, and the dilution factor, q_{H_2S}, can be simplified to:

$$q_{H_2S} = \frac{m\text{DIC}}{m\text{DIC} + 2m\text{H}_2\text{S}}$$

If a reduced carbon substrate is used (redox state of –IV for hydrocarbons such as methane, etc.) the reaction will be something like:

$$CH_4 + SO_4^{2-} \rightarrow HS^- + HCO_3^- + H_2O$$

where 1 mole of HS^- indicates the production of only 1 mole of DIC. In this case, the dilution factor for the process of sulphate reduction alone can be expressed as:

$$q_{H_2S} = \frac{m\text{DIC}}{m\text{DIC} + m\text{H}_2\text{S}}$$

In these equations H_2S represents the sum of all reduced sulphur species (i.e. H_2S and HS^-). Furthermore, sulphate reduction is an alkaline reaction, and generally results in the precipitation of $CaCO_3$. The common ion effect for Ca^{2+} during dissolution of gypsum also causes calcite to precipitate.

Sulphate reduction affects not only $^{14}C_{DIC}$ but also $\delta^{13}C_{DIC}$ due to the addition of ^{13}C-depleted DIC from the organics, for which $\delta^{13}C_{org}$ is generally about –25 to –30‰. The ^{13}C content of the final DIC after carbonate dissolution and sulphate reduction can be calculated from:

$$\delta^{13}C_{DIC-final} = \frac{\delta^{13}C_{DIC-rech} \cdot m\text{DIC}_{rech} + \delta^{13}C_{org} \cdot m\text{H}_2\text{S}}{m\text{DIC}_{rech} + m\text{H}_2\text{S}}$$

The calculated values for $\delta^{13}C_{DIC-final}$ should match closely with values for $\delta^{13}C$ measured for the sample DIC. If not, other geochemical processes such as sulphide precipitation, or additional carbonate reactions may have occurred.

Incorporation of geogenic CO_2

Deep crustal or mantle sources CO_2 of volcanogenic or metamorphic origin are not uncommon in deeply circulating groundwaters. Geothermal waters in volcanic settings or along fault zones associated with plate boundaries will often have elevated P_{CO_2} values that reflect a subsurface source. This mantle-derived CO_2 has been observed in geothermal waters and fumeroles from seismically active belts and has been found to have a $\delta^{13}C$ value of about –6‰ (Barnes et al., 1978; Marty and Jambon, 1987).

Another geogenic source of CO_2 is from the thermal metamorphism of limestone. Where rising magma interacts with carbonate strata, decarbonation of $CaCO_3$ takes place, with the preferential loss of heavy CO_2. Such reactions can take place at temperatures as low as 600°C, although the presence of silica and production of wollastonite [$CaSiO_3$] lowers the reaction temperature to less than 300°C. Metamorphic CO_2 is several permil enriched above the carbonate precursor, with values typically between 5 and 10‰.

A ^{14}C correction based on carbon mass balance is unlikely to be useful because of the poor constraint on the molar contribution of geogenic CO_2. However, a $\delta^{13}C$ mixing model may be appropriate if the value for the geogenic CO_2 is known. Accordingly, a geogenic dilution factor q_{geo} would be determined as:

$$q_{geo} = \frac{\delta^{13}C_{DIC-meas} - \delta^{13}C_{geo}}{\delta^{13}C_{rech} - \delta^{13}C_{geo}}$$

The $\delta^{13}C$ value for DIC acquired in the recharge area ($\delta^{13}C_{rech}$) is determined according to the $\delta^{13}C$ mixing model described above.

Methanogenesis

Methanogenic activity in groundwaters is another process that contributes additional carbon to the DIC pool. Both the ^{14}C activity and concentration of this additional carbon must be quantified. Chapter 5 discusses the various processes that can contribute methane to groundwater. Here we will only consider the common case where methane is produced by biological activity.

In the absence of more efficient electron acceptors such as O_2, NO_3^- and SO_4^{2-}, bacteria will thrive on substrates of organic carbon, producing both CO_2 and CH_4 as reaction by-products. In fresh waters, the favoured reaction pathway is the reduction of CO_2 to CH_4 using H_2. This can be expressed as a series of simplified reactions involving CO_2 and H_2 production by fermentative bacteria, followed by CH_4 production by methanogens:

$$2CH_2O + 2H_2O \xrightarrow{\text{fermentation}} 2CO_2 + 4H_2 \xrightarrow{\text{methanogenesis}} CH_4 + CO_2 + 2H_2O$$

The overall reaction produces equal concentrations of CO_2 and CH_4. The effect on $\delta^{13}C$ is dramatic, as CO_2 reduction is accompanied by a very strong fractionation. Whiticar (1986) shows that there is in the order of 60 to 90‰ enrichment of the DIC relative to the methane produced in such settings. For an initial organic carbon substrate with $\delta^{13}C = -25$‰ and an enrichment factor of $\varepsilon^{13}C_{CO_2-CH_4} = 70$‰, the methane and DIC products would have $\delta^{13}C$ values of –60‰ and +10‰, respectively. Hence, the measured $\delta^{13}C_{DIC}$ provides good evidence for methanogenesis in groundwaters.

The calcite solubility must also be considered. The production of CO_2 indicates that this is an acid-generating reaction and so calcite will dissolve, if available. This adds additional carbonate from the aquifer matrix to the DIC pool, which must be considered in the ^{14}C mass balance. The dissolution of aquifer carbonate follows the reaction:

$$2CH_2O + H_2O + CaCO_3 \rightarrow CH_4 + 2HCO_3^- + Ca^{2+}$$

For each mole of methane produced, two moles of HCO_3^- are added to the DIC pool, both of which would be ^{14}C-free if the source of CH_2O is old.

Developing an algorithm to correct for these effects is sketchy at best, given the uncertainties in measuring the production of CH_4; a gas with low solubility and high diffusivity. However, an attempt at calculating a dilution factor for methanogenesis q_{CH_4} is a useful exercise as it serves to constrain modelled ^{14}C ages. Note that if the CH_2O substrate in the reaction above is derived from soils in the recharge area, then there is no correction to be made, as this carbon will have a ^{14}C activity close to that of the atmosphere and soil CO_2. Dilution due to addition of carbonic acid and calcite dissolution would be corrected for by the traditional models such as the chemical mass balance model (CBM) or the $\delta^{13}C$ model. If the source of organic carbon is from within the aquifer itself, and can be shown to be ^{14}C-free, then a correction factor calculated as follows may approximate the ^{14}C dilution from methanogenesis:

$$q_{CH_4} = \frac{mDIC_{meas} - 2mCH_4}{mDIC_{meas}}$$

The contribution of CH_4 can be quantified using $\delta^{13}C$ as well, considering that production of 1 mole of CH_4 yields 2 moles of HCO_3^- after carbonate dissolution. If we assume values for the DIC and $\delta^{13}C_{DIC}$ in the recharge area (as discussed for the $\delta^{13}C$ mixing model), as well as values for $\delta^{13}C_{CH_2O}$ and $\delta^{13}C_{carb}$ we can produce a $\delta^{13}C$ mixing model for methanogenesis that doesn't require a measurement of the amount of CH_4 produced. The $\delta^{13}C$ composition of DIC produced by methanogenesis ($\delta^{13}C_{DIC-CH_4}$) and carbonate dissolution, according to the reaction above, can be approximated as:

$$\delta^{13}C_{DIC-CH_4} = \frac{\delta^{13}C_{carb} + \delta^{13}C_{CH_2O} + \frac{1}{2}\varepsilon^{13}C_{CO_2-CH_4}}{2}$$

The ^{14}C dilution factor can then be calculated according to:

$$q_{CH_4} = \frac{\delta^{13}C_{DIC-meas} - \delta^{13}C_{DIC-CH_4}}{\delta^{13}C_{DIC-rech} - \delta^{13}C_{DIC-CH_4}}$$

The $\delta^{13}C_{DIC\text{-}rech}$ value represents the ^{13}C composition of DIC in the recharge area prior to methanogenesis but after open and closed carbonate dissolution. It must be measured in recharge area groundwaters or calculated according to the approach used for the $\delta^{13}C$ mixing model discussed above.

For typical values of $\delta^{13}C_{CH_2O} \approx -25‰$, $\varepsilon^{13}C_{CO_2\text{-}CH_4} \approx 60‰$ and $\delta^{13}C_{carb} \approx 0‰$, values for $\delta^{13}C_{DIC\text{-}CH_4}$ would be on the order of 5‰. This correction factor would have to be used in conjunction with the chemical mass balance model to correct the DIC pool for dilution from dissolution of carbonate prior to methanogenesis. The $\delta^{13}C$ mixing model would not work for this calculation of carbonate dissolution in the recharge area because of the additions of ^{13}C-enriched DIC from methanogenesis.

Where the organic substrate is not ^{14}C-free, its ^{14}C activity must be taken into consideration. The corrected dilution factor q for this source of DIC can be approximated by the algorithm:

$$q_{CH_4\text{-}corrected} = 1 - q_{CH_4} \cdot (1 - a^{14}C_{CH_2O}/100)$$

where $a^{14}C_{CH_2O}$ is the ^{14}C activity of the organic substrate in pmC.

Dilution factors for multiple processes

When correcting ^{14}C for discrete geochemical processes, a process-specific dilution factor, q, is determined. In the case where more than one process has occurred, the overall dilution factor is the product of the different q-factors. Using the example of carbonate dissolution (q determined from models such as CMB or ALK) followed by dolomite dissolution and sulphate reduction, the net dilution factor would be:

$$q_{net} = q_{carb} \cdot q_{dol} \cdot q_{H_2S}$$

It must be stressed that simple models such as this may improve the reliability of age estimates, but must be used with caution, and with an understanding of their sensitivity to input parameters. It is essential that groundwater ages are modelled for the full possible range of values for input parameters, as well as for the range of errors on measured parameters (e.g. H_2S concentrations, DIC in the recharge area, or the error on ^{14}C activity measurement).

Revisiting the groundwaters in southern Oman

With these new insights, we can take another look at the groundwaters in southern Oman, for which the conventional correction models were presented above (Table 8-4 and Fig. 8-7). Recall that these calculations did not take sulphate-reduction into account (Table 8-3). Here, we will examine these groundwaters again with further correction for this additional source of ^{14}C dilution.

First, some background on the hydrogeological setting in southern Oman. Precipitation in the Sultanate of Oman is chronically deficit, and water supplies are met by groundwater, augmented with some costly, desalinized seawater. A major carbonate aquifer discovered beneath the interior Najd plateau in the 1980s added significantly to the country's water supply inventory. The Upper Umm Er Radhuma aquifer outcrops in the Dhofar Mountains, but hosts flowing artesian groundwaters throughout most of the interior (Fig. 8-11).

Environmental Isotopes in Hydrogeology 223

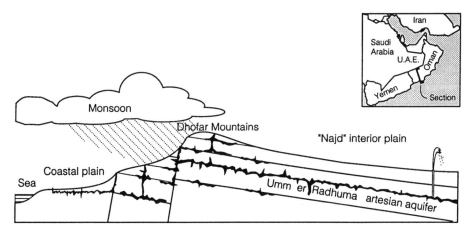

Fig. 8-11 Schematic geological cross section through the coastal plain, Dhofar Mountains and "Najd" interior plain. (10 × vertical exaggeration). The Umm er Radhuma formation hosts an extensive fissured limestone aquifer with artesian groundwater (from Clark et al., 1987).

The stable isotope data demonstrate that these artesian groundwaters are unrelated to the modern precipitation in the recharge area, and so are paleogroundwaters recharged under an earlier climate (Fig. 8-12). The modern monsoon has a distinct isotopic character, very close to that of local seawater, and recharges the coastal groundwaters (see Fig. 3-11). The Upper Umm er Radhuma aquifer, however, receives no recharge from the modern monsoon.

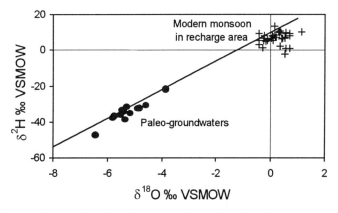

Fig. 8-12 Stable isotope signature of modern and paleogroundwaters in southern Oman.

The carbonate-dissolution correction models applied to these groundwaters (Fig. 8-7) suggest recharge during a period between about 15 ka and 30 ka, although there is a lot of variation. However, a paleoclimatic reconstruction based on travertines in Oman indicate a switch from arid to pluvial conditions only after 15 ka, and a continuation of the pluvial conditions until *ca* 4000 yr ago (Clark and Fontes, 1991). These groundwater ages do not fit this paleoclimate reconstruction, and suggests that the models have not corrected for all the ^{14}C dilution processes.

The presence of HS^- in these waters (Table 8-3) tells us that sulphate reduction is taking place. The fissures in this aquifer are also lined with secondary calcite, as further evidence for sulphate reduction. A correction factor based on the algorithm developed earlier, and using CH_4 as an analogue for the organic carbon source, produces dilution factors ranging from $q_{H_2S} = 0.67$ to 0.99. For these calculations, the contribution to the DIC pool from sulphate reduction was calculated from the Rayleigh enrichment in ^{34}S (Fig. 6-7). Thermocatalytic hydrocarbon, found elsewhere in the Umm er Radhuma aquifer, was believed to be the source of organic carbon, and was assigned a $\delta^{13}C$ value of –26‰. The results are compared with ages from the $\delta^{13}C$ dilution correction model (Fig. 8-13) (Clark et al., 1996). The error bars presented for each age are based on a sensitivity analysis using the range of reasonable input parameters and errors on measured parameters, as recommended above. These revised ages now match well with the paleoclimatic reconstruction, indicating recharge during the early Holocene pluvial period.

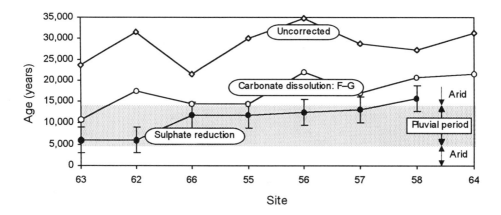

Fig. 8-13 ^{14}C ages for groundwaters in southern Oman, corrected for carbonate dissolution and sulphate reduction.

Modelling ^{14}C ages with NETPATH

The complications discussed above can be modelled by tracing the geochemical and isotopic evolution using an integrated mass-balance approach. The Water Resources Division of the United States Geological Survey (USGS) has developed an interactive geochemical model NETPATH for determining groundwater ages, based upon the geochemistry along a flow path and the boundary or initial conditions set by the modeller (Plummer et al., 1994; available at <http://h2o.usgs.gov/software>). The NETPATH geochemical code is used to interpret net geochemical and isotopic mass-balance reactions between initial and final wells along a flow path. The method uses a mass-balance approach (chemical and isotopic) rather than thermodynamic equilibrium. It is best suited for regional, confined flow systems where changes between sampling points can be observed.

NETPATH is particularly useful in flow systems with complicated geochemistry, such as sulphate reduction and methanogenesis. Constraints (controlling parameters) include C, S, Ca, Mg, Na, Cl, ^{34}S, redox; with controlling phases such as dolomite, calcite, gypsum, aquifer organic matter, CO_2, cation exchange, halite and pyrite.

One of the first applications of NETPATH in a geochemically complicated system is that of the methanogenic groundwaters of the Alliston aquifer of southern Ontario (Aravena et al., 1995). Here, high DOC levels characterize groundwaters from confined glacio-fluvial outwash sediments and paleozoic bedrock aquifers. Methane concentrations in bedrock wells vary between 0 and 5 mmoles/L (60 mg-C/L) and in the confined alluvial aquifer complex vary up to 3.5 mmoles (40 mg-C/L). The source of the methane was identified as *in situ* biogenic production, and not thermocatalytic methane from the bedrock, on the basis of the δ^2H and $\delta^{13}C$ (Fig. 5-10).

Biogenic methanogenesis generates a complementary source of DIC from aquifer-derived DOC, and represents a significant proportion of the aqueous carbon pool in the Alliston aquifer. Both the DOC and the CH_4 had ^{14}C activities of 2 to 15 pmC and were shown to originate from peat, buried within these glacial sediments at the termination of a warm interstadial period at about 35 ka. In this case, modelling the carbon geochemistry and isotope data with NETPATH showed that the reaction path involved incongruent dissolution of dolomite, ion exchange, methanogenesis and oxidation of sedimentary organic carbon. Corrected groundwater ages indicated late Pleistocene recharge to the aquifer (Aravena et al., 1995).

^{14}C Dating with Dissolved Organic Carbon (DOC)

The development of AMS (accelerator mass spectrometry) for the precise measurement of very small amounts of carbon (<5 mg C) now allows measurement of ^{14}C in DOC. Like DIC, ^{14}C-active DOC is derived from the groundwater in the soil zone. Unlike DIC, however, it is unaffected by dilution during the host of carbonate reactions experienced by groundwater, and so provides an independent method to date groundwater.

From the discussion of DOC in Chapter 5, soil-derived organics are dominantly humic and fulvic acids. However, subsurface sources of humic substances, such as buried peat or brown coal, are not uncommon (e.g. Aravena, 1993; Geyer et al., 1993), and it is essential to establish a pedogenic rather than geogenic origin. The intial ^{14}C activity of the soil DOC may also be less than 100 pmC. As soils develop over thousands of years, a component of their organic carbon (some of the humic compounds) can be sub-modern. For this reason, it is the fulvic acid fraction which is generally derived from the most recently decomposed vegetation, that is used for dating.

Dating groundwaters with DOC is not without methodological difficulties. Its concentration in groundwater is typically below 1 mg-C/L, which makes sampling difficult. DOC is usually stripped from 100 L or more of groundwater, using ion exchange resins, and then eluted in the laboratory and fractionated into humic (HA) and fulvic acid (FA) components. The FA is then analysed by AMS.

The initial ^{14}C activity in fulvic acid ($a_o^{14}C_{FA}$)

The FA fraction of DOC is more labile than HA, but can also be derived from older sources of organic carbon in the subsurface. Like DIC, an initial ^{14}C activity that is less than 100 pmC must be used in the decay equation. Data are presented in Fig. 8-14 from five German sites (Geyer, 1993) and three Canadian sites (Wassenaar et al., 1991), where FA was collected from shallow, tritium-bearing groundwaters. The $a^{14}C_{FA}$ measurements range between 100 pmC to as low as 38 pmC, with most in the 75 to 100 pmC range. Recall that if groundwaters contain tritium, any atmospherically-derived carbon fractions should have 100 to ~130 pmC from "bomb" ^{14}C, although a lag of several years to decades for the formation of humic substances from dead vegetation can be expected.

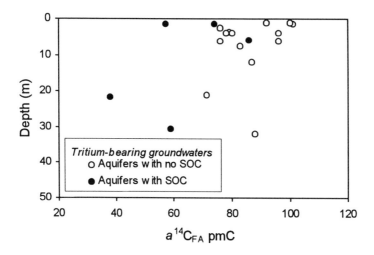

Fig. 8-14 The ^{14}C activity of FA in modern, tritium-bearing groundwater from Quaternary sediments. Sedimentary organic carbon (SOC) in some aquifers is a source of at least part of the FA. The aquifer derived FA has lower ^{14}C activities (data from Wassenaar et al., 1991; Geyer et al., 1993)

Clearly, FA is a mixture of variable-aged organic carbon sources in the soil. Some is also derived from sedimentary organic carbon (SOC) in the aquifer. Where redox conditions evolve through NO_3^- reduction, SO_4^{2-} reduction and fermentation, bacteria may preferentially use the younger (labile) FA — a process that would reduce the ^{14}C content of the residual FA sampled in the groundwater. Just as we do for ^{14}C dating with DIC, this less-than-modern initial ^{14}C activity in the FA ($a_o^{14}C_{FA}$) must be taken into account in the decay equation.

Advantages and disadvantages of DOC

Clearly, ^{14}C dating with soil-derived DOC holds some advantage over DIC in that it experiences less dilution through reaction in the subsurface. AMS measurements have a resolution of up to 10 half-lifes. With an initial ^{14}C activity closer to 100 pmC rather than 50 pmC, the groundwater dating range is extended by at least one half-life (5730 years). Further, DIC in very old groundwaters can experience further dilution through exchange or dissolution/re-precipitation reactions. DOC in paleogroundwaters will not be so affected since old organic matter in aquifers contributes little to the fulvic acid pool, although it does release humic acids. Its high initial $a^{14}C$ and greater long-term stability extends the potential DOC dating range perhaps beyond 30,000 years.

However, problems with DOC dating arise in aquifers with advanced redox evolution, and where subsurface humic sources are available. Microbial activity can result in a disproportionate loss of ^{14}C-active carbon. In methanogenic aquifers, subsurface SOC (sedimentary organic carbon) can be dissolved by bacteriological activity, contributing to a dilution of DOC. Aravena and Wassenaar (1993) show that for methanogenic aquifers, corrections can be made for DOC dilution by SOC sources, but require an assessment of the $a^{14}C$ of the SOC. While in many aquifers this SOC may be very old and ^{14}C-free, aquifers of glacio-fluvial origin from Canadian and European settings can have remarkably high contents of SOC that still have measurable ^{14}C activities.

The following case studies serve to emphasize some aspects of DOC dating. Note that the expression "DOC dating" is misleading. Reasonable results have only been obtained on the FA component. Other components of the DOC pool are of uncertain origin and cannot be used.

Case studies for ^{14}C dating with DOC and DIC

The Milk River aquifer

The simple geometry, uncomplicated geochemistry, and well-defined hydraulic setting of the Milk River sandstone aquifer in western Canada have attracted teams of isotope hydrogeologists over the past two decades to test their tools (e.g. Drimmie et al., 1991; Ivanovich et al., 1992; Hendry and Schwartz, 1988; 1990). The Cretaceous sandstone of the Milk River Formation forms a 15,000-km^2 artesian aquifer that outcrops in the invitingly named Sweetgrass Hills of northern Montana, and dips northward into southern Alberta (Fig. 8-15). The aquifer is bounded above and below by low permeability shales which contain saline formation waters with high Cl$^-$ and enriched ^{18}O contents. The δ^{18}O values increase down-dip along the flow path, accompanied by a similar Cl$^-$ enrichment. The stable isotopes provide evidence for modern recharge, paleogroundwaters and diffusional mixing with the adjacent shale beds (Fig. 8-16).

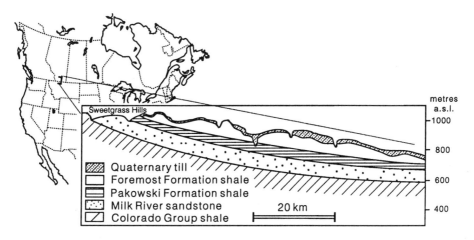

Fig. 8-15 The Milk River aquifer (modified from Hendry and Schwartz, 1990).

The stable isotope distribution in the Milk River aquifer (Fig. 8-16) shows three principal groupings of groundwaters. Group A data represent late-glacial to Holocene recharge. The Sweetgrass hills are at the southern limit of ice expansion and parts of them remained ice-free (Dyke and Prest, 1987). The more enriched Group B samples are ^{14}C-free, and may have recharge during the last (warmer) interglacial period. Group C waters are from the most distal part of the aquifer. Their elevated δ^{18}O and Cl$^-$ result from mixing with formation waters diffusing from the confining shales (Hendry and Schwartz, 1988), discussed in Chapter 6.

Murphy et al. (1989) and Wassenaar et al. (1991) analysed two DOC fractions, for comparison with ^{14}C ages based on DIC (Table 8-5). Murphy et al. sampled groundwaters in wells from the western area, both proximal to outcrop areas, and further north along the flow path. Wassenaar et al. sampled just two wells along a single, well-defined flow path in the eastern part of the aquifer system.

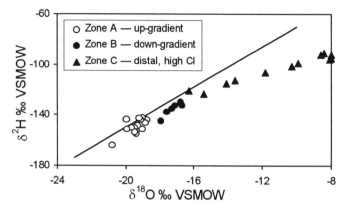

Fig. 8-16 $\delta^{18}O$ vs. δ^2H composition of groundwaters from the Milk River aquifer in southern Alberta. Zone A groundwaters are found within 20 km of the recharge area; Zone B groundwaters were sampled further along the northward flow direction; Zone C (higher Cl⁻) groundwaters were sampled from the most distal parts of the aquifer (modified from Drimmie et al., 1991). A steady enrichment in isotopes is observed in the direction of flow in the aquifer.

Table 8-5 Carbon isotopes and chemistry for groundwaters of the Milk River aquifer, southern Alberta, Canada

Well	← DIC →			DOC	HMW DOC (FA)		LMW DOC		CH_4	SO_4^{2-}
	mg/L	^{14}C	$\delta^{13}C$	mg-C/L	^{14}C	$\delta^{13}C$	^{14}C	$\delta^{13}C$	mg/L	mg/L
Murphy et al. (1989) — western groundwaters										
5	625	28.1	−11.8	2.8	20.0 ± 0.5	−25.8	0.4 ± 0.2	−40.0	0.1	481
7	695	18.6	−6.8	10.6	37.6 ± 1.1	−26.0	1.5 ± 0.1	−40.2	0.5	1282
8	785	3.0	−12.6	3.1	7.2 ± 0.3	−25.7	63.5 ± 0.7	−20.2	2.0	1565
4	865	0.4	+5.5	16.7	0.7 ± 0.2	−24.3	0.8 ± 0.1	−35.3	27.5	7.4
11	845	0.0	−16.4	2.3	7.3 ± 0.3	−25.8	0.6 ± 0.2	−36.5	33.7	0.0
Wassenaar et al. (1991) — eastern flow path										
85	647	9.2	−12.4	2.0	30.6	−30.0	46.8	−11.0	—	216.1
52	769	1.7	−12.5	1.4	6.5	−30.0	7.4	−11.6	—	330.5

Both studies identify the principal geochemical processes which have reduced the ^{14}C content of the DIC along both flow paths. Cation exchange reduces Ca^{2+} concentrations and provokes calcite dissolution. Methanogenesis takes place, and is believed to follow a CO_2 reduction pathway (recall from Chapter 5), which accounts for the enriched $\delta^{13}C_{DIC}$ in Well 4. Elsewhere, oxidation of methane (Well 11) accounts for the depletion in $^{13}C_{DIC}$ (−16.35‰) which is likely mediated by sulphate reducing bacteria. Groundwater age estimates were made (Table 8-6) using the PHREEQE reaction path geochemical model (Parkhurst et al., 1980) and the WATEQ-ISOTOP model (Reardon and Fritz, 1978).

Table 8-6 Groundwater age estimates from DIC and DOC radiocarbon modelling for the Milk River aquifer

Well	^{14}C age (ka) DIC	^{14}C age (ka) HMW DOC	^{14}C age (ka) LMW DOC
Murphy et al. (1989)			
5	6.5 to 9.6	13.3	45.7
7	5.3 to 7.4	8.1	34.7
8	19.0 to 22.8	21.8	3.8
11	undefined	21.7	42.3
Wassenaar et al. (1991)			
85	15.0 to 17.6	9.8	6.3
52	29.0 to 30.7	22.6	21.5

Detailed characterization of the DOC by column separation and nuclear magnetic resonance spectrometry (Murphy et al., 1989), and by ^{14}C and ^{13}C measurements provide the following observations.

1. Two principal DOC fractions were isolated: a high molecular weight fraction (HMW), isolated on XAD-8® resin represented 18% of the total DOC. The HMW fraction was taken to be dominantly fulvic acid (FA), with only a minor humic acid (HA) component. A low molecular weight fraction (LMW) isolated on Silicalite® was identified as short-chain aliphatic compounds and represented 8 to 32% of the DOC.

2. The HMW fraction is accepted as being largely derived from soils in the recharge area, and hence should be useful for ^{14}C dating.

3. Much of the LMW compounds has a structure close to kerogen, which would suggest an origin in the aquifer and/or the confining shale beds. If so, this would not be suitable for dating.

4. The ^{14}C of the HMW (fulvic acid) fraction in both studies had activities as high or considerably higher than the associated DIC, indicating that this DOC fraction was indeed "datable."

5. The ^{14}C activities of the LMW fraction of the DOC in the western region, except for one anomalous sample, were less than 1 pmC, indicating that this DOC fraction is indeed dominated by aquifer-derived kerogen and not datable. However, the same LMW fraction from the eastern flow path had ^{14}C activities almost as high as the HMW fulvic acids. Wassenaar et al. conclude that along their flow path, it is soil derived, and that contributions from aquifer kerogen are only important further down-gradient or in the western region.

6. Aquifer-derived kerogen identified in the LMW fraction is also believed to contribute a minor amount to the HMW DOC in the western groundwaters. This dilution then acts to overestimate the HMW ages.

Comparison of DIC and DOC age estimates (Table 8-6) indicates that the HMW (fulvic acid) component of DOC gives the youngest ages. Further, these are without any correction for the minor reduction of $a^{14}C$ in the recharge environment ($a^{14}C_{FA-DOC} < 100$ pmC) due to the participation of "old" soil carbon during the formation of FA (Fig. 8-14). The FA ages also provide the best match with groundwater ages that have been constrained using hydraulic parameters.

The conclusion of both studies is then that the fulvic acid component is indeed derived from the soils in the recharge area. Further, FA seems to move conservatively through this sandstone aquifer (no significant bacterial or sorption reactions).

The Gorleben study, Germany

The Gorleben site in Germany is a site of active investigation on the suitability of salt domes in Germany for the long-term containment of high-level nuclear waste. Since subsurface salt erosion — as documented by salt springs — is a potential problem, detailed hydrogeological and geochemical investigations were undertaken in the area. The Gorleben aquifer comprises fluvial Quaternary sediments overlying the sedimentary strata hosting the salt formations. A component of the hydrogeological investigations was to date the Gorleben aquifer groundwaters with tritium and radiocarbon.

The initial conclusions, based on DIC-^{14}C ages corrected for geochemical processes, point to the presence of very old groundwater, with ages approaching 30,000 years in parts of the system. This required independent verification, and so the DOC component of the groundwaters, was then examined. The HMW fraction was sampled and separated into fulvic and humic acids for analysis of their ^{13}C and ^{14}C compositions (Fig. 8-17) (Artinger et al., 1996; Geyer et al., 1993).

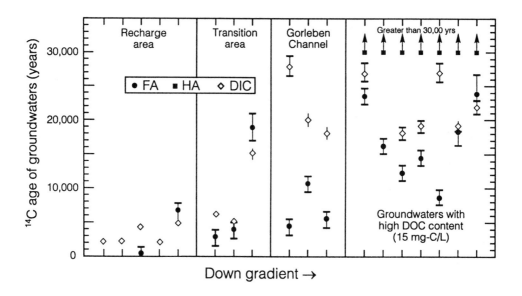

Fig. 8-17 Comparison of ^{14}C ages derived from the fulvic acid (FA) and humic acid (HA) components of DOC with corrected DIC ^{14}C ages (modified from Artiger, 1996).

The results show that at most sites, the FA fraction is considerably younger than the DIC, which is clearly too old. Findings also show that in this aquifer system, humic acids cannot be used for dating. They are almost entirely derived from lignite found in abundance in the aquifer sediments. Consequently, their ^{14}C ages are substantially above those determined with DIC and FA-DOC. In samples with high DOC contents, the HA-DOC dates approach the dating limit, which in this area is assumed to be close to 30 ka. (This dating limit is determined by corrections due to unavoidable contamination with foreign organic carbon during preparation procedures.)

The FA data appear to be reliable at least for ages up to about 10 ka, assuming a $a_o{}^{14}C_{FA}$ to be between 70 and 80 pmC, as derived from tritium-bearing groundwaters in recharge environments (Geyer et al., 1993). Higher $^{14}C_{FA}$ ages are not quite as certain because they are always found in groundwater with very high (>>100 mg-C/L) DOC contents. The fulvic acids separated from such high DOC waters have molecular weights above 8000 amu, (atomic mass units) whereas those with $^{14}C_{FA}$ ages below 10 ka have much lower molecular weights. The former "heavy" fulvic acids are well within the range of molecular weights known for humic acids. One may thus suspect, that in samples with very high DOC contents, sufficient fulvic acid was released from the lignites to dilute the soil-derived fulvics or that a portion of the humic acids is carried into the fulvic acid fraction during separation.

The measured ^{14}C-DIC concentrations were corrected with the CMB model connected to the PHREEQE geochemical code. The model ages thus obtained are for most cases greater than the FA ages although for ages below ~5 ka there is good agreement. For ages above 10 ka, the comparison is difficult — as indicated above.

In summary, several wells at Gorleben deliver water with "corrected" ^{14}C$_{DIC}$ ages > 15 ka, yet have ^{14}C$_{FA}$ ages that are as low as 4 ka. In this case, there can be little doubt that the age of this sample is closer to 4 than to 15 ka. This interpretation is supported by δ^{18}O and δ^2H values, which are similar to modern meteoric waters, rather than depleted Pleistocene water. It remains to be determined what processes are responsible for the low ^{14}C-DIC concentrations. A simple two-component mixing of water with young ^{14}C$_{FA}$ age and another with old ^{14}C$_{DIC}$ age is unreasonable. Geochemical processes must be implicated.

The relevance of this study is not only the important hydrogeological information obtained from DOC data, but also shows that the simple term "DOC" dating is misleading. A primary separation of DOC into at least the humic and fulvic acids is necessary. Dating should then focus on fulvic acids, since these are the dominant DOC compounds leaving the soil zone. The study also shows that, at least in high DOC environments, the fulvic acid fraction can be "contaminated" either through release of FA from sedimentary organic carbon or humic acid fractions which are carried with the fulvic acid through the preparation system. Improved organic geochemical methods may in future provide a tool to correct for subsurface contributions to the fulvic acid.

Chlorine-36 and Very Old Groundwater

The long half-life of ^{36}Cl (301,000 years) and generally simple chemistry of Cl$^-$ makes this radioisotope an interesting tool for dating very old groundwater, with application throughout the Quaternary period. Interest in ^{36}Cl applications has broadened over the past decade to include recharge studies, groundwater infiltration rates (see Chapter 4; Walker et al., 1992) and rates of erosion.

Groundwater dating is based on two fundamental methods: (i) decay of cosmogenic and epigenic ^{36}Cl over long periods of time in the subsurface, or (ii) in-growth of hypogenic ^{36}Cl produced radiogenically in the subsurface. In either case, an essential parameter is its initial concentration in recharging groundwaters. Andrews and Fontes (1992) provide a good review of ^{36}Cl applications and problems for dating old groundwaters.

Units of expression for ^{36}Cl data

A word about the expression of ^{36}Cl data and units is important. The low concentrations of ^{36}Cl and high precision of AMS (accelerator mass spectrometry) measurements allow concentrations to be expressed as an atomic ratio. Thus, although β^- counting is theoretically possible and has been done in the past (Rozani and Tamers, 1966) AMS technology now permits direct measurement of concentration rather than activity. The specific concentration of ^{36}Cl is measured on total Cl extracted from water or mineral samples, and so is expressed as atoms of ^{36}Cl per total Cl [moles Cl $\times$ Avogadro's number ($6.022 \cdot 10^{23}$)]:

$$R^{36}Cl = \frac{\text{atoms } ^{36}Cl}{Cl}$$

By convention, this very small number is multiplied by 10^{15} for ease of expression (giving values typically between 0 and several thousand. In groundwaters, the concentration of ^{36}Cl per litre, represented as $A^{36}Cl$, is an important expression as it is independent of the Cl^- content. $A^{36}Cl$ is calculated as:

$$A^{36}Cl = \frac{\text{atoms } ^{36}Cl}{L} = \frac{^{36}Cl}{Cl} \cdot mCl^- \cdot 6.022 \cdot 10^{23}$$

where mCl^- = moles Cl^-/L = (mg/L $\cdot$ 10^{-3})/35.5, and $^{36}Cl/Cl = R^{36}Cl$. $A^{36}Cl$ is usually expressed $\times 10^{-7}$ to give values between 0 and 100.

Together, these two expressions for ^{36}Cl concentration provide insights into the origin and behaviour of ^{36}Cl in groundwaters. $R^{36}Cl$ (or simply ^{36}R) will not change during concentration by evaporation, whereas $A^{36}Cl$ will increase. Conversely, $A^{36}Cl$ (or ^{36}A) will not necessarily change with increases in Cl^- in groundwaters by leaching or evaporite dissolution (^{36}Cl-free additions of Cl^-), whereas $R^{36}Cl$ will decrease proportionally.

Cosmogenic production of ^{36}Cl

Groundwaters in recharge areas derive cosmogenic ^{36}Cl from two sources: atmospheric production and epigenic or surface production. Atmospheric ^{36}Cl is produced in the upper stratosphere through the bombardment of argon gas by cosmic radiation, according to the following two reactions (Andrews and Fontes, 1992):

$$^{40}Ar + p \rightarrow \, ^{36}Cl + n + \alpha \quad (67\%)$$

$$^{36}Ar + n \rightarrow \, ^{36}Cl + p \quad (33\%)$$

where: n = neutron, α = alpha particle and p = proton

Atmospheric residence time is minimal and ^{36}Cl, together with stable Cl^-, is washed to the surface by precipitation or arrives as dry fallout. Common ^{35}Cl in the atmosphere can also be irradiated by cosmic flux to produce ^{36}Cl and gamma radiation:

$$^{35}Cl + n \rightarrow \, ^{36}Cl + \gamma$$

The atmospheric production of ^{36}Cl is highly affected by the geomagnetic latitude (Fig. 8-18) due to the shielding role played by the magnetosphere against cosmic radiation.

This atmospheric signal in recharge waters can be complicated by epigenic (surface) production. Cosmic radiation can penetrate the upper several metres of the crust, which allows for activation and spallation of Cl, K, Ca and Ar in minerals and soil moisture to produce ^{36}Cl according to the following nuclear reactions:

$$^{35}Cl + n \rightarrow \, ^{36}Cl + \gamma$$

$$^{39}K + n \rightarrow \, ^{36}Cl + n + \alpha$$

$$^{40}Ca + n \rightarrow \, ^{36}Cl + p + \alpha$$

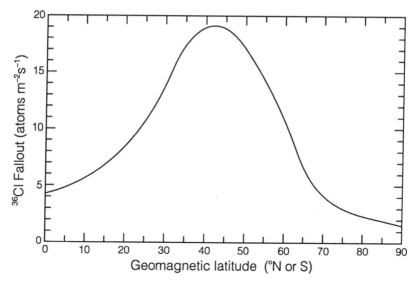

Fig. 8-18 Fallout of ^{36}Cl produced by natural cosmic radiation as a function of geomagnetic latitude (from Andrews and Fontes, 1992).

The first reaction is the most efficient due to the large activation cross-section of ^{35}Cl (Andrews and Fontes, 1992). Hence, high Cl environments (arid regions) will have higher epigenic ^{36}Cl production rates.

Subsequently, ^{36}Cl will enter the hydrological cycle through rainfall or as dry fallout and then begins a decay cycle according to the following decay mechanisms:

$^{36}_{17}Cl \rightarrow {}^{36}_{18}Ar$ beta emission (98%)

$^{36}_{17}Cl \rightarrow {}^{36}_{16}S$ electron capture (2%)

If the system remains closed to other sources of ^{36}Cl, losses should be due to radioactive decay and will reflect the subsurface residence time of the groundwater. The latitudinal variation and coastal proximity (source of Cl⁻ in rain from sea spray) will produce a geographical distribution of the specific ^{36}Cl concentration (^{36}Cl/Cl⁻) in precipitation. Bentley et al. (1986) calculate this atmospheric input for the United States (Fig. 8-19), although the contribution from epigenic production is not considered.

Observed and predicted ^{36}R values in sub-modern (pre-bomb) groundwater compare remarkably well (Bentley et al., 1986; Table 8-7). The atmospheric fallout rate used, however, was greater than what it is now believed to be (Fig. 8-18), which implies a discrepancy that must be attributable to epigenic production. Andrews et al. (1994) use the ^{36}Cl content in Holocene groundwater recharged at the geomagnetic latitude 56.2°N to calculate a cosmogenic fallout rate of 30.6 atoms ^{36}Cl m^{-2}s^{-1}, which is considerably higher than that of modern fallout (Fig. 8-18). This discrepancy also suggests that epigenic production is important.

Provided that the input of cosmogenic ^{36}Cl from both fallout and epigenic production can be determined, decay of this signal over long times in the subsurface can provide an indication of age. AMS precision allows measurement down to about 2% of average input levels (about 5 half-lives), dating by decay should be possible for the past 1.5 Ma.

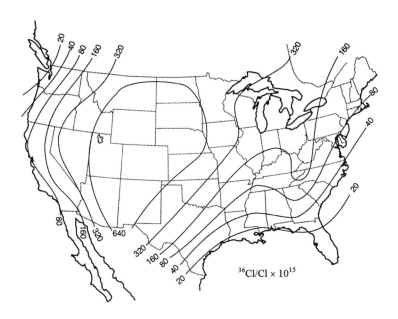

Table 8-7 Comparison of measured and calculated $^{36}Cl/Cl$ ratios in shallow, sub-modern groundwaters (Bentley et al., 1986)

Region	Average $^{36}Cl/Cl$ (x 10^{15})	Predicted $^{36}Cl/Cl$ (x 10^{15})
Southeast Arizona	365 ± 78	400
Madrid Basin	254 ± 78	250
Southeast Texas	32 ± 3	30
Southern Alberta	453 ± 111	500
Central New Mexico	707 ± 37	>640

This optimistic note that dating should be possible to 1.5 Ma is correct for groundwaters in which the subsurface production of ^{36}Cl and mixing of Cl from different sources do not play a role. As will be shown with the following case studies, this is seldom the case.

Subsurface production of ^{36}Cl

Below a depth of several tens of metres, most cosmic radiation has been attenuated. Here, decay (and minor fission) of natural radionuclides (principally U, Th and K) irradiate light elements producing an *in situ* neutron flux. This acts upon ^{35}Cl in the reaction above to produce hypogenic ^{36}Cl. *In situ* production will depend on the concentration of natural radionuclides, and will vary between aquifers and geological settings. With time, hypogenic ^{36}Cl will accumulate and decay until the rate of production is matched by the rate of decay. This is the point of secular equilibrium, and is usually reached after about 5 half-lifes. In this case, an estimate of subsurface residence time is indicated by the degree of in-growth toward secular equilibrium (Fig. 8-20).

Calculation of the fixed rate of subsurface production is an important yet difficult parameter to determine. It depends upon the U and Th content of the host rock, its light element composition,

Cl⁻ content in the groundwater and porosity. Andrews et al. (1986) show the calculation of neutron fluxes and ^{36}Cl production using the example of the Stripa granite. Bentley et al. (1986) calculate typical ^{36}R values at secular equilibrium for given neutron fluxes in various environments (Table 8-8). Given a fixed initial ^{36}R in recharge groundwaters, and a fixed rate of subsurface production, groundwater age t (yr) is derived from measured ^{36}Cl/Cl ratios according to the equation (from Bentley et al., 1986):

$$t = \frac{1}{\lambda^{36}Cl} \ln \frac{A_t(R_t - R_{se})}{A_o(R_o - R_{se})}$$

Here, R_t is the measured ^{36}Cl/Cl ratio after time t, R_{se} is the ratio at secular equilibrium (for the given aquifer), R_o is the initial cosmogenic ratio, λ^{36}Cl is the decay constant for ^{36}Cl (ln2/t$_{½}$ = 2.303 · 10^{-6}), A_t is the measured Cl⁻ concentration (atoms/L) and A_o is the initial Cl⁻ concentration.

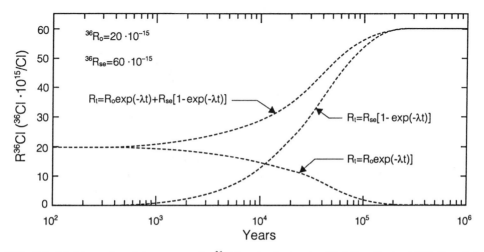

Fig. 8-20 Calculated ingrowth and decay curves for ^{36}Cl in groundwaters (modified from Fontes, 1985) Here, R is an abbreviation for ^{36}R or R^{36}Cl.

Table 8-8 Neutron flux from U and Th and secular equilibrium ^{36}Cl/Cl ratios for various rock types (Bentley et al., 1986)

Rock type	Total production (neutrons kg^{-1} yr^{-1})	Equilibrium ^{36}Cl/Cl at infinite time (× 10^{15}) ratio
Granite	15,837	30.1
Sandstone	2,884	4.68
Shale	11,789	12.5
Limestone	2,482	10.9

Example of the Great Artesian Basin, Australia

This approach was applied to the Great Artesian Basin in Australia (Fig. 8-21). This regional sandstone aquifer outcrops along its northeastern margin and flow is essentially westward. Tritium and ^{14}C are found only in the recharge region, implying that groundwater ages throughout most of the basin are greater than at least 50,000 years. Measurements of ^{36}Cl/Cl ratios indeed show a decrease in a westerly direction. If no subsurface additions of Cl⁻ are evident, then this decrease should be related to decay. Using the above equation to account for subsurface production in the sandstone, isochrons can be drawn for the groundwaters.

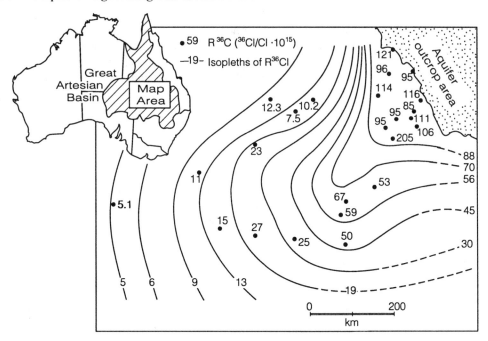

Fig. 8-21 Chlorine-36 in the Hooray sandstone aquifer of the Great Artesian Basin, Australia. The decrease in $R^{36}Cl$ along the flow direction suggests decay and, therefore, increasing groundwater age (modified from Bentley et al., 1986).

In this study, subsequent measurements of ^{36}Cl in the recharge environment as well as salinity distribution in the aquifer itself have changed some interpretations of these data from the Great Artesian Basin. Andrews and Fontes (1992) show that there is a subsurface source of Cl^- that affects the age-flow distance relation seen in Fig. 8-21. These effects become apparent on ^{36}R-Cl and ^{36}A-Cl diagrams.

Fig. 8-22 shows that in the recharge area (identified on the basis of high ^{14}C and ^{3}H activities), groundwaters with similar Cl^- concentrations have similar $^{36}Cl/Cl$ ratios. This indicates evaporative concentration of ^{36}Cl and Cl in the recharge environment. The subsurface ratios, however, show mixing between two groundwaters: (i) a non-evaporated end-member from the recharge environment (high ^{36}R) and (ii) a subsurface end-member with high Cl^- and with a low ^{36}R in secular equilibrium with production in the sandstone ($^{36}R_{se} \sim 5 \cdot 10^{-15}$). In this case, low $^{36}Cl/Cl$ ratios along the flow path don't reflect groundwater age, but rather chloride mixing. Nor does the secular equilibrium seen in distal groundwaters necessarily reflect age. Rather, this appears to be activation of *in situ* Cl and not Cl^- carried from the recharge area.

From Fig. 8-23, we see that the distal groundwaters have lost their high initial ^{36}Cl activities (^{36}A) due to either decay or mixing with waters which have reached secular equilibrium. In either case, these distal waters are at least 1 half-life old (>300 ka). Andrews and Fontes (1992) conclude that (i) the Great Artesian Basin now receives evaporated recharge with high and variable ^{36}R, (ii) ^{36}R changes due to Cl additions from the aquifer, and (iii) there has been a climatic shift from wet to arid conditions since the distal groundwaters were recharged.

Subsurface additions to the Cl^- and ^{36}Cl reservoir has been observed elsewhere. Andrews et al. (1994) show that in the Triassic Bunter (East Midlands) sandstone (Fig. 8-8), subsurface ^{36}Cl dominates in the most distal part of the aquifer. There, chloride from residual marine brines is gained by the main groundwater flow system. This is similar to groundwaters in the Milk River sandstone (Fig. 8-15) which gain Cl^- by diffusion from the adjacent aquitards. In both cases, the

^{36}Cl brought with the *in situ* source of Cl⁻ reflects the residence time of the brine, not the groundwater, and dating is not possible. However, in upper parts of the aquifer where Pleistocene water is present, ^{36}Cl can be used to determine ^{36}Cl fallout in the past.

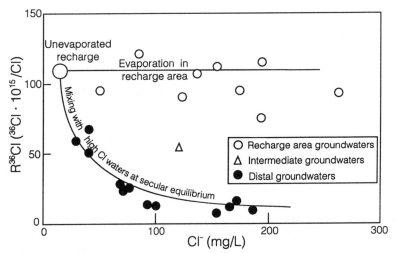

Fig. 8-22 The ^{36}Cl/Cl ratio (^{36}R) plotted against Cl⁻ concentration for groundwaters in the Great Artesian Basin (Hooray sandstone). Groundwaters in recharge area have gained Cl⁻ through evaporation. The mixing line is between unevaporated recharge and high Cl⁻ groundwaters at secular equilibrium with in situ neutron flux. Distal groundwaters have gained Cl⁻ from the aquifer, and plot on the mixing line.

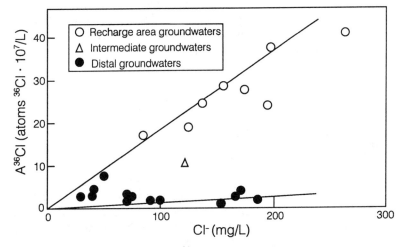

Fig. 8-23 The ^{36}Cl concentration in atoms per litre (A^{36}Cl) plotted against Cl⁻ concentration for groundwaters in the Great Artesian Basin (Hooray sandstone). Groundwaters from the recharge area, have greater ^{36}Cl concentrations than groundwaters from the intermediate or distal zone (Andrews and Fontes, 1992).

Summary of ^{36}Cl in groundwater dating

The use of ^{36}Cl as a groundwater dating tool is clearly not without its problems. Dating either relies on the decay of surface derived ^{36}Cl (atmospheric fallout plus epigenic production) or the activation of surface-derived Cl⁻ towards secular equilibrium. Unfortunately, very old groundwaters often gain Cl⁻ and ^{36}Cl by diffusion from the aquifer matrix or adjacent aquitards. In such cases, the ^{36}Cl ages are then of the chloride, and not the water. If decay of an established cosmogenic signal can be measured, and/or subsurface in-growth of ^{36}Cl from the recharge environment can be determined, then "ages" can be calculated.

The Uranium Decay Series

The half-life of certain nuclides in the ^{238}U decay series (Fig. 8-24) offer interesting possibilities for measuring the age of groundwaters. The other two actinide decay schemes (^{235}U and ^{232}Th), together with the ^{238}U series, are implicated in another dating method based on the accumulation of ^{4}He (from α-decay) in groundwater. Within the ^{238}U decay series the degree of disequilibrium between the activities of certain parent–daughter radionuclide pairs may be used to estimate age. As groundwaters in the recharge area leach or etch uranium and its daughters from the rock, the activity ratios in solution are altered, providing an initial activity in solution that will evolve, providing a measurement of time.

The ^{234}U/^{238}U activity ratio is the most useful for groundwater because of its high solubility and because the long half-lifes of these nuclides allow dating of groundwaters tens to hundreds of thousands of years old. The decay of ^{238}U produces two short-lived daughters (^{234}Th and ^{234}Pa), making ^{234}U the first long-lived daughter to accumulate in this decay series. Other activity ratios such as ^{230}Th/^{234}U or ^{226}Ra/^{230}Th are influenced by geochemical processes and difficult to interpret. They may provide, however, additional information on the geochemistry of the system under investigation (Pearson et al., 1991).

$^{234}U/^{238}U$ disequilibrium

The evolution of U and ^{234}U/^{238}U is linked to redox evolution in groundwater. Under oxidizing conditions typical of recharge areas, the oxidized U^{6+} species (UO$_2^{2+}$) is favoured and uranium concentrations are high (mg/L range). At greater depths and distances along aquifers, the less-soluble reduced U^{4+} species (UO$_2$) dominates. In strongly reducing groundwaters, the uranium content is usually below 0.1 ppb. Complexation with carbonate anions also increases uranium solubility.

If a rock mass has been undisturbed for a considerable period of time, the decay of parent and in-growth of daughter radionuclides will produce a state of "secular equilibrium" where all activity ratios in the decay series will be equal to 1. Thus, the activity of ^{238}U (a^{238}U) equals the activity of its (great-grand) daughter ^{234}U (a^{234}U) and the activity ratio a^{234}U/a^{238}U = 1. Note that this is not a concentration ratio, which would be a much smaller fraction, equal to the half-life ratio for an activity ratio of 1 (secular equilibrium). The m^{234}U/m^{238}U concentration ratio at secular equilibrium is then $5.51 \cdot 10^{-5}$, and so the amount of ^{234}U is very little. It is the short half-life of ^{234}U that gives it a higher radioactivity. Ratios are always expressed as activity ratios (i.e. ^{234}U/^{238}U $\equiv a^{234}$U/a^{238}U).

As groundwaters gain their uranium content from the aquifer matrix, one would expect that their ^{234}U/^{238}U would be equal to 1. However, groundwaters generally have ratios greater than 1 due to the higher solubility of ^{234}U, which is situated in crystal lattices damaged by decay of ^{238}U, and due to the recoil of ^{234}Th (grandparent of ^{234}U) from fracture surfaces into solution following the α – decay of ^{238}U (Andrews et al., 1982). Auto-oxidation of ^{234}U to U^{6+} during decay also enhances its solubility. Factors important to this process include the concentration and distribution of uranium within the rock matrix and on fracture surfaces, the chemical state of the uranium in the solid phase and the geochemical properties of the groundwater.

Groundwaters in the recharge area are generally oxidized and aggressive to mineral weathering reactions. It is here that most of the uranium is taken into solution. The ^{234}U/^{238}U activity ratio of recharging groundwaters is invariably greater than 1. At the Stripa site (an underground laboratory for fractured-rock hydraulics testing in Sweden) shallow groundwaters had $^{234}U/^{238}U$ activity ratios of 2 to 3, and exceeding 10 at the 330 m level (Andrews et al., 1982). In the Triassic Bunter sandstone of the English East Midlands, Andrews et al. (1983) measure activity ratios between 2.78 and 7.0. In this case, dissolution of dolomitic zones within the aquifer was identified as the source of uranium.

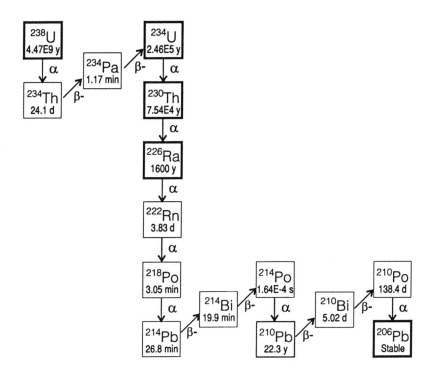

Fig. 8-24 The ^{238}U decay series, with principal decay pathway, decay mechanism and half-life of radionuclides (s—second, min—minute, d—day, y—year). Long-lived nuclides in bold boxes.

As groundwaters circulate to greater depth and redox evolves to low Eh values, uranium solubility decreases and uranium will precipitate on fracture walls. Although ^{234}U leaching now no longer takes place, the $^{234}U/^{238}U$ activity ratio can still increase due to enhanced ^{234}Th recoil (ejection into solution during decay) from the precipitating ^{238}U. At the same time, ^{234}U will decay, thus lowering the $^{234}U/^{238}U$ activity ratio. The relative importance of these two opposing processes is controlled by the fracture surface area per volume of interstitial water (S), which is largely related to the width of the fracture. S for millimetre-wide fracture apertures may be less than 1 to 2 cm^2 cm^{-3}, whereas it will be >10,000 for micron-wide fractures.

For a closed system in which the $^{234}U/^{238}U$ is permitted to grow, Andrews et al. (1982) present a model which estimates the time required for a certain value to be reached. It requires a knowledge of the recoil escape probability, the density of the rock and the rock-water contact surface, as well as the ratio between uranium contents of the rock available to exchange and in solution. This information and especially the ratio of leachable uranium in the rock to that in solution ($U_{rock}/U_{gw} \sim 1000$) is usually not available to the hydrogeologist yet $^{234}U/^{238}U$ can easily be measured. In Fig. 8-25, this model was used to demonstrate the influence of fracture opening and uranium content in fracture coatings on $^{234}U/^{238}U$ for groundwaters in granite.

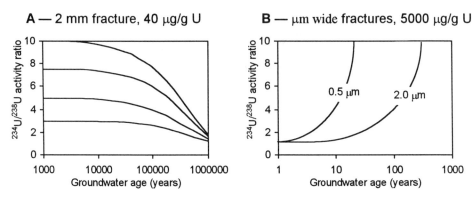

Fig. 8-25 Relationships of $^{234}U/^{238}U$ to age for groundwater with 10 µg/kg U in a typical granitic rock. The two diagrams show the influence of fracture aperture and uranium content in the fracture. In **A**, fractures are 2 mm wide, with $S = 10^7$ cm^2/cm^3, and fracture surfaces have a uniform uranium content of 40 µg/g. Four examples with different initial $^{234}U/^{238}U$ activity ratios are shown. In **B**, the uranium content is much greater (5000 µg/g) and fracture apertures are much smaller (S = 10,000 and 40,000 cm^2/cm^3) (Andrews et al., 1982, with kind permission from Elsevier Science Ltd., The Boulevard, Langford Lane, Kidlington, OX 51 GB, U.K.).

If a groundwater with elevated $^{234}U/^{238}U$ (i.e. >1) moves into an aquifer in which no uranium is present and/or reducing conditions exist then $^{234}U/^{238}U$ will decrease as the ^{234}U decays ($t_{½}$ = 246,000 years). The $^{234}U/^{238}U$ activity ratio after time, t, is determined as:

$$\left[\frac{^{234}U}{^{238}U}\right]_t = 1 + \left(\left[\frac{^{234}U}{^{238}U}\right]_o - 1\right) \cdot e^{-t \cdot \lambda_{234}}$$

This is the simple case where no further contribution to ^{234}U in the groundwater occurs (i.e. no recoil of ^{234}Th). Decaying systems of this type have been found in deep aquifers with reducing conditions and very low solubility of uranium.

Clearly, the use of uranium isotopes to determine groundwater age requires a very clear understanding of the porosity configuration as well as the distribution of uranium and geochemical conditions throughout the aquifer. For the Stripa granite and the Bunter sandstone studies, spatial variations proved to be more important than decay and in neither case was it possible to estimate groundwater ages. On the other hand, it was shown that the uranium series data are useful to monitor changes to groundwater flow regimes over time. This is valuable information where the potential migration of radionuclides from radioactive waste repositories is being investigated. In such cases, measurement of additional radionuclides of the ^{238}U and ^{232}Th decay series may also be warranted (e.g. Pearson et al., 1991).

Dating with ^{226}Ra and ^{222}Rn

Two other isotopes of the ^{238}U decay series that can yield age information are ^{226}Ra and its daughter ^{222}Rn. Radium-226 has a half-life of 1620 years whereas ^{222}Rn is a short-lived isotope ($t_{½}$ = 3.8 days). Both are produced in the ^{238}U decay chain (Fig. 8-24). Their geochemistry differs from other nuclides in these series. Radium is a divalent alkali-earth cation that behaves like Ca^{2+} and Ba^{2+} in solution. Radon is inert and transport is essentially diffusion controlled.

Groundwaters gain ^{226}Ra through two principal mechanisms including dissolution (rock-etch) of the rock matrix and alpha-recoil of ^{226}Ra into solution on decay of ^{230}Th. The solubility of Ra^{2+}

increases with salinity due to complexation (Herczeg et al., 1988), but can co-precipitate with Ca^{2+} or Ba^{2+} in carbonate saturated waters. Ingrowth of ^{226}Ra approaches a secular equilibrium activity controlled by its production rate and half-life, and becomes constant after about 8000 years (Andrews et al., 1989). High concentrations are then a reflection of long residence time. Data from the Stripa granite in Sweden shows that ^{226}Ra activities increase from about 0.02 Bq/kg in shallow groundwaters to 0.3 at intermediate depths and reaching over 5 Bq/kg at the deepest levels (400 to 900 m) of the site. Deriving quantitative age information requires accurate estimation of dissolution rates, but offers good possibilities for dating in the 1000 to 5000 year period (Hillaire-Marcel et al. 1997).

The high radioactivity of ^{222}Rn gained medical interest years ago for its alleged curative powers (recall the famous radon spas all over Europe) although now is recognized as a carcinogen (it is associated with lung cancer in miners working underground in uranium mines). It is a very soluble noble gas and so reacts little with the rock matrix. This conservative behaviour and short half-life, as well as its relative ease to sample and measure, make it a radionuclide of hydrogeological interest.

The presence of radon in groundwater is a clear indication that its parent ^{226}Ra is not far removed. Thus, it is an indicator for actively circulating groundwater and can identify groundwater discharges in surface systems that are otherwise free of radon. With information on the distribution of uranium and radium in rock and fracture minerals, it is possible to derive hydraulic information on rock permeabilities and fluid movements. A collection of case studies has been assembled by Graves (1988).

Whereas most studies look at the discharge of radon, its in-growth has been used to estimate residence times for artificial groundwater recharge. The water supplies of the city of Dortmund in Germany rely on artificially recharged groundwater with water from the Ruhr River as the primary source. Artificial tracing of drinking water supplies is not permitted, and so a method developed by Hoehn and Gunten (1989) was adopted. The in-growth of ^{222}Rn in the artificially recharged groundwater is measured. The activity of ^{222}Rn will reach a secular equilibrium after about 5 half-lifes i.e. after 15 days — provided the input from ^{226}Ra decay is constant in time. In the Dortmund study, it was indeed possible to identify water with different residence times, to show mixing of recharge water with water which infiltrated from a nearby lake, and recognize contributions from underground leaky pipes (Hoehn et al., 1992).

^{4}He and old groundwater

The ingrowth of helium from radioactive decay in crustal waters offers a qualitative measure of time. The uranium decay series is dominated by α (plus β^-) decay (Fig. 8-24). During the decay of ^{238}U to ^{206}Pb, 32 atomic mass units are lost, which is accounted for by 8 α particles. These are 2n, 2p, nuclei that pick up two electrons and become ^4He atoms. The ^{232}Th decay series also produces α particles. Another source of crustal helium is through the fission of ^6Li by neutrons in high U and Th rocks:

$$^6Li + n \rightarrow \alpha + {}^3H \rightarrow {}^4He + {}^3He + \beta^-$$

As the reaction shows, both ^3He and ^4He are produced. According to the rates and reactions for subsurface production, both the total He and the $^3He/^4He$ ratio in groundwater will evolve (Table 8-9). Both measurements can be used to estimate groundwater residence time.

The concentration and ^{3}He/^{4}He ratio of atmospheric helium (Table 8-9) represents a dynamic steady-state of He diffusion from the mantle through spreading ridges plus crustal ^{4}He, and loss through planetary helium escape (Nicolet, 1957). Mantle helium is primordial, incorporated during planetary accretion, and is the largest reservoir of ^{3}He (Kurtz and Jenkins, 1981).

By contrast, crustal helium is highly enriched in ^{4}He and greatly exceeds the He concentration in air-saturated water. The ^{3}He/^{4}He ratio is often expressed as a fraction of that in air; and thus, R/R$_{air}$ for crustal fluids = 0.007 to 0.022 (Andrews, 1985). Rocks rich in U, Th and K can have ^{3}He/^{4}He < 10^{-10} with complementary increases in total He (Clarke and Kugler, 1973).

Table 8-9 Helium abundances and isotope ratios in different reservoirs (R = ^{3}He/^{4}He)

Source	He concentration	^{3}He/^{4}He	R/R$_{air}$
Atmosphere	5.24 ppmv	$1.38 \cdot 10^{-6}$	1
Surface water	$4.5 \cdot 10^{-8}$ cm^3 STP g^{-1}	$\sim 1 \cdot 10^{-6}$	~1
Crustal fluids	10^{-7} to 10^{-4} cm^3 STP g^{-1}	$\sim 10^{-8}$ to $<10^{-10}$	0.007 to 0.022
Mantle He	up to $2.7 \cdot 10^{-5}$ cm^3 STP g^{-1}	1 to $3 \cdot 10^{-5}$	7 to 21

Dating groundwaters by He in-growth requires that some details of the host aquifer be known. Corrected for atmospheric helium gained during recharge (Table 7-10), measured ^{4}He should then reflect the groundwater residence time. The helium production after time t can be described by (Andrews et al., 1982):

$$[He] = \rho \, \phi^{-1} \, t \, (1.19 \cdot 10^{-13} \, [U] + 2.88 \cdot 10^{-14} \, [Th])$$

where: [He] is the groundwater helium content in cm^3 STP g^{-1} water
ρ is the rock density in g cm^{-3}
ϕ is the fractional porosity of the rock
[U] and [Th] are the uranium and thorium contents of the rock matrix in ppm

Andrews et al. (1983) compare different dating methods in two aquifers. One is our now-familiar case study of the Triassic Bunter sandstone aquifer in the East Midlands, U.K (Fig. 8-8). The other is the Blumau aquifer in SE Austria, a Tertiary fluvial sand confined by overlying clays. The helium dating considered only the ^{4}He contents. Both aquifers show consistent increase in He contents with age for samples with measurable radiocarbon ages (Fig. 8-26). However, these He contents are greater than the He excesses calculated from the above equation. A helium age of 78,000 years was calculated for the older Blumau sand aquifer, which cannot be supported by hydrological models. Even greater He excesses were calculated for the Bunter sandstone, giving ^{4}He ages up to 10 times greater than the ^{14}C ages. In both cases, the ^{4}He excesses were attributed to diffusion of helium into the aquifer from adjacent strata.

No ^{3}He measurements were made in these studies, although such data would have most likely documented that indeed radiogenic helium was added to the groundwater. Recall that the ^{3}He/^{4}He ratio is strongly dependent on the lithium contents of the aquifer rocks and, therefore, can potentially indicate where the helium was produced, if different rock types are present (Andrews, 1985; 1987). As these two aquifers show, diffusion of He from other strata must be taken into account. Nevertheless, the systematic increase in He along the flow gradient, whether from decay within the aquifer or diffusion from adjacent strata, provides a useful tool to estimate groundwater age. Empirical or semi-quantitative corrections for diffusion can extend the He-dating range well beyond that of ^{14}C.

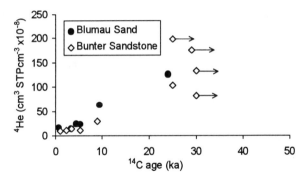

Fig. 8-26 He contents in two confined aquifers (Blumau fluvial sand in southeastern Austria and Bunter sandstone in the East Midlands, England) correlated with their corrected radiocarbon ages. Arrows indicate "greater than" ages (data from Andrews et al., 1983).

A cautionary note should be made with respect to helium accumulation in regions of tectonic activity. Active faults and other crustal discontinuities can act as conduits for migration of helium, which is a highly diffusive gas. In geothermal groundwaters from the Cordillera of western Canada, He concentrations up to $1.760 \cdot 10^{-5}$ cm^3 STP g^{-1} H$_2$O were measured in thermal waters with modern ^{14}C activities and measurable ^{3}H. The low ^{3}He/^{4}He ratio of $7.8 \cdot 10^{-8}$ (R/R$_{air}$ = 0.06) signifies also a strong radiogenic He source. These strong inputs of crustal ^{4}He to a young groundwater implicated mixing with He migrating along a regional thrust fault underlying the flow system (Phillips, 1994).

Problems

1. You have been sampling groundwaters from a Cretaceous limestone aquifer in southern Jordan (LMWL: δ^2H = 7.6 δ^{18}O + 16) and have produced the following data. How can these data be used to constrain the age of these groundwaters. Give the general age range provided by the methods that you suggest.

$\delta^{18}O$	δ^2H	^3H TU	^{14}C pmC	$\delta^{18}O$	δ^2H	^3H TU	^{14}C pmC
−5.9	−28	9.4	40.1	−5.7	−34	<0.8	14.1
−6.1	−31	6.5	45.3	−6.1	−40	<0.8	12.2
−6.2	−33	2.5	39.6	−5.9	−37	<0.8	9.5
−5.7	−28	6.7	43.9	−6.0	−35	<0.8	8.9
−5.4	−26	8.3	51.1	−6.4	−38	<0.8	9.6
−5.9	−31	9.6	46.4	−6.1	−37	<0.8	7.1
−5.5	−28	5.8	44.7	−6.3	−40	<0.8	12.8
−5.5	−30	6.9	45.8	−5.8	−35	<0.8	10.5
−6.4	−33	8.1	52.2	−6.1	−39	<0.8	10.2
−5.9	−30	7.4	48.3	−5.9	−37	<0.8	7.9
−5.1	−22	8.3	41.8	−6.1	−37	<0.8	6.5
−5.0	−24	4.2	43.6	−6.4	−40	<0.8	8.3

2. What would be the isotopic composition ($\delta^{13}C_{DIC}$) of a groundwater that has recharged under open system conditions in a soil with $\delta^{13}C_{CO_2}$ = −23‰ and with pH 6.8 at a temperature of 10°C?

3. Calculate the $\delta^{13}C$ of the $CO_{2(g)}$ in equilibrium with a water whose DIC (2.1 ppm) has a $\delta^{13}C = -1.5‰$ and a pH of 6.8 at 15°C. Determine the appropriate enrichment factors from equations in Table 1-5. What is a likely source of this CO_2?

4. Results of analyses for tritium and $a^{14}C_{DIC}$ in a groundwater sample submitted to you for interpretation are 10.8 TU and 18.5 pmC. How would you proceed and what additional data or observations would you request for a meaningful interpretation?

5. The following analysis was carried out on a groundwater from a carbonate aquifer. Assume that the Ca + Mg accurately reflect the total amount of calcite dissolved and that Ca + Mg was modified by evaporite dissolution and ion exchange (i.e. Na and Cl should be balanced, differences reflect ion exchange) (concentrations in mg/L):

pH	T°C	HCO_3^-	SO_4^{2-}	Cl^-	Na^+	Ca^{2+}	Mg^{2+}	$\delta^{13}C_{DIC}$	$\delta^{13}C_{soil-CO2}$	$\delta^{13}C_{carb}$	$^{14}C_{DIC}$
7.91	25	119	1.2	8.9	17.0	25	2.8	−13.5‰	−21‰	0‰	55 pmC

(i) Calculate the uncorrected age of this groundwater, and the corrected age using the ALK and STAT models, and compare the results.

(ii) Could these results be explained as an open system for CO_2 uptake and calcite dissolution, and if so, what is the "age" of the water?

(iii) Assume that a pH of 6.25 and DIC of 0.96 mmoles (at $P_{CO_2} = 10^{-1.8}$) were measured for the infiltrating groundwaters in the recharge area, prior to closed system dissolution of calcite. Calculate the age of these groundwaters using the $\delta^{13}C$ mixing model.

(iv) What is the corrected age using the chemical mass balance models (CMB-Chem and CMB-Alk) again presuming that carbonate dissolution took place under closed system conditions?

(v) Test the age correction for this groundwater sample using the Fontes-Garnier model of carbonate dissolution and matrix exchange.

6. The following analysis (mg/L) is for a groundwater sampled from a Tertiary carbonate aquifer under exploration for a supply of water for petroleum drilling. Describe the evidence for any geochemical processes that have likely affected the activity of $^{14}C_{DIC}$.

PH	T°C	Ca^{2+}	Mg^{2+}	Na^+	K^+	HCO_3^-	Cl^-	SO_4^{2-}	HS^-	DOC
8.1	25	95	28	285	3.5	129	524	8.2	65	8

log Si_{cal}	$\delta^{34}S_{SO4}$	$\delta^{13}C_{DIC}$	^{14}C pmC
0.81	46.2	−20.7	3.45

(i) Determine a corrected age for this water using the $\delta^{13}C$ model and the Fontes Garnier model and compare your results with the uncorrected age.

(ii) Apply a correction for sulphate reduction, assuming that the DOC in the sample has a redox state of 0 (i.e. DOC of the form CH_2O) and is ^{14}C-free.

Chapter 9
Water–Rock Interaction

The isotopic composition of groundwater is in most cases controlled by meteorological processes. There exist, however, some extreme geological environments where reaction between groundwater and the aquifer matrix or subsurface gases can modify the water's "meteoric" signature. Some typical processes are shown in Fig. 9-1. Effects of such water-rock interaction dominate at high temperature and over geological time scales, but can also be observed in shallow groundwater flow systems at low temperature. The alteration of the $\delta^{18}O$–δ^2H composition of groundwater gives insights into its subsurface history and geochemical reactions. In this chapter the isotope exchange processes in extreme hydrogeologic systems are explored.

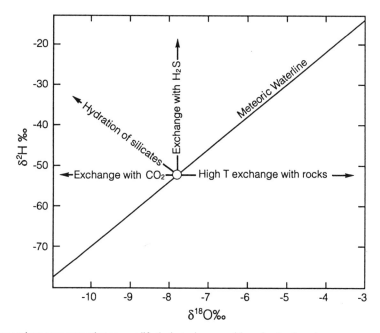

Fig. 9-1 Isotope exchange processes that can modify the isotopic composition of meteoric waters.

Mechanisms of Isotope Exchange

In the literature, reference is often made to the process of isotope exchange with the aquifer matrix to account for alteration of the stable isotope content of water. So what is isotope exchange and how is it achieved? The exchange of an isotope between water and a mineral is restricted to ^{18}O or ^2H, although exchange of other isotopes such as ^{13}C can occur between minerals, solutes and gases. To have exchange, the two phases must be in contact and be reacting, such as ^{18}O between water and calcite:

$$CaCO_2{}^{18}O + H_2O \leftrightarrow CaCO_3 + H_2{}^{18}O$$

between quartz or chalcedony and water:

$$SiO^{18}O + H_2O \leftrightarrow SiO_2 + H_2{}^{18}O$$

exchange with feldspar:

$$CaAl_2Si_2O_7{}^{18}O + H_2O \leftrightarrow CaAl_2Si_2O_8 + H_2{}^{18}O$$

exchange with clay hydration waters (chlorite in this example):

$$14(Mg,Fe)_5Al_2Si_3O_{10}(OH)_7(^{18}OH) + H_2O \leftrightarrow 14(Mg,Fe)_5Al_2Si_3O_{10}(OH)_8 + H_2^{18}O$$

Isotope exchange is achieved by various mechanisms. These have been discussed in Chapter 1 in the section on isotope fractionation. Dissolution and re-precipitation of the mineral is one mechanism, where ^{18}O (or ^2H if present in the mineral or gas) is mixed between the two reservoirs. Mineral alteration, such as the weathering of feldspars to clay minerals, allows the exchange of oxygen between the alteration fluid and the silicate structure. Both oxygen and hydrogen exchange with clay hydration waters.

A third isotope exchange process is the direct exchange of ^{18}O (or ^2H) between water and the mineral crystal lattice. The kinetics of this diffusion-controlled reaction are temperature dependent and mineral-specific. Oxygen exchange with hornblende takes place only at high temperature (Dodson, 1973), but occurs at decreasing temperatures for quartz > muscovite > biotite and finally feldspar. The loose lattice structure of feldspar allows ^{18}O to exchange, even at temperatures below 100°C (O'Neil and Taylor, 1967).

As an isotope exchange reaction proceeds between water and another phase, the two reservoirs will move towards isotopic equilibrium, for the ambient temperature. In a closed system with no significant flux in or out, the relative sizes of the reacting reservoirs become an important consideration. Where water-rock ratios are low, isotope exchange can impart a measurable effect on the water.

High Temperature Systems

Geothermal waters were in the past considered to be juvenile or magma-derived fluids that had found their way to the earth's surface. Isotope hydrology has since demonstrated that these are largely meteorically-derived fluids. Nonetheless, magmatic and mantle-derived fluids must represent a component of deep crustal fluids and even geothermal waters in some localities. Their identification and interaction with crustal rocks can be observed through ^{18}O and ^2H systematics.

Oxygen is the most abundant element in the Earth's crust. Fig. 9-2 shows the range in $\delta^{18}O$ found in various water and rock reservoirs. This figure shows that ^{18}O is preferentially partitioned into the rock reservoirs. Although hydrogen has a much lower crustal abundance, it is present in several rock forming minerals and, of course, in water. The range of δ^2H in various water and rock reservoirs is shown in Fig. 9-3. Unlike ^{18}O, the greatest enrichments of ^2H are observed in water, while minerals and rocks are preferentially depleted. These contrasting behaviors are important in the isotopic evolution of water in high temperature systems.

Magmatic water and primary silicates

As magma cools, water is continuously exchanging ^{18}O and ^2H with the newly formed minerals. At these high (>800°C) temperatures, exchange reactions are fast, and little to no fractionation occurs. Given the high concentration of oxygen in the rock as compared with that in the fluid reservoir, the fluid takes on the $\delta^{18}O$ composition of the rock. In most settings, this is on the order of +6 to +9‰ VSMOW. Not so for ^2H, which is poorly represented among the main rock-forming minerals. Feldspars, the principal mineral of granites to gabbros and basalts contain no hydrogen. Only a few minor rock-forming minerals such as biotite [K(Mg,Fe)$_3$(AlSi$_3$O$_{10}$)(OH)$_2$] and hornblende [NaCa$_2$(Mg,Fe,Al)$_5$Si$_6$Al$_2$O$_{22}$(OH)$_2$] contain H. As hydrogen represents a only minor component of the mineral, and these ferromagnesian minerals represent less than 10% of granites and other primary silicates, the δ^2H composition of the rock itself can evolve during crystallization, along with the fluid.

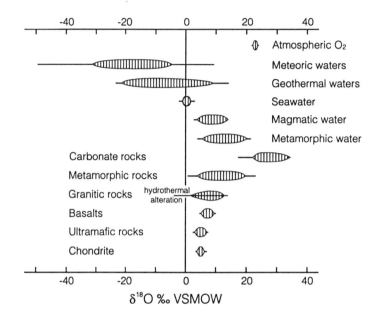

Fig. 9-2 The range of δ^{18}O in various crustal rock and water types.

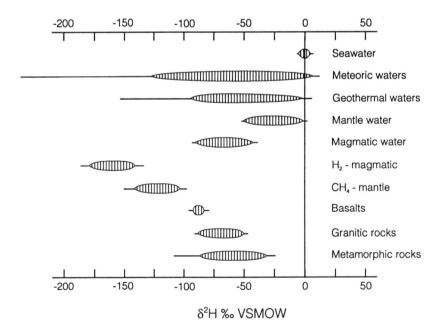

Fig. 9-3 The δ^2H composition of the major crustal reservoirs of hydrogen.

The $\delta^{18}O$ composition of granites, basalts and metamorphic rocks (Fig. 9-2) are enriched above seawater, with values up to 20‰. By contrast, the range of δ^2H compositions for these rocks (Fig. 9-3) is considerably depleted, due to the preferential fractionation of ^2H into H_2O. For example, the principal rock-forming minerals bearing H include hornblende and biotite, which are depleted by 10 to 20‰ at their initial point of crystallization. Fractionation between H_2O and these minerals increases to almost 50‰ at 400°C, by which point exchange with water is kinetically impeded.

For oxygen, there is little fractionation between water and silicate minerals at magmatic temperatures, although there is rapid exchange. By consequence, the $\delta^{18}O$ composition of magmatic or mantle-derived water would be the same as that of the host rock. Although no mantle or magmatic water has been sampled, its isotopic composition would be similar to that of the bulk rock (Fig. 9-4).

As a crystalline rock mass cools, the isotopic composition of its fluid will evolve. The water-rock ratio is low in most crystalline settings below several hundred metres depth and it is the rock reservoir that dominates isotopic exchange. The mineral-water fractionation factors for ^2H and ^{18}O increase with decreasing temperature, but are opposite in sign. At 300°C, $\varepsilon^{18}O_{min-H_2O}$ is over 5‰, and increases to 20‰ below 100°C. By contrast, $\varepsilon^2H_{min-H_2O}$ is about –65‰ at 300°C and –150‰ below 100°C. The theoretical evolution of $\delta^{18}O$ and δ^2H in a magma fluid is shown in Fig. 9-4, assuming an infinitely low water-rock ratio and no evolution of the isotopic composition of the rock mass itself. In reality, δ^2H_{rock} will evolve to lower values due to the relatively small reservoir of H in rocks, an effect which then constrains the enrichment of ^2H in the residual water. This evolution for crustal fluids was interpreted from the isotopic composition of chlorite and epidote in fracture minerals from the western Bohemian Massif in Germany by Simon and Hoefs (1993). These fluids are thought to be the precursors to the brines sampled from 4000 m depth in the German KTB (Continental Drilling Project) borehole.

The rate of exchange between water and primary silicates is an important consideration, and as pointed out above, requires geological time periods to proceed at low temperatures. Cole and Ohmoto (1986) is a good reference for further reading on this subject.

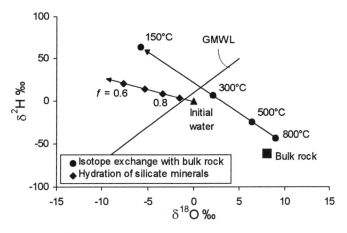

Fig. 9-4 Theoretical evolution of the isotopic composition of water during exchange with feldspar and hydration of primary silicate minerals (discussed below). Buffering by rock with an infinitely small water to rock ratio is assumed. Isotope exchange (●) modelled for equilibration with rock mass at successively lower temperatures. No consideration is given for minor evolution of the δ^2H (or $\delta^{18}O$) composition of the rock mass. The evolution of a crustal fluid during low temperature hydration of primary silicates (♦), discussed below, is modelled for Rayleigh distillation of seawater, down to residual fractions f of 0.6.

The ^{18}O shift in geothermal waters

Craig, in 1963, first noted that geothermal waters showed positive $\delta^{18}O$ shifts of varying degrees from the meteoric water line (MWL). He attributed it to high temperature (up to 300°C) exchange of ^{18}O between the fluids and reservoir rocks. No effect on the ^2H content of thermal waters is evident, due to the low representation of H in most rock-forming minerals. Truesdell and Hulston (1980) show a classic diagram (Fig. 9-5) of the geothermal effect for groundwaters from some of the more well-studied geothermal systems in the world.

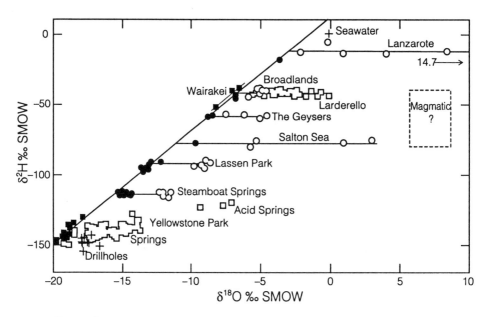

Fig. 9-5 The $\delta^{18}O$ and δ^2H composition of geothermal waters from various regions. Meteoric waters from each locality are shown in filled symbols. Open symbols demonstrate the strong ^{18}O-shift due to isotopic exchange (from Truesdell and Hulston, 1980). Lanzarote (Arana and Panichi, 1974) — carbonate; Broadlands, New Zealand (Giggenbach, 1971) — crystalline reservoir, T = 273°C; Larderello, Italy (Panichi et al., 1977) — volcanic, T = 180–260°C; The Geysers, California (Truesdell and Hulston, 1980) — crystalline reservoir, T = 250°C; Salton Sea, California (Craig, 1966) — carbonates, T = 300°C; Lassen (Craig, 1963) — crystalline; Steamboat Springs, Colorado (Craig, 1963) — crystalline; Yellowstone, Wyoming (Truesdell and Hulston, 1980) — volcanic. Note that these early data predate VSMOW.

One could suggest that this shift is due to mixing with juvenile or magmatic waters. However, in this case, values for δ^2H would be expected to shift as well. White (1974) estimated the isotopic composition of magmatic water to be $\delta^{18}O$ = 6 to 9‰ and δ^2H = –40 to –80‰, and mixing lines would then be expected to converge towards this composition. This is not the case for many geothermal waters, and it has been concluded that juvenile waters do not contribute to these geothermal systems.

Groundwaters of meteoric origin are generally far from isotopic equilibrium with the minerals of the host aquifer rocks, given that the water is highly depleted in ^{18}O compared to the minerals (Fig. 9-2). Isotope exchange will slowly move the system towards equilibrium. In most aquifers, the temperature is too low and circulation rates too high for any quantitatively significant alteration of the groundwater. However, in geothermal systems above at least 100°C, the increased temperature has two main effects. At higher temperature, the equilibrium fractionation for ^{18}O between minerals and water is reduced, thus increasing the initial degree of disequilibrium. The rate of isotope exchange is also increased at elevated temperatures. The ^{18}O-

shift observed in geothermal waters reflects isotopic evolution towards mineral-water equilibrium, at reservoir temperatures.

The initial isotopic difference between the rocks and the recharge water ($\Delta^{18}O_{rock-water}$), and the water-rock ratio, are also important factors — as in any mixing relationship. The ^{18}O contents of crystalline and carbonate rocks are considerably higher than in most meteoric waters (Fig. 9-2). Coupled with a low water-rock ratio, even minor exchange will impart a measurable shift to the water.

Let's take a look at reactions in a geothermal reservoir hosted by silicate and carbonate rock. Silicate minerals have $\delta^{18}O$ values in the range of ~8 to 12‰ VSMOW. Equilibrium fractionation in these exchange reactions decreases to less than 10‰ at high temperature (Fig. 9-6). Thus, a rock with a bulk $\delta^{18}O$ of 10‰ and a meteoric water of –15‰ has an isotopic disequilibrium of 15‰. As exchange proceeds, the $\delta^{18}O$ of the water becomes enriched while the feldspars and secondary silica phases (such as chalcedony) become depleted.

In a carbonate reservoir at high temperature, the degree of disequilibrium for meteoric waters is extreme. Carbonate minerals are highly enriched in ^{18}O, with values close to 29‰ VSMOW (0‰ VPDB) (Fig. 9-2). Consider that at ambient temperatures (during the formation of marine carbonate) $\varepsilon^{18}O_{CaCO_3-H_2O}$ = 29‰, whereas at 250°C it is reduced to 8‰. Thus, meteoric waters with $\delta^{18}O$ = –15‰ in a 250°C carbonate reservoir with $\delta^{18}O_{CaCO_3}$ = 29‰ VSMOW would be depleted from equilibrium by 29 +15 – 8 = 36‰. Small wonder that the largest geothermal shifts seen above are for the carbonate dominated systems (e.g. Lanzarote and Salton Sea geothermal systems, Fig. 9-5).

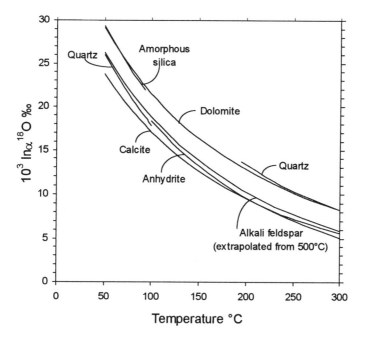

Fig. 9-6 ^{18}O fractionation as a function of temperature for exchange with amorphous silica (Kita et al., 1985) quartz (0-100°: Kawabe, 1978; >195°: Shiro and Sakai, 1972), anhydrite (Chiba et al., 1981), calcite (O'Neil, Clayton and Mayeda, 1969), and dolomite (Northrop and Clayton, 1966).

Other important factors include the rate of circulation of meteoric waters through the geothermal reservoir, the porosity of the system, and the availability of fresh rock surfaces for alteration. Geothermal waters in low porosity systems with low water-rock ratios can have a significant shift providing that flow is not highly channelled along fractures with altered surfaces. By contrast, geothermal waters in systems with high water-rock ratios will have a diminished geothermal shift.

Andesitic volcanism and geothermal waters

A strictly horizontal $\delta^{18}O$ shift (no δ^2H enrichment) in geothermal waters is not always the case. Giggenbach (1992) re-evaluated stable isotope data from geothermal systems associated with circum-Pacific andesitic volcanism. He shows that geothermal waters in these settings are mixtures of local meteoric waters and magmatic water that has been recycled from the upper mantle ($\delta^{18}O = 8 \pm 2‰$ and $\delta^2H = -20 \pm 10‰$). The isotopic composition of this "andesitic" water evolves from the dehydration of the descending oceanic crust. Exchange of water between clays and a decreasing pore water reservoir generates a residual seawater with enriched δ^2H. Enrichment of ^{18}O is attributed to exchange with oxygen in minerals at high temperature in the upper mantle.

An example of geothermal waters with a possible "andesitic" component is found at Mount Meager, in the Coastal Mountains of western Canada. Here, geothermal waters discharge at temperatures up to 60°C on the periphery of a Quaternary volcanic complex associated with subduction along the coast. The isotope hydrology of hotsprings and thermal drillhole waters indicate recharge mainly by meteoric waters recharging locally on the volcanic complex (Fig. 9-7). The deep (200°C) geothermal water (MC1), however, falls on a mixing line between local meteoric waters and Giggenbach's andesitic water (inset in Fig. 9-7). The ratio suggests that MC1 water is a mixture of local meteoric groundwater and about 22% andesitic water from the upper mantle.

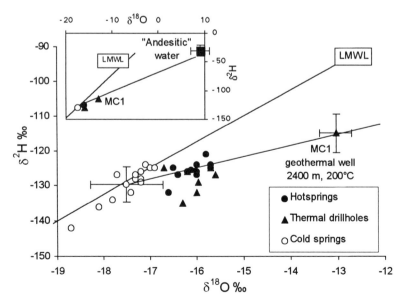

Fig. 9-7 Stable isotope diagram for waters at the Meager Creek geothermal area (data from Clark et al., 1982: Ghomshei and Clark, 1992). Andesitic water composition from Giggenbach (1992). Local meteoric water line based on precipitation and cold spring data from Mount Meager.

Subsurface steam separation

A second subsurface process that can affect the isotopic composition of thermal waters is steam loss in the subsurface. During ascent, >100°C a fraction of a geothermal water will flash to steam as the pressure diminishes. Fractionation during boiling imparts an enrichment on the residual liquid. Although the fractionation between water and steam is greatest at low temperature, it is significant at geothermal temperatures. For ^{18}O it is 5‰ at 100°C, and a little over 2‰ at 200°C (Table 1-4 and associated equations). Steam separation can take place as single step, multi-step or continuous processes (Truesdell et al., 1977), all of which affect the isotopic composition of the residual water. Arnason (1977) calculates the change in ^{18}O or ^{2}H contents of the liquid phase according to the following equation, which assumes that the steam and water remain in contact (closed system equilibrium conditions):

$$\delta_{w-low\,T} - \delta_{w-high\,T} = (1000 + \delta_{w-low\,T}) \cdot (1-[f + (1-f)/\alpha_{v-w}])$$

where high T and low T are the initial and steam-separation temperatures, respectively, and α_{v-w} is the vapour-water fractionation factor for the temperature of separation.

Depending on the degree of steam separation, which is a function of the initial and final temperature, the isotopic composition of the remaining thermal water may be enriched by up to several permil in ^{18}O. The isotopic composition of the geothermal waters in MC1 (Fig. 9-7) was calculated in this manner, from isotope measurements of post-flash (100°C) waters (Ghomshei and Clark, 1993).

Geothermometry

Maximum subsurface temperatures experienced by geothermal waters can be recorded by ionic and stable isotope ratios in solutes and the water itself. Such information is important in geothermal resource evaluation. It also reflects on depth of groundwater circulation, based on an understanding of regional tectonics and geothermal gradients. Geochemical and isotopic geothermometers developed over the past two decades rely on the assumptions that the two species or compounds coexisted and have equilibrated within the geothermal reservoir, that temperature is the main control on their ratio, and that re-equilibration has not occurred during ascent and discharge.

Cationic relationships: Studies of Na, K and Ca in aqueous systems suggest that cation concentrations are controlled by temperature-dependent equilibrium reactions with feldspars, mica and calcite. Several semi-empirical equations for the temperature have been determined on the basis of cation ratios (in ppm):

$$T°C = \frac{1217}{log(Na/K) + 1.483} - 273 \qquad \text{(Ellis and Mahon, 1967)}$$

$$T°C = \frac{1647}{log(Na/K) + \beta\,[log(\sqrt{Ca}/Na) + 2.06] + 2.47} - 273$$

where $\beta = 1/3$ for Na waters and $\beta = 4/3$ for Ca waters (Fournier and Truesdell, 1973)

At temperatures less than about 200°C, the solubility of magnesium silicate increases and Mg plays a role in the controlling reactions, which must be accounted for (Fournier and Potter, 1979). In lower temperature or high salinity systems, other cations relationships including Na/Li and Mg/Li are important, e.g.:

$$T°C = \frac{2200}{\log(\sqrt{Mg/Li}) + 5.47} - 273 \qquad \text{Kharaka and Mariner (1987)}$$

The principal drawback to these chemical geothermometers is that given time or low water-rock ratios, re-equilibration can occur through exchange reactions at lower temperatures during ascent.

Silica solubility: The increased solubility of quartz and its polymorphs at elevated temperatures has been used extensively as an indicator of geothermal temperatures (Truesdell and Hulston, 1980; Fournier and Potter, 1982). In systems above about 180 to 190°C, equilibrium with quartz has been found to control the silica concentration, whereas at lower temperatures, chalcedony is the controlling phase (Arnasson, 1976). These relationships are graphically presented in Fig. 9-6. Temperature can be derived from the following relationships for equilibrium with these silica polymorphs from 0 to 250°C, where Si concentrations are in ppm (from Fournier, 1981):

Quartz (no steam loss) $T°C = 1309/(5.19 - \log Si) - 273$

Quartz (max. steam loss) $T°C = 1522/(5.75 - \log Si) - 273$

Chalcedony $T°C = 1032/(4.69 - \log Si) - 273$

Amorphous silica $T°C = 731/(4.52 - \log Si) - 273$

$\Delta^{18}O$ in sulphate-water exchange: Although several isotope exchange reactions can be used as indicators of subsurface temperatures, the exchange of ^{18}O between dissolved sulphate and water is often the most useful because equilibration is rapid at reservoir temperatures greater than about 200°C and pH less than 7, conditions that favour SO_4^{2-}–H_2O exchange. Below reservoir temperatures of 150 to 200°C, the abiotic exchange half-time increases exponentially from several years to some 10^6 years at 100°C and pH 7 (Chiba and Sakai, 1985). Only during bacterial sulphate reduction can accelerated ^{18}O exchange take place. Recall from Chapter 6 that during this process $\delta^{18}O_{SO4}$ increases asymptotically towards an equilibrium value with the water at the ambient temperature. If sulphate reduction in the discharge area is not important, then a measure of the reservoir temperature is often preserved. These two equations provide similar estimates:

$$10^3 \ln\alpha^{18}O_{SO4-H2O} = 3.251 \cdot 10^6 \, T^{-2} - 5.1 \qquad \text{(Mitzutani and Rafter, 1969)}$$

$$10^3 \ln\alpha^{18}O_{SO4-H2O} = 2.88 \cdot 10^6 \, T^{-2} - 4.1 \qquad \text{(McKenzie and Truesdell, 1977)}$$

$\Delta^{34}S$ in SO_4^{2-}-H_2S exchange: In geothermal systems where both dissolved sulphate and sulphide are present, the large temperature-dependent fractionation of ^{34}S can be a useful geothermometer. Exchange is fast under acidic conditions, although exchange at near-neutral pH can take hundreds of years (Truesdell and Hulston, 1980) and so ascending fluids can preserve their temperature record. The exchange follows the following reaction:

$$10^3 \ln\alpha_{SO4-H2S} = 5.07 \cdot 10^6 \, T^{-2} + 6.33 \qquad \text{(Robinson, 1973)}$$

Δ¹³C in CO₂-CH₄ exchange: In groundwaters where methane coexists with dissolved inorganic carbon, ^{13}C will exchange through the reaction:

$$CO_2 + 4 H_2 \leftrightarrow CH_4 + 2H_2O$$

The temperature of equilibration can be calculated from the relationship of Bottinga (1969):

$$10^3 \ln\alpha^{13}C_{CO_2-CH_4} = 2.28 \, (10^6 \, T^{-2})^2 + 15.176 \, (10^6 \, T^{-2}) - 8.38$$

or of Richet et al. (1977):

$$10^3 \ln\alpha^{13}C_{CO_2-CH_4} = -0.62 \cdot 10^9 \, T^{-3} + 6.616 \cdot 10^6 \, T^{-2} + 6.04 \cdot 10^3 \, T^{-1} - 3.08$$

At high temperature, isotopic equilibrium is achieved quickly and is preserved in gases discharging from geothermal systems. However, re-equilibration often takes place as geothermal waters move into shallower zones at lower temperatures during ascent. In low temperature systems, isotopic equilibrium between CO_2 and CH_4 can be achieved during bacterial activity (Truesdell and Hulston, 1980).

Δ²H in H₂-H₂O exchange: The H_2-H_2O pair is one of the few that give consistent agreement. Equilibration is faster for this exchange reaction than other geothermometers. Consequently, calculated temperatures are often closer to discharge temperatures than reservoir temperatures. Calculations can be made using a combination of the temperature equations for ^2H fractionation between H_2O vapour and water, and between H_2 and H_2O vapour (Table 1, inside front cover).

Low Temperature Water-Rock Interaction

The "geothermal shift" observed for $\delta^{18}O$ in geothermal groundwaters is a trend towards isotopic equilibrium between ^{18}O-depleted meteoric waters and ^{18}O-enriched rocks at high temperature. In lower temperature environments, however, the reverse can be observed. Increased fractionation between water and minerals results in ^2H-enriched and ^{18}O-depleted waters that plot above the meteoric water line. Retrograde exchange between water and primary silicate minerals, and hydration of primary silicates, are the two dominant exchange reactions.

Two conditions are required if significant shifts are to be observed: (i) very low water-rock ratios where the rock reservoir dominates in exchange reactions, and (ii) geological time scales, because rates of exchange and hydration are slow at low temperature. The crystalline rocks of most shield terrains host brines that meet these criteria, and exhibit these trends.

Hydration of primary silicate minerals

Below temperatures of about 300°C, the hydration of silicate minerals becomes an important reaction affecting the isotopic evolution of water in crystalline rocks. Reactions such as the alteration of feldspar (anorthite) to clay (kaolinite):

$$CaAl_2Si_2O_8 + 3H_2O \rightarrow Ca^{2+} + Al_2Si_2O_5(OH)_4 + 2OH^-$$

result in a significant uptake of water. Fractionation of ^{18}O occurs between the fluid and the silicate structure as well as between the fluid and mineral hydration water. For ^2H, fractionation occurs only between the fluid and the clay hydration waters. Consequently, progressive hydration can result in a dramatic evolution due to Rayleigh distillation of the residual fluid.

The contribution from the ^{18}O-enriched primary silicates (5 to 10‰; Fig. 9-2) during alteration will also play a role, acting to buffer the δ^{18}O of the fluid towards an equilibrium value for exchange with the rock. Isotopic fractionation between the fluid and authigenic clay minerals partitions ^{18}O into the mineral, resulting in a ^{18}O-depletion in the residual fluid ($\varepsilon^{18}O_{clay-H_2O}$ < 0‰). The fractionation between oxygen in the silicate structure of clay minerals and the residual fluid varies between about 10‰ at 150°C and 25‰ at 25°C:

$$10^3 \ln\alpha^{18}O_{kaolinite-H_2O} = 2.5 \cdot 10^6 \, T^{-2} - 2.87 \qquad \text{(Land and Dutton, 1978)}$$

$$10^3 \ln\alpha^{18}O_{smectite-H_2O} = 2.67 \cdot 10^6 \, T^{-2} - 4.82 \qquad \text{(Yeh and Savin, 1977)}$$

Fractionation of ^{18}O between clay hydration waters and residual fluid is similar and on the order of 15 to 30‰ at 25°C (Table 9-1).

The reverse occurs for hydrogen, where ^{2}H preferentially fractionates into the water. This imparts an enrichment in ^{2}H in the residual fluid, leaving the clay mineral up to 60‰ depleted in ^{2}H ($\varepsilon^2 H_{clay-H_2O}$ = −30 to −60‰; Table 9-1).

Under conditions of low water-rock ratios, progressive water loss will result in a Rayleigh distillation of the residual water. From other systems (see the example of rainout in Chapter 2) we know that at low residual fractions f the water will have evolved to extreme isotope values. Fig. 9-4 shows this process for the evolution of seawater. In this case, fractionation factors of $\varepsilon^{18}O_{clay-H_2O}$ 10‰ and $\varepsilon^2 H_{clay-H_2O}$ = −40‰ were used to simulate reaction at elevated temperatures.

Table 9-1 Fractionation of ^{18}O and ^{2}H between interlayer water in clay minerals and water at low temperature

Clay	$\varepsilon^{18}O_{min-wat}$ 25°C	$\varepsilon^2 H_{min-wat}$ 25°C	Reference
Montmorillonite	27	−62	Savin and Epstein, 1970
Kaolinite	27	−31	Savin and Epstein, 1970
Smectite	27	−62	Savin and Epstein, 1970;
		−30*	Lawrence and Taylor, 1972
Gibbsite	18	−16	Lawrence and Taylor, 1972
Illite	23 @ 22°		James and Baker, 1976

* Under weathering conditions.

The example of shield brines

The occurrence of high salinity Ca-Na-Cl brines in deep crystalline environments has been recognized for over a century, although it is only in the past two decades that they have gained the interest of hydrogeologists. Nuclear agencies became interested in learning more about these fluids, and how they may affect corrosion and transport of radioactive waste that may be buried in such deep crystalline environments. Inflows at mines have been the principal source of information for brine studies in the Canadian Shield (e.g. Fritz and Reardon, 1979, Fritz and Frape, 1982). Subsequent studies done on the Fennoscandian Shield and the German continental deep drill hole (KTB) (Fritz and Lodemann, 1990) all confirm the basic observations that (i) Ca-Na-Cl brines are present at depths below active groundwater flow systems in most crystalline rock environments, and (ii) these brines are enriched in deuterium with δ^2H values usually > −35‰ and δ^{18}O < 0‰ such that they all plot above the meteoric water line (Fig. 9-8). Considerable debate has ensued over the origin and age of these groundwaters, and how they have gained their salinity.

In assessing their origin, the most notable observation is their marked enrichment in δ^2H. Further, a strong correlation for stable isotope-salinity in data from the principal Canadian Shield mining camps suggest mixing between local precipitation and a ^{2}H−enriched brine of relatively uniform isotopic composition (δ^{18}O = −10 to −7‰ VSMOW; δ^2H = −20 to 0‰ VSMOW) (Frape and Fritz, 1982). Many of these brines contain measurable tritium, indicating that they have a component of recent recharge. This confirms that the trend lines observed in Fig. 9-8 are mixing lines rather than lines of δ^{18}O and δ^2H evolution. Fritz and Reardon (1979) attributed the unusual isotopic composition of Shield brines to evolution during hydration of primary silicates under closed system conditions, although low-temperature exchange with feldspars may be in part responsible for the negative shift in δ^{18}O (Fig. 9-4).

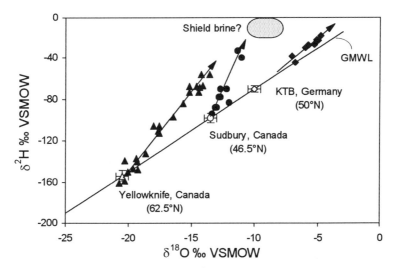

Fig. 9-8 Deuterium and ^{18}O composition of brines and shallow groundwaters in the Canadian Shield (Sudbury Basin and Yellowknife mining camps), and from the Bohemian Massif (KTB deep continental drillhole, Germany). Local meteoric waters shown in open triangle. Canadian Shield brine is a hypothetical rock-equilibrated fluid. Salinity (not shown) increases in samples with increasing deviation from the meteoric water line along trend lines shown, indicating mixing between local meteoric waters and a rock-equilibrated brine (Data from Douglas, 1997; Frape and Fritz, 1982; Fritz and Lodemann, 1990.)

If these brines are in complete equilibrium with the host rocks, then all vestiges of their original isotopic signature, whether meteoric, oceanic or magmatic, are lost, and the question of their origin remains unanswered. This may also apply to the dissolved constituents (Fritz et al., 1994). Considering their high salinities, one may suspect that these fluids are allochthonous, although an *in situ* origin through leaching of Cl⁻ from halide-bearing minerals and fluid inclusions within the rock mass is possible. An allochthonous fluid is certainly the case at the KTB (Lodemann et al., 1997) where strontium isotope data for open vein calcites are not in equilibrium with present-day brines. A relevant observation for Shield brines is that they must be exceedingly old to be so extensively altered at low temperature.

Whether there has been a direct or indirect marine precursor is still open to discussion. Additional isotopes now useful in studies of brine origin and mixing include ^{6}Li, ^{11}B and ^{81}Br.

Analysis by inductively coupled plasma mass spectrometry (ICP-MS) has made the two lighter isotopes more accessible to research. For example, Bottomley et al. (1994) use the strong correlation between $\delta^{11}B$ and salinity to infer a Paleozoic seawater origin for Canadian Shield brines. Subsequent work (Bottomley et al., 1997) supports a marine origin by showing that the δ^6Li composition of deep brines from Yellowknife, Canada is very close to the marine value of –32‰ LSVEC yet highly depleted from the host rocks which have δ^6Li closer to 0‰ LSVEC.

Low-temperature exchange in sedimentary formations

Another environment in which extensive water-rock interaction has modified the original isotopic composition of the pore water is found in deep sedimentary basins. The Na-Cl salinity of basin brines led to early interpretations that these fluids originated as seawater that experienced salinity increases through evaporation or evaporite dissolution. Clayton et al. (1966) published some of the earliest isotopic measurements of sedimentary basin brines (Fig. 9-9). They are characterized by a trend of ^{18}O-enrichment with minor ^2H-enrichment, that originates with local meteoric waters. However, the most saline brines are generally more enriched than an evaporated seawater. Fig. 2-24 shows that the brine created through the evaporation of seawater eventually becomes more depleted in ^{18}O and ^2H due to the formation of hydration sheaths around ions.

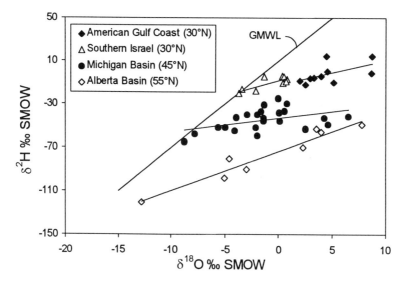

Fig. 9-9 The $\delta^{18}O$ - δ^2H relation for formation brines from selected sedimentary basins (Michgan Basin, Alberta Basin, American Gulf Coast — Clayton et al., 1966; Southern Israel — Fleischer et al., 1977). Note that most of these data were produced prior to the synthesis of VSMOW.

The characteristic deviation from the meteoric water line for sedimentary basin brines is attributed to a number of processes including ^{18}O exchange with carbonate minerals at elevated temperatures (Clayton et al., 1966), ^2H exchange with hydrocarbon, H_2S and hydrated minerals, dewatering of clays during compaction, hyperfiltration (see below), hydration of anhydrite (Bath et al. (1987) and mixing with meteoric waters (Longstaffe, 1983; Bein and Dutton, 1993; Musgrove and Banner, 1993).

From Fig. 9-9, the strongest effect is the positive shift ^{18}O. Consider that marine limestone is precipitated in equilibrium with seawater at about 25°C ($\varepsilon^{18}O_{CaCO_3-H_2O}$ = 28.4‰). Its $\delta^{18}O$ value would be 28.4‰ VSMOW or –2.4‰ VPDB. Groundwaters of meteoric origin ($\delta^{18}O_{H_2O}$ of say –10‰ VSMOW) circulating through these formations are then 10‰ below isotopic equilibrium. At formation temperatures of 50 to 100°C, $\varepsilon^{18}O_{CaCO_3-H_2O}$ is reduced to 20‰ or less. The disequilibrium now becomes even greater (28.4 – 20 + 10 = 18.4‰). Oxygen exchange between formation carbonate and formation fluids will slowly shift the $\delta^{18}O$ composition of the fluid towards that of the rock. Although water-rock ratios in sedimentary formations may be up to several percent, the much larger rock reservoir will ultimately control the $\delta^{18}O$ composition of the fluid.

Dehydration of gypsum to anhydrite at elevated temperatures will also impart an ^{18}O enrichment on formation fluids. Experiments by Fontes (1965) and by Gonfiantini and Fontes (1967) show that at low temperature, the crystallization water in gypsum [$CaSO_4 \cdot 2H_2O$] is enriched by 3 to 4‰ in ^{18}O with respect to pore water. Like clays, the ^2H enrichment factor is reversed. For gypsum hydration waters, ^2H is depleted by about 15‰ (Fig. 9-10).

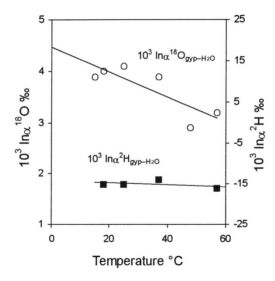

Fig. 9-10 Fractionation of ^{18}O and ^2H between the water of crystallization and the crystallizing solution for gypsum ($CaCO_3 \cdot 2H_2O$) (from Fontes, 1965; Fontes and Gonfiantini, 1967b).

The relatively modest δ^2H enrichment in basin brines (Fig. 9-9) can then be partially explained by exchange with or dehydration of gypsum and clay minerals that were precipitated from seawater. These hydration waters are depleted in ^2H with respect to seawater, but will be relatively enriched with respect to equilibrium with circulating meteoric waters (Table 9-1; Fig. 9-10). Further, this disequilibrium will be greater at higher temperature, where fractionation is reduced. Where water-rock ratios are low, isotope exchange in shale and evaporite strata will enrich the pore waters in ^2H.

Deuterium enrichment can also arise through exchange with H_2S. Hydrocarbon-bearing strata that host formation waters with high sulphate contents can produce H_2S. The strong partitioning of ^2H into the H_2O reservoir ($\alpha^2H_{H_2O-H_2S}$ = 2.2 @ 50°C) produces H_2S depleted by over 500‰. With sufficient *in situ* production of H_2S, this can impart a positive shift in $\delta^{18}O_{H_2O}$.

Hyperfiltration of isotopes

Enrichment trends for isotopes in deep groundwaters can, in certain cases, be attributed to hyperfiltration. This is a valid explanation for enrichments observed in low permeability formations. When groundwater moves by advection along steep hydraulic gradients through shales or clays, the matrix can act like a reverse-osmosis membrane, retarding the movement of larger molecules. The filtrate would then be depleted in heavier isotopes. Phillips et al. (1986) proposed hyperfiltration to explain enrichments in Cl⁻ and $\delta^{18}O$ observed down-gradient in the Milk River aquifer of northern Montana and southern Alberta. Hitchon and Friedman (1969) account for $\delta^{18}O$ and δ^2H enrichments in the western Canadian sedimentary basin brines with hyperfiltration.

Experimental work by Coplen and Hanshaw (1973) produced depletions of 0.8‰ for $\delta^{18}O$ and 2.5‰ for δ^2H in NaCl solutions forced through smectite clay. Fritz, Hinz and Grossman (1987) show an opposite trend for ^{13}C in DIC during hyperfiltration, where an enrichment of 1.5‰ in the filtrate was measured. This example demonstrates the case where reaction plays a role. Concentration of the DIC on the up-gradient side of the smectite columns alters the distribution of DIC species within the bulk solution, producing a ^{13}C enriched "polarization" layer. Diffusion along this new concentration gradient enriches the filtrate in ^{13}C. These authors cite a similar example for ^{37}Cl during hyperfiltration.

Caution must be exercised in employing hyperfiltration in hydrogeological studies, given the specific flow and aquitard conditions required. Hendry and Schwartz (1988), for example, measured Cl⁻ and $\delta^{18}O$ in the pore waters of the confining formations adjacent to the Milk River aquifer and found that matrix diffusion of these species into the aquifer was a more likely explanation for their enrichments along the flow path.

Strontium Isotopes in Water and Rock

Strontium is a divalent cation that readily substitutes for Ca^{2+} in carbonates, sulphates, feldspars and other rock-forming minerals. Like Ca^{2+}, it participates in water-rock reactions, and is a minor component of most groundwaters. Strontium isotopes ($^{87}Sr/^{86}Sr$) have proven to be a useful indicator of water-rock interaction, and as a tracer for groundwater movement and the origin of salinity.

The strontium isotope ratio for carbonate rocks has been measured throughout the Phanerozoic (Fig. 9-11). This $^{87}Sr/^{86}Sr$ curve reflects the relative contributions of strontium to the ocean from continental weathering and from hydrothermal activity along mid-oceanic ridges (Veizer, 1989). The primordial $^{87}Sr/^{86}Sr$ ratio of 0.699, derived from meteorites, has been steadily increasing due to the decay of ^{87}Rb. Modern seawater has $^{87}Sr/^{86}Sr = 0.709$ (Faure, 1986), which is intermediate between ^{87}Sr-depleted ocean basalts (~0.703) and ^{87}Sr-enriched continental rocks (0.710 to 0.740).

Most of the Phanerozoic is characterized by a general decrease in $^{87}Sr/^{86}Sr$ (Fig. 9-11) due to increasing activity along mid-ocean spreading ridges. The late Cenozoic marine sediments experienced a dramatic increase in ^{87}Sr due to climate cooling and increased rates of continental weathering by glaciation. This rapid and steady increase through the Pliocene and Pleistocene provides a high-resolution dating tool for sedimentary rocks. The strong variation in $^{87}Sr/^{86}Sr$ through Phanerozoic time and between rock types provides for strong contrasts between differing geological terrains.

The abundance of ^{87}Sr, the daughter of ^{87}Rb, is directly linked to the geochemistry of potassium, for which Rb^+ will readily substitute. K-rich rocks will have high ^{87}Rb and ^{87}Sr contents, and this is reflected in the $^{87}Sr/^{86}Sr$ ratio of water with which they have equilibrated. Thus, groundwaters that have geochemically evolved in differing geological terrains will have contrasting strontium isotope ratios. For example, Yang et al. (1996) use this phenomenon to derive from $^{87}Sr/^{86}Sr$ ratios in the St. Lawrence river a pattern of contributions from various tributaries and landscapes.

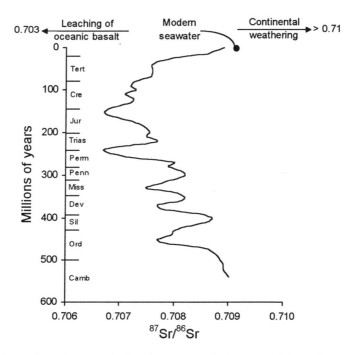

Fig. 9-11 Strontium isotopes in sedimentary rocks throughout Phanerozoic time (modified from Veizer, 1989).

A similar approach can be taken if subsurface water-rock interaction influences the geochemistry of saline groundwaters in crystalline and basinal settings or even where normal groundwater has residence times long enough to see the influence of strontium leaching — which may go very fast. Bullen et al. (1996) report strontium isotope data from a very young groundwater system in crystalline terrain. They show that it is possible to recognize water that has been in contact with K- and Sr-poor feldspars and distinguish it from water flowing through K-rich terrain where biotite and K-feldspars dominate the strontium geochemistry. In this way, different groundwater bodies can be distinguished, mixing relationships examined, and allochthonous vs. autochthonous sources of salinity be identified.

An interesting use of $^{87}Sr/^{86}Sr$ data for a hydrogeological investigation was presented by Scholtis et al. (1996) who investigated the suitability of slightly metamorphosed Cretaceous marl for the containment of radioactive waste. Strontium isotope analyses show a clear differentiation of groundwaters with differing origins. More importantly, the fracture carbonates that precipitated from these waters have the same $^{87}Sr/^{86}Sr$ composition as the water. Therefore, it was possible to confirm that little or no cross-formational water movement took place, an important observation in a repository project.

Isotope Exchange in Gas - Water Reactions

In most normal groundwater systems, dissolved gases have little or no influence on the isotopic composition of groundwater. Rather, the isotopic composition of a gas phase is generally controlled by exchange with the water. Remember that water has a concentration of some 55.6 moles H_2O per litre. Only in very unusual circumstances would gas-water ratios be high enough to affect the fluid, and be considered in the interpretation of groundwater isotope data (Fig. 9-1). Examples may be found in volcanic or metamorphic terrains with strong CO_2 production, in sedimentary strata with extensive sulphate reduction taking place, or in landfills (Chapter 6).

Deuterium shift — exchange with H_2S

Hydrogen sulphide (H_2S and HS^-) can be generated in groundwaters where an organic carbon substrate is available to support bacterial sulphate reduction (see Chapter 6). It can also be produced abiotically in sedimentary basins during the maturation of hydrocarbon. H_2S readily exchanges hydrogen with water, and is analogous to the carbonate system where speciation is controlled by pH (Fig. 6-3):

$$H_2S \leftrightarrow H^+ + HS^- \leftrightarrow 2H^+ + S^{2-}$$

Because exchange takes place quickly (reaction half-time is in the order of minutes, depending on temperature and pH), kinetic reactions are not important and the fractionation factor is easily measured. It is also one of the largest fractionations observed in isotope hydrogeology. At 25°C, the fractionation factor is (Table 1 — front cover):

$$\alpha^2 H_{H_2O\text{-}H_2S} = 2.358$$

Remember that most α-values are close to 1. This large fractionation is used commercially as the primary stage used to produce heavy water (^2H_2O) in the Canadian Deuterium Uranium nuclear reactors (CANDU). In expressing such a large fractionation in permil notation, we have to be careful of the expression used. Using the thermodynamic expression we get:

$$10^3 \ln \alpha^2 H_{H_2O\text{-}H_2S} = 858‰$$

whereas using the enrichment factor yields:

$$\varepsilon^2 H_{H_2O\text{-}H_2S} = (\alpha - 1) \cdot 10^3 = 1358‰$$

and the measured values of δ^2H for H_2S in equilibrium with H_2O gives yet a third value for the isotope separation (using $\delta^2H_{H_2O} = -70‰$ VSMOW):

$$\alpha^2 H_{H_2O\text{-}H_2S} = 2.37 = \frac{1000 + \delta^2 H_{H_2O}}{1000 + \delta^2 H_{H_2S}} \qquad \text{(recall from Chapter 1)}$$

$$\delta^2 H_{H_2S} = -606‰ \text{ VSMOW}$$

and so, $\Delta^2 H_{H_2O\text{-}H_2S} = \delta^2 H_{H_2O} - \delta^2 H_{H_2S} = 536‰$

Despite such strong fractionation, it still takes a substantial amount of sulphate reduction in groundwaters to modify their original ^{2}H contents. A mass-balance calculation shows that the production of 0.1 mole of H_2S would enrich the water by 1‰ in ^{2}H:

$$mH_2O \cdot \delta_o^2H_{H2O} = (mH_2O - mH_2S) \cdot \delta_f^2H_{H2O} + mH_2S \cdot \delta^2H_{H2S}$$

where: $\delta_o^2H_{H2O}$ = initial value for water = –70‰
$\delta_f^2H_{H2O}$ = final value for water = –69‰
$\delta^2H_{H2S} = \delta_f^2H_{H2O} - \Delta^2H_{H2O-H2S} = -69 - 536 = -537$‰
mH_2O = 55.6 moles-H_2O /L

To produce a shift in δ^2H_{H2O} of 1‰, it would require the generation of 0.104 moles or 3.5 g/L H_2S. A similar result can be determined using a Rayleigh calculation, according to the equation:

$$\delta_f^2H_{H2O} \cong \delta_o^2H_{H2O} + 10^3 \cdot (\alpha^2H_{H2S-H2O} - 1) \cdot \ln f$$

where the fraction of water reacted (1–f) is equal to mole fraction of H_2S produced (mH_2S/mH_2O), which in this case is 0.0017 or 0.1 moles/L of H_2S. Note that $\alpha^2H_{H2S-H2O}$ is the inverse of the value given above (1/2.358 = 0.424). For a readily discernible (say 5‰) shift in δ^2H_{H2O}, about 17.6 grams of sulphide per litre would be produced, requiring reaction of about 50 g/L SO_4^{2-} (70 g of gypsum per litre) in solution.

Groundwaters circulating through strata with clay units or other sources of sedimentary organic carbon, and where residual high SO_4^{2-} pore waters, evaporite sequences, or other sources of sulphate are available, will likely have an H_2S component. However, in most cases, dissolved sulphide concentrations are in the order of a few millimoles or less, and their effect on the isotopic composition of water is negligible.

^{18}O exchange between H_2O and CO_2

Another gas-water exchange reaction often implicated in the modification of this isotopic composition of water is that of ^{18}O between CO_2 and H_2O (Fig. 9-1). However, like exchange with H_2S, it is rarely observed in nature. At 25°C, the accepted value for the fractionation factor $\alpha^{18}O_{CO2-H2O}$ is 1.0412 (Friedman and O'Neil, 1977) and isotopic equilibrium is achieved in a matter of a few hours (equilibration with CO_2 is the basis of measuring $\delta^{18}O$ in waters). Thus, in systems with high ratios of CO_2 to water, one could expect a negative shift in $\delta^{18}O_{H2O}$. However, a negative ^{18}O-shift is seldom seen. This may be due to the high volumes of CO_2 required for a measurable shift in $\delta^{18}O$, as well as to the nature of subsurface CO_2 sources which are normally enriched in ^{18}O to begin with.

CO_2 generated by the high-temperature decomposition of limestone (contact metamorphism) is a principal cause for natural, high P_{CO2} groundwaters. The decarbonation of $CaCO_3$ produces CO_2 that is enriched in ^{18}O (Shieh and Taylor, 1969; Valley, 1986). Bottinga (1968) calculated the isotope enrichment factor $\varepsilon^{18}O_{CO2-CaCO3}$ in this reaction to be ~ 6‰ at 600°C. Thus, metamorphic CO_2 from carbonate with a $\delta^{18}O$ of about 30‰ VSMOW may be as enriched as 36‰ VSMOW. In isotopic equilibrium with H_2O of $\delta^{18}O$ = say –10‰, this CO_2 would then have a $\delta^{18}O$ value of +31.2 (–10 + 41.2 ‰). The net difference between the initial $\delta^{18}O_{CO2}$ and the equilibrium value is less than 5‰, hardly enough to significantly modify the water. It is not surprising that this type of shift in the isotopic composition of meteoric waters is seldom observed.

High pH Groundwaters — The Effect of Cement Reactions

Isotopic fractionation between hydroxide [OH$^-$] and water for both ^{18}O and ^{2}H are enormous, but seldom observed because most hydroxide-water reactions take place only at high pH. The exchange of hydroxide with minerals such as hydroxyapatite [Ca$_5$(PO$_4$)$_3$OH] is insufficient to affect the isotopic composition of water. However, portland cement type reactions can have an effect. As cements are used in many underground structures and in sealing piezometer installations, it is perhaps one of the more important exchange reactions affecting the isotopic composition of water.

Dry portland cement is a mixture of oxides, dominated by CaO and SiO. Mixed with water, a series of Ca-Si hydrates forms, the crystallization of which gives cement its strength. Calcium oxide will also hydrate to form portlandite [Ca(OH)$_2$]:

$$CaO + H_2O \rightarrow Ca(OH)_2$$

Portlandite is very soluble, and generates high pH pore waters in cement grout. The dissociation constant for portlandite is $10^{-5.02}$ at 25°C, and the pH of a portlandite-saturated solution is 12.52 (aOH$^-$ = $10^{-1.48}$). If strong isotope fractionation occurs between OH$^-$ and H$_2$O, then there could be an effect on the isotopic composition of cement pore waters.

Green and Taube (1963) determined the water-hydroxide fractionation factor:

$$10^3 \ln\alpha^{18}O_{H_2O\text{-}OH} = 39.2‰ \quad @\ 15°C$$

which they extrapolate to ~40‰ at 25°C. Although this is a large fractionation, we can show that shifting pH to 12.5 will not produce a measurable effect on the δ^{18}O of the water. Consider that the concentration ratio aOH$^-$/aH$_2$O at pH 12.5 is $10^{-1.5}$/55.6 = 0.000575, so the activity of water is 0.9994. Using the simplified Rayleigh relationship, with f = 0.9994, we can determine the δ^{18}O shift in H$_2$O between a neutral pH and the alkaline pH:

$$\begin{aligned}\delta^{18}O_{pH12.5} &= \delta^{18}O_{pH7} - \varepsilon^{18}O_{H_2O\text{-}OH^-} \cdot \ln f \\ &= \delta^{18}O_{pH7} + (-40 \cdot -0.000575) \\ &= \delta^{18}O_{pH7} + 0.023‰\end{aligned}$$

which is less than the measurement error. If the solution had an extraordinary pH of 13, this would only produce a 0.07‰ shift.

If we consider that the pore waters in hydrated cement and cement-like rocks are the residual waters remaining after hydration of the solid cement grout, then f becomes a considerably larger number, and will be based on the water-mineral ratio. These calculations assume no further fractionation between free hydroxide and the hydroxide mineral. Using a molar ratio of say 3:1, f becomes 0.67 and the difference between initial and final $\delta^{18}O_{H_2O}$ becomes 16‰.

There is, then, the potential for modifying the isotopic composition of groundwater within or near cement structures. Dakin et al. (1983) found that groundwaters sampled from piezometers sealed with cement grout were enriched by 1 to 2‰ above associated springs and other groundwaters. In a series of laboratory experiments, they verified that cement pore waters were indeed ^{18}O-enriched.

The fractionation factor for ^{2}H in hydroxide systems is even more dramatic. Heinzinger and Weston (1964) measured an incredible –1400‰ for $10^3\ln\alpha$ between OH$^-$ and H$_2$O at 13.5°C. This gives a fractionation factor of:

$$\alpha^2H_{H_2O\text{-}OH} = 4.055$$

and thus according to:

$$\alpha^2H_{H_2O\text{-}OH} = \frac{1000 + \delta^2H_{H_2O}}{1000 + \delta^2H_{OH}} = 4.055$$

the δ^2H for hydroxide in equilibrium with a water with say δ^2H = –70 would be –771‰. Such a strong depletion in the hydroxide phase would enrich the residual waters following hydration, assuming no additional mineral-OH$^-$ fractionation occurred. This example is presented to show the strong isotope effects associated with the hydration of cement minerals. Although these conditions are seldom found naturally, one should consider these reactions when looking at data collected from wells or piezometers where contamination from cement seals or cemented casings is possible.

Problems

1. Is the strong δ^2H enrichment trend observed in Shield brines the result of primary water-rock interaction or mixing? Explain.

2. Take the case where the hydration of primary silicate minerals occurs under closed system conditions with low water-rock ratios at 25°C. Create a diagram to show the evolution of the δ^{18}O and δ^2H composition of the residual water, as it is consumed to 50% of its original volume. Assume an original water composition of δ^{18}O = –10‰ and δ^2H = –70‰ and that there is no contribution from ^{18}O in the primary silicates. Use fractionation factors from Table 1 (front cover) for ^{18}O and ^{2}H fractionation between smectite and water.

   Now do the same calculations for the residual pore water during hydration of anhydrite to gypsum. Use the fractionation factors from the equations developed for data in Fig. 9-10 at 25°C and at 60°C.

3. Equilibrate a meteoric water (say with δ^{18}O = –15‰ VSMOW) with marine limestone (δ^{18}O = 0‰ VPDB) at 80°C and determine its final δ^{18}O composition, assuming an infinitely small water-rock ratio.

   If you increase the water-rock (volume) ratio to 0.33 (25% porosity), what would you expect the δ^{18}O of the water to be. What about the limestone?

4. Redox reactions in a hydrothermal system (T = 150°C) cause the coexistence of HS$^-$, H$_2$S, and SO$_4^{2-}$. Write a mass balance equation which would permit you to determine the δ^{34}S value of the original sulphur source, and apply it to a system with the following concentrations.

HS$^-$	0.013 moles/L	$\varepsilon^{34}S_{HS\text{-}H2S}$ = –2‰ at 150°C	
H$_2$S	0.001 moles/L	δ^{34}S = –30‰ CDT	
SO$_4^{2-}$	0.02 moles/L	δ^{34}S = 19‰ CDT	

- Comment on the possible origin of this sulphate.

- Given that the first dissociation constant for H_2S is $10^{-7.0}$, what would be the pH of this groundwater.

5. This problem builds on elements of isotope hydrogeology that have been gained from all previous chapters. It serves as a case study that integrates the use of isotope techniques to determine groundwater provenance, subsurface history, and groundwater age. It requires, however, access to the website for this book to find the location map and data set. They can be found at <**www.science.uottawa.ca/~eih/Ch6/prob6.htm**>. These data have been produced for groundwaters and precipitation sampled from sites given on the adjoining map. Your task is to evaluate these data and provide a brief but informative report with respect to the recharge origin and subsurface history and mean residence times of these groundwaters. In your evaluation, include relevant graphs as interpretive tools and to support your observations. In your report, note any assumptions you use in your interpretations.

Chapter 10
Field Methods for Sampling

This chapter was prepared in collaboration with R.J. Drimmie, Department of Earth Sciences, University of Waterloo, Ontario, Canada.

Hydrogeological investigations and interpretations are no better than the quality of the results obtained from samples collected in the field. This chapter describes the equipment required and procedures to be followed for field measurements and sampling. It also presents the more common analytical procedures used in various laboratories, which dictate the size of sample needed for analysis (including repeat measurement) and any required preservation.

The procedures to collect samples for environmental isotopes and geochemical analyses can be straight forward (e.g. $\delta^{18}O$ and δ^2H in water) or very complex (e.g. helium isotopes). Analytical methods are also continually improving, and different laboratories will have different requirements and constraints on sample size. *It is important to contact the analytical laboratory before field work commences in order to assure that your sampling protocol meets their requirements.* Links to the websites for various laboratories can be found at the Isogeochem website:<**http://beluga.uvm.edu/geowww/isogeochem.htm**>. Another excellent reference site is found at Syracuse University <**www.geochemistry.syr.edu/cheatham/InstrPages.html**>. These and other links can also be accessed from the site for this book, <**www.science.uottawa.ca/~eih**>.

Some general comments on sampling:

- *Chemicals:* Most chemicals used for preservation should be reagent grade purity, to avoid contamination (such as traces of carbonate in NaOH for sampling $^{14}C_{DIC}$). In some cases, such as the acid rinse used to clean residual carbonate from a ^{14}C sampling carboy, the chemicals need not be reagent grade. Consider the isotopes and chemistry to be measured.

- *Bottle Labels:* Repeated sampling for different isotope analyses at a given site will soon produce a large number of bottles with the same site number. Give careful consideration to your labelling system. It serves to help you remember what you sampled (considering that some samples get archived and pulled out for later analysis) and it serves the laboratory. Laboratory personnel will want to know what to analyse, and how you have treated the sample. Surprises such as unidentified brine samples, or samples preserved with HgCl, won't gain you any friends in the lab. Use a unique field code with sequential numbering for sample identification, with the analysis included, for example:

Field code:	SB8 - ^{13}C	
Date:	96/08/24	*Time*: 17:22
Site:	Hollyburn	
Well name:	Main Lodge No. 1	
Analysis:	$\delta^{13}C$ of DIC	
Field Alkalinity:	70 mg/L as HCO_3^-	
E.C.:	392 µS/cm	
Filtered:	0.45 µ	
Preservation:	NaN_2	

Don't use an all-inclusive field code that has all site information. Keep it short to avoid the errors (and tedium) of transcribing. Use a narrow-tip *waterproof* marker or pencil (graphite doesn't fade) on a sticky label. The label should then be totally covered by clear tape. Print out customized labels before the field work from a table using a word processor and sheets of adhesive paper.

Other information to include in your notes, but not necessary for the lab are:

Sample site: (well, piezometer, spring, rain gauge, river, etc.)

Sampling depth: (depth of well, piezometer, bailer or pump intake, surface)

Sample method: (well pump, bailer, suction pump, grab sample, etc.)

Preparation: (well pumped 45 minutes, piezometer developed with N_2 purge)

Condition of sample site: (abandoned well, newly installed piezometer, etc.)

Environmental conditions: (precipitation, temperature)

Other hydrological data: (runoff, recent use of well, or other relevant details)

- **Sample Containers:** Water samples must be collected and stored in well-sealed bottles. The bottle will vary according to the size of sample required (and supplies available). The best are glass bottles with a tight seal, although these are more likely to leak if the cap is not taped to prevent loosening. Gases diffuse much more slowly through glass bottles than plastic, although glass is less robust for shipping. For this reason, high-density linear polyethylene (HDPE), linear polyethylene (LPE) or polypropylene (PP) bottles are recommended. They are produced with tight-fitting screw caps (without rubber seals that tend to host colonies of bacteria).

Exceptions are noted for individual sampling protocols below, and sometimes any available container will do just fine. In some third world countries, well rinsed product bottles (e.g. shampoo) are sold in markets, and are fine for $\delta^{18}O$–δ^2H samples. Glass "soda pop" bottles with screw caps are excellent for gas samples, as they are designed to retain well over an atmosphere of CO_2. Test the integrity of the cap after taking the sample by inverting the bottle and squeezing. No water should come out.

If sampling is done for ^{18}O, ^2H and tritium alone, then changes in chemistry of the water or minor biological activities in the sample container during storage can be ignored. The situation is different for certain dissolved constituents (notably DIC and DOC) since both chemical reactions and biological activity can affect the sample. In this case, the sampling equipment should be extremely clean, biologically inactive and the samples will have to be preserved. This is best done with a very minor amount of Na-azide (NaN_2). Some investigators use mercuric chloride (Hg_2Cl_2). We do not recommend it because of difficulties with disposal.

Sample containers for gas analyses will vary with the type of gas to be collected. A description is provided in the section on gas analyses. Important to note is also that shipment by air freight requires adherence to safety standards. For this reason, alkaline and acid solutions or certain chemicals cannot be transported by air.

Table 10-1 provides a summary of the basic requirements to be met in terms of sample size and storage.

Table 10-1 Sample size, preservation and storage time. Note that laboratory capabilities with small sample sizes vary considerably. This table is a guideline only, and laboratory requirements should be discussed before sampling. For a list of laboratories, see <http://beluga.uvm.edu/geowww/isogeochem.htm>

Isotope	Method of analysis	Analytical precision	Sample size	Field measurements and preservation	Storage life
WATER					
^{18}O	IRMS (CO_2 equil.)	±0.1‰	10 mL	b-pl	>1 year
^{2}H	IRMS (Zn red. to H_2)	±1‰	10 mL	b-pl	>1 year
^{3}H	Direct LSC	±8TU	20 mL	b-pl	Decays, $t_{½}$=12.43 yr
	Enriched LSC	±0.8 TU	250 mL	b-pl	
	Propane synthesis	±0.1 TU	1000 mL	b-gl	
	^{3}He-ingrowth, IRMS	±0.1 TU	50 mL	b-gl	
DISSOLVED INORGANIC CARBON (DIC: H_2CO_3, HCO_3^-, CO_3^{2-})					
^{13}C	IRMS (acid. to CO_2)	±0.15‰	10 mg HCO_3^-	pH, filt, b-gl, NaN_2, 4°C	Months
	IRMS (acid. to CO_2)	±0.15‰	25 mg $BaCO_3$	pH, filt, fp-(NaOH, $BaCl_2$), b-pl	~1 year
^{14}C	LSC (conv. to C_6H_6)	±0.3 pmC	0.5-3 g C	fp-(NaOH, $BaCl_2$), b-pl	~1 year
	LSC on Carbasorb®	±5 pmC	1-3 g C	Strip into Carbasorb® solution	Unlimited
	GPC (acid. to CO_2)	<± 0.3 pmC	3-5 g C	fp-(NaOH, $BaCl_2$), b-pl	~1 year
	AMS (acid. to CO_2)	<± 0.3 pmC	5 mg C	b-pl, 4°C	Months
DISSOLVED ORGANIC CARBON (DOC)					
^{13}C	Oxid. to CO_2 - IRMS	±0.5‰	20 mg C	pH, filt, b-gl, NaN_2 or HCl	1 month
^{14}C	AMS (comb. to CO_2)	±0.5 pmC	5 mg C	Ion exchange resin	< 1 month on resin
DISSOLVED SULPHATE (SO_4^{2-})					
^{34}S	IRMS (comb. to SO_2)	±0.3‰	20 mg $BaSO_4$	Filt, fp-($BaCl_2$)	Unlimited
^{18}O	IRMS (conv. to CO_2)	±0.5‰	0.1g SO_4	Filt, fp-($BaCl_2$)	Unlimited
DISSOLVED SULPHIDE (H_2S, HS^-)					
^{34}S	Comb. to SO_2 - IRMS	±0.3‰	25 mg CdS	Filt, fp-(ZnAc or CdAc)	Unlimited
DISSOLVED NITRATE (NO_3^-) AND NH_4^+					
^{15}N	Conv. to N_2 - IRMS	±0.2‰	4 mg N	Acid to pH 2 with HCl	~3 months
^{18}O	Conv. to CO_2 - IRMS	±0.5‰	25 mg NO_3	Acid to pH 2 with HCl	~3 months
HALIDES					
^{37}Cl	IRMS (conv. to CH_3Cl)	±0.1‰	1-10 mg Cl^-	b-pl	Unlimited
^{36}Cl	AMS (conv. to AgCl)	±10^{-15}	1-10 mg Cl^-	b-pl	Unlimited
^{129}I	AMS (conv. to AgI)	±10^{-13}	2-10 mg I^-	b-pl	Unlimited
DISSOLVED URANIUM					
$\dfrac{^{234}U}{^{238}U}$	α spectrometry	±0.05	1-5 µg U	Filt, acid. with HCl	Months
	TIMS (U-coated Fe)	±0.01	<1 µg U	Filt, acid. with HCl	Months
DISSOLVED GASES					
He	IRMS ($^{3}He/^{4}He$)	±1· 10^{-8}	50 mL H_2O	Copper tube sampling device	Months
^{39}Ar	GPC	±3 dpm cm^{-3}	15 m^3 H_2O	Gas cylinder	Unlimited
^{85}Kr	GPC	±3 dpm cm^{-3}	100 L H_2O	Gas cylinder	Decays, $t_{½}$ = 10.72 yr
CH_4	IRMS (^{13}C, ^{2}H)	±0.1‰	10 mmol CH_4	Glass syringe, septum bottle	Diffuses, 1 month
H_2	IRMS (^{2}H)	±1‰	10 mmol H_2	Glass syringe, septum bottle	Diffuses, days
CARBONATE MINERALS					
^{13}C	IRMS (acid. to CO_2)	±0.1	<10 mg	^{13}C and ^{18}O measured for single extraction	—
^{18}O	IRMS (acid. to CO_2)	±0.1	<10 mg	—	—
SULPHATE and SULPHIDE MINERALS					
^{34}S	IRMS (comb. to SO_2)	±0.3	25 mg $CaSO_4$ or FeS_2	—	—
^{18}O	IRMS (conv. to CO_2)	±0.5	25 mg $CaSO_4$	—	—

Notes to Table 10-1:
IRMS: isotope ratio mass spectrometry (sample conversion to gas).
AMS: accelerator mass spectrometry.
LSC: liquid scintillation counting.
GPC: gas proportional counting (mainly for inert gases).
TIMS: thermal ionization mass spectrometry.
b-pl: plastic bottle, preferably high density polyethylene (HDPE) or polypropylene (PP) with tight sealing cap.
b-gl: glass bottle, with tight sealing cap (plastic, not rubber cap liner or cone).
pH: generally required for interpretation of ^{13}C and DIC data.
Filt: filter to 0.45µ, preferably in field. Filtering is not usually required for developed wells and springs which run clear. Filtering should be avoided for large water samples for ^{14}C, or for dissolved sulphide samples due to the potential for contamination by atmospheric CO_2 and O_2.
fp-(reagents): precipitation of sample from water in field using specified reagents.
NaN$_2$: sodium azide, antibacterial agent.
4°C: refrigerate until analysis. Recommended to reduce biological activity, and minimize diffusion of gases through plastic container walls.

Groundwater

When sampling groundwater, consider the following points:

- Does the sample represent *in situ* groundwater, or is it contaminated by other fluids or by contact with the atmosphere? Is the well or piezometer contaminated by drilling fluid or has evaporation or exchange with the atmosphere occurred?

- How well does the sample represent the aquifer or groundwater being investigated: i.e. are there spatial or temporal variations that are not represented by the sampling program? Taking one sample in the summer will not characterize very well the isotope content of groundwater within the recharge area of an aquifer, nor will one sample from a site in a regional aquifer characterize its isotopic composition.

Sample sites

Natural springs often represent ideal sources for sampling groundwater and its dissolved constituents as they are continually flushed. Care must be taken to ensure that sampling is done as close as possible to the point of discharge in order to minimize the effects of atmospheric contamination and degassing. Springs discharging from alluvial materials can allow atmospheric contamination of groundwaters before they can be sampled. This is why some groundwaters can contain geochemically forbidden combinations, such as H_2S and O_2.

Wells and piezometers provide the only means of acquiring groundwater samples from discrete aquifer horizons. It is important that drilling and well completion have been carried out with the hydrogeological considerations of sampling in mind, i.e. all drilling fluid has been removed from the screened zone. In order to ensure maximum sample quality, it is recommended to remove several well volumes prior to sampling. Note that some drill additives can sorb to borehole walls or penetrate the aquifer under the hydraulic pressures of drilling and therefore, removal of these additives can become virtually impossible. This is especially true for organic additives, in which case DOC sampling should not be attempted. If a reliable sample is needed, then the drilling and sampling protocol must be designed accordingly.

Domestic wells and water distribution systems may provide the only sources of groundwater samples in many studies with no budget for drilling. A vast groundwater monitoring network in the form of domestic, industrial and municipal wells may be at your disposal. Groundwater investigations are often carried out in areas with a tradition of groundwater exploitation, such as agricultural watersheds or arid and semi-arid regions. In such cases, your sampling points are pre-installed and ready to go, and most well-owners are generally pleased to participate in your study. Such a network of wells is less than ideal because you have no control over the horizons tapped by a given well, or over the completion — many will be open hole, possibly tapping several horizons. Nonetheless, with a careful review of water well completion reports (these are required by law for most regions), the most suitable wells can be selected for sampling.

Sampling private or municipal wells presents the problem of access as most well heads are sealed, and the nearest point of sampling may be well downstream of the well intake. Consider the impacts of sampling from, for example, the tap on the side of a house. The plumbing system for the house should have no impact on the $\delta^{18}O$ and δ^2H, but may affect pH and ^{13}C in the carbonate system. A water softener will also affect chemistry, and trace elements will certainly be affected by the plumbing system.

In general, the following points should be considered for well sampling:

(1) Get the well completion date. Examine the well log of geology, hydrostratigraphic units and well completion details. Determine the type and depth of pump installed.

(2) Identify the zone of water intake to be sampled and attempt to exclude contributions from other zones.

(3) Static (observation) wells should be flushed before collecting samples. During sampling, the pump or bailer intake should be as close to the screened zone as possible.

(4) Water supply wells in use need no flushing, although samples should be collected from a point near the well head. If they can only be sampled from a tap somewhere downstream of the distribution system, find out which types of water treatment and storage systems are installed.

Getting water from the well

Since prehistoric times, our ingenuity to get water from a well has evolved from hauling it up in goat skin bladders or with an Archimedes' screw to early suction pumps and modern high rate turbine pumps. Most domestic water wells have electric submersible pumps installed that provide positive-pressure water with no atmospheric contact. In shallow-dug wells, mechanical lifting is still practiced. Samples collected with such devices, usually from young groundwater systems, are perfectly acceptable for most analyses, provided exchange with the atmosphere can be minimized.

Positive pressure systems should be used in investigations involving DIC, H_2S, rare gases, or methane, to minimize degassing before sampling. Suction pumps can be used without concern for analysis of ^{18}O and ^2H in water, or isotopes in sulphate, nitrate or chloride. However, negative pressure pumping causes CO_2 and other dissolved gases to escape from solution, which affects pH, DIC, ^{13}C and other isotope analyses. Pressure is maintained at or above atmospheric level by submersible pumps, gas-driven pneumatic pumps and even bailers. Some hand-held and hand-powered pumps can even deliver samples at the formation pressure, well above atmospheric pressure, and should be considered for sites where organic or noble gases are to be sampled. Within national nuclear waste repository studies, very sophisticated sampling devices for positive pressure collection have been constructed and, in some countries, the designs are available for copying and general use.

Deuterium and oxygen-18

Sampling for ^{2}H and ^{18}O in water is simple, since neither is measurably affected by chemical and biological processes. Evaporation through leaky bottle caps or partially filled bottles may pose a problem, as it may enrich the residual water. Samples that contain volatile organic compounds which would escape during analysis will have to be marked and pretreated in the laboratory. Such gases are problematic in automated, on-line analytical systems.

SAMPLING SYNOPSIS:	^{18}O and ^{2}H in groundwater
VOLUME:	100 mL, minimum 10 mL.
EQUIPMENT:	No special requirements.
CONTAINER:	Plastic screw cap bottle (HDPE, LPE, PP).
PRESERVATION:	Tightly sealed to prevent evaporation. No filtering required.
ANALYSIS	Both isotopes are measure on isotope ratio mass spectrometers (IRMS)
	δ^2H: This analysis is done on hydrogen gas obtained on-line or off-line (in glass "breakseal" tubes) through high-temperature reduction of water (3 µL) on metal — either specially prepared zinc, chromium or uranium. Hydrogen-bearing compounds such as organics will also react. An isotope exchange technique that involves exchange of hydrogen gas with the water sample in the presence of platinum powder as a catalyst is in preparation by the IAEA.
	δ^{18}O: Analyses are done on carbon dioxide that has equilibrated with the water sample at a constant temperature (usually 25°C). Between 1 and 3 mL water are added to reaction vessels, then evacuated by pumping through capillary tubes (nonfractionating) connected to the mass spectrometer.

Tritium

The sampling procedure for ^{3}H is as simple as that for ^{2}H and ^{18}O in water. For direct analysis, one 125 mL bottle can be used for all three isotope determinations, with no filtering or special preservation. The electrolytic enrichment method usually requires 250 mL to allow a repeat measurement. More precautions are taken when low detection limits are required. Helium-3 ingrowth and gas proportional counting allow detection to ^{3}H/H $< 10^{-19}$. In this case it is best to use tight-sealing glass bottles and, for maximum avoidance of atmospheric contamination, the bottles should be flushed with argon.

One potentially serious source of contamination includes the tritium-emitting luminous dials that can be found on some watches and compasses. Watches are a particular problem because the tritium emitted is partially absorbed into the body and the biological half-life is 10 days.

Furthermore, samples that contain ^{222}Rn (also a β^- emitter) must be degassed before analyses.

SAMPLING SYNOPSIS: Tritium

VOLUME:
- Direct tritium (±8 TU)   50 mL
- Enriched tritium (±0.8 TU)   250 mL
- Ultra low level (ULL: ±0.1 TU) 500 mL, flushed with argon
- ^{3}He ingrowth (±0.1 TU)   50 mL, flushed with argon

EQUIPMENT: Thief sampler (bailer) or pump, minimize contact with air.

PRESERVATION: Tightly sealed to avoid contact with air. No filtering required.

TECHNIQUE:
- Rinse bottle completely with sample, fill completely and cap.
- ULL — Glass bottle should be perfectly dry and, if possible, flushed with high purity dry argon to remove adsorbed moisture. Fill bottle to brim once, no rinsing and cap tightly.

ANALYSIS
- A variety of measurement techniques have been developed for tritium determinations, the most simple one being Liquid Scintillation Counting (LSC) of β^- decay. Liquid scintillation counting can be also done on samples enriched by electrolytic reduction of the sample volume, which concentrates the heavy isotopes, ^{2}H and ^{3}H, in the residual water.
- Ultra Low Level counting (ULL) is done with gas proportional counters usually on propane or H$_2$ prepared from electrolytically enriched samples. Again, dissolved organic hydrogen compounds will affect the analysis.
- Tritium can also be measured by the ^{3}He in-growth method where the water sample (~ 250 mL) is degassed on a vacuum line, sealed in glass and stored for several months. The ^{3}He ingrown from ^{3}H decay is then measured by mass spectrometry (see Chapter 1 or 7 for decay of ^{3}H).

Carbon-13 in DIC

Carbon-13 analyses on dissolved inorganic carbon are a requirement for the interpretation of radiocarbon data but are also important for the interpretation of geochemical and microbiological processes in groundwater. Sampling can be done in association with radiocarbon samples (see below), or on its own for ^{13}C studies.

Sampling for ^{13}C$_{DIC}$ is most simply done with dense plastic or glass bottles. Absolutely tight caps are required and biological activities in the sample bottles must be prevented. The latter is achieved if some Na-azide (NaN$_2$) is added to the sample. Surface waters that contain low levels of DIC (equilibrium with the atmosphere) and higher levels of DOC are more prone to modification by microbiological activity than groundwaters, which normally have high DIC and low DOC.

Traditionally, DIC was sampled for ^{13}C measurements by precipitation from solution in the field as an alkaline BaCO$_3$ slurry, in the same way that conventional ^{14}C analyses are sampled (see below). However, for the small samples needed for δ^{13}C measurements, this method is cumbersome and prone to contamination by atmospheric CO$_2$. A far better way is to submit water samples, preserved against bacteriological activity, to the lab where direct acidification under vacuum is done to extract the CO$_2$.

SAMPLING SYNOPSIS: ^{13}C in DIC

VOLUME

Water volume depends on the concentration of DIC. A field alkalinity measurement will give you this. About 5 mg HCO$_3^-$ is required for an analysis. The following guide for sample volumes is conservative, and should allow for repeat analyses.

Alkalinity (mg/L HCO$_3^-$)	Volume of sample (mL)
10	500
50	100
100	50
200	25

EQUIPMENT

- Bottles: use amber glass or heavy plastic bottles with tight caps, preferably with a silicon septum insert to permit extraction with a syringe in the laboratory. Alternatively, use a double valve steel or glass cylinder of sufficient volume to hold water sample with at least 5 mg of carbon.
- Thief samplers or down-hole pumps can be used, if degassing does not occur during sampling. Exchange with air has to be avoided.

PRESERVATION

- None, if the samples are kept dark and cool to be analyzed within at most a few days after sampling.
- Na-azide should be added if the samples are kept for more than a week or two, or if you anticipate high levels of DOC (>1 to 2 mg-C/L).

TECHNIQUE

- Fill clean and rinsed bottles with a hose at the bottom of the bottle to avoid splashing but allow overflow. If a flow-through, double-valved sampler is used, flush with at least two volumes before closing.
- If necessary, add to sample bottles before filling a minor amount of Na-azide (less than the tip of a spatula).
- Close and keep cool and out of light.

ANALYSIS

Samples are injected by syringe into pre-evacuated glass vessels containing acid (usually H$_3$PO$_4$ in sidearm) that are mounted on a vacuum line. CO$_2$ so released is frozen in a liquid N$_2$ trap, cryogenically purified and sealed in a glass breakseal, or fed directly to mass spectrometer.

Radiocarbon in DIC

Only about 10^{-12} of all carbon in the biosphere and atmosphere is radiocarbon, and so substantial sample sizes are required to measure its activity with conventional β^--decay counting equipment. As the activity of ^{14}C is extremely low, 0.5 to 3 grams carbon are usually needed for measurement by gas proportional counting (GPC) or liquid scintillation counting (LSC). Much smaller samples (less than 10 mg C) can be measured by accelerator mass spectrometry (AMS). Radiocarbon measurements are expressed as "percent modern carbon" or "pmC," where "modern" refers to wood grown in 1890 in a pollution-free environment.

Sampling for LSC or GPC: The necessary amount of carbonate must be stripped from up to 25 L of water. Table 10-2 shows the yield of DIC (g C) from a 25L jug for various alkalinities. The DIC is stripped from solution as a $BaCO_3$ slurry under high pH conditions. Some laboratories recommend stripping DIC as $SrCO_3$. Sr^{2+} is not toxic like Ba^{2+}, but it is lighter and does not settle out as fast.

While sampling, some precautions have to be taken. The sampling jug should have a spigot at the base to drain the water after precipitation. Splashing during filling must be avoided to prevent exchange with atmospheric CO_2. If the water is presumed to be very old, these precautions are mandatory. For younger waters, errors due to less careful sampling are often negligible, especially for groundwater with $P_{CO_2} >> 10^{-3.5}$ atm, as such waters degas during discharge. In any case, allow the jug to overflow by at least half its volume when filling.

The water is made alkaline to convert all DIC species to carbonate (CO_3^{2-}), which precipitates with the dissolved Ba^{2+} added as $BaCl_2 \cdot 2H_2O$. The distribution of carbonate species is strongly pH dependent and CO_3^{2-} dominates above pH 10.3 (see Chapter 5). This pH is achieved through addition of a strong base (generally NaOH) and ensures total precipitation of DIC.

The reactions follow as:

Raising the pH to >11: $HCO_3^- + NaOH \rightarrow CO_3^{2-} + Na^+ + H_2O$

Precipitation of $BaCO_3$: $CO_3^{2-} + BaCl_2 \rightarrow BaCO_3 + 2Cl^-$

Sodium hydroxide (NaOH) is added in concentrated form and must be carbonate-free. It is commercially available and can be used undiluted as a 10 N solution, but it can also be prepared in the laboratory by adding sufficient solid NaOH pellets to distilled water to obtain an approximately 15 N solution. At this concentration, sodium carbonate is no longer soluble and will settle out as a white precipitate. The viscosity of such NaOH solutions is very high and it will take at least 1 week before all of the precipitate has settled. Aliquots are decanted in small, tight plastic bottles to be used for at most 1 to 2 days. Note: if left open to the atmosphere, it quickly becomes contaminated with $^{14}CO_2$. Usually 30 mL of 10 N NaOH solution will be required to raise the pH above 11.

The amount of barium chloride used should be kept to a minimum. It is toxic both as dust and in dissolved form. As sulphate will be stripped from solution as $BaSO_4$, the amount of barium used must be increased accordingly. Table 10-2 shows the minimum amount of barium chloride required as a function of sample alkalinity and sulphate content. An excess of about 50% should be used.

NOTE: NaOH is highly corrosive and extreme care has to be taken in its handling. It is not accepted on commercial flights, and must be shipped in advance and packaged according to airline protocols. If for any reason carbonate-free NaOH cannot be obtained or preserved, the precipitation of the carbonate sample must proceed without it. This may result in an incomplete precipitation of the dissolved carbon but would not affect the ^{14}C content. Such samples should be analysed for ^{13}C only in order to normalize the ^{14}C. Separate samples for "interpretable" $\delta^{13}C$ must be collected.

Table 10-2 Carbon-14 sampling, water volume and BaCl$_2$·2H$_2$O requirements for a standard 25-L container

SO$_4^{2-}$ content →	50 mg/L	100 mg/L	200 mg/L	500 mg/L	1000 mg/L	Grams C from 25-L jug
Alkalinity (mg/L HCO$_3^-$)	Minimum BaCl$_2$·2H$_2$O required (grams per 25-L jug)					
10	4	7	14	33	65	0.05
25	6	9	15	34	66	0.12
50	8	11	18	37	69	0.25
100	13	16	23	42	74	0.49
150	18	21	28	47	79	0.74
200	23	26	33	52	84	0.98
250	28	31	38	57	89	1.23
500	53	56	63	82	114	2.46

The NaOH should be added before the BaCl$_2$·2H$_2$O since many groundwaters have a very high equilibrium P$_{CO_2}$ (10^{-2} atm or greater) and, if solid barium chloride is added first, then the "salting-out effect" (due to increased ionic strength) will cause a significant and almost immediate loss of carbon dioxide gas from the sample. However, it is imperative that both NaOH and barium chloride be added rapidly one after the other because any alkaline solution exposed to air will absorb atmospheric carbon dioxide.

After the chemicals are added, the jug is tightly closed and rolled to mix its contents. Thereafter, >4 hours settling time should be allowed. When draining the supernatant water, the incoming air should be stripped of its CO$_2$ to avoid atmospheric contamination. A CO$_2$ trap consisting of Ascarite® (CO$_2$-absorbing compound) can be placed at the air intake on top. This can be made by filling a >50 mL syringe with a layer of glass wool, then Dryerite® to trap moisture, ~ 20 mL Ascarite, then more Dryerite and a cap of glass wool to hold it all in. Put the syringe through a rubber bung that fits the top of the jug. The excess Ba^{2+} can be precipitated from the water and the hydroxide neutralized by adding some sulphuric acid, to be environmentally friendly.

With the clear water drained off, a 1-L bottle is fitted with a rubber bung to the top of the jug and the sample swilled into it. Be sure to estimate correctly the amount of sediment in the jug before swilling it into a bottle. The bottle is tightly sealed and ready for shipment to the laboratory.

SAMPLING SYNOPSIS: DIC–^{14}C for conventional analyses (LSC, GPC)

VOLUME: 0.5 to 3 g C stripped from 25 to 50 L water as BaCO$_3$.

EQUIPMENT:
- 25-L (or 50-L) jug for precipitation of sample, pre-rinsed with acid and deionized water.
- 1 L sample shipment bottle.
- Barium chloride (BaCl$_2$·2H$_2$O reagent grade).
- 10 N NaOH (carbonate free).
- Ascarite trap to fit precipitation jug.
- Rubber bung to fit jug with large center hole to fit over neck of the transfer bottle.

TECHNIQUE:
- Make field measurements of alkalinity and SO$_4^{2-}$ (Table 10-2).
- Fill jug with sample water, prevent splashing, allow overflow.

	• QUICKLY add the necessary amount of NaOH (about 20-30 mL of 10 N) and the premeasured amount of $BaCl_2 \cdot 2H_2O$. • Roll to mix. Allow to settle for a minimum of 4+ hours. • Drain, if possible, with Ascarite® trap. This water should not be drained off near a water source. • Transfer slurry to shipment bottle, Cap and allow further settling, then pour off excess water.
ANALYSIS	Conventional measurement is done in liquid scintillation counters (LSC) on samples converted to benzene (C_6H_6), or by gas proportional counting on CO_2 or C_2H_2 produced from the $BaCO_3$. Accuracy for both is about ±0.3 pmC. For survey purposes, direct counting on CO_2 absorbed by Carbasorb® is easier and requires less carbon, but has an accuracy of ±5 pmC.

Sampling for AMS: Considering the rather involved procedure to strip sufficient DIC from groundwater for conventional analysis, it is of little wonder that accelerator mass spectrometry (AMS) has gained so much interest. For AMS, only about 5 mg carbon is required. This reduces the sample volume for typical groundwaters from up to 100 L for LSC to <500 mL for AMS. This tremendous advantage in sample size allows extraction in the laboratory by the same method as that for $\delta^{13}C_{DIC}$ (see above). It also minimizes contamination by reducing the amount of sample "handling" in the field. We can also consider analysing samples that are not accessible by conventional methods, such as brines with <10 mg/L HCO_3^-, and high sulphate discharging at low flow rates from deep geologic settings, or porewaters in low-permeability sediments.

SAMPLING SYNOPSIS:	**DIC-^{14}C for Accelerator Mass Spectrometry (AMS)**

VOLUME:	About 5 mg C are required for standard analysis.
EQUIPMENT:	• Same as for $^{13}C_{DIC}$ extraction. • Container: acid- and deionized water- rinsed glass bottle (dried) with a silicon septum in cap (250 mL to 1 L), heavy-wall plastic or glass bottle, double-valve steel or glass cylinder. • NaN_2 for waters with >1 to 2 mg-C/L DOC.
TECHNIQUE:	• Samples must be collected without air contact and kept dark and cool. • If DOC is present in significant quantities, biological activity should be killed with NaN_2.
ANALYSIS	DIC is extracted from solution by acidification under vacuum, in the same manner as for $\delta^{13}C$ (above). The purified CO_2 is sealed in a glass breakseal for conversion to black (elemental) carbon at the AMS facility. Accelerator mass spectrometry is basically an atom counting method. The carbon, mounted on Fe-filaments in the source of the mass spectrometer, is ionized and accelerated through a magnetic field to separate the three main isotopes. Accuracy is slight less than for high-precision gas counting, yet is more than adequate for hydrogeological purposes.

Carbon-13 and ^{14}C in DOC

Dissolved organic carbon has gained interest in hydrogeological studies because of its importance in the geochemical evolution of groundwater, and because DOC from the soil zone can be used for dating. For groundwater dating purposes, it seems that fulvic acids most closely represent the soil component of DOC, and must be separated from higher weight humic acids.

Sampling and preparation of DOC is rather complex and requires some experience. Summaries of the field and analytical methods are provided by Artinger et al. (1995), Aravena (1993), Wassenaar et al. (1991) and Geyer et al. (1993). Humic acid (HA) can be precipitated from a water sample by acidification with HCl to pH 2, whereas the residual fulvic acid (FA) — still in solution — can be separated on an XAD-8 resin followed on-line by a silicalite column to trap at least a portion of the low molecular weight DOC (LMW-DOC). The sum of these three fractions does not necessarily account for all DOC, but collection of >80 % is possible in most waters. Since DOC concentrations in groundwater in general are <<5 mg/L, very large volumes of water have to be treated in the field. For this reason, enrichment techniques are being developed, with the most promising being enrichment by reverse osmosis (Kim et al., 1995). This technique has been field tested and permits rapid processing of the necessary volumes of water in the field and easy treatment and separation in the laboratory.

The separated and cleaned acids are combusted to CO_2, which is either purified for $\delta^{13}C$ determinations or further preparation for ^{14}C-AMS analyses. The ^{13}C determinations done on AMS for ^{14}C correction purposes are not to be used for detailed studies.

Sulphur-34 and ^{18}O in aqueous sulphur compounds

Sampling sulphate in water with no sulphide (H_2S and HS^-) is straightforward. Samples are precipitated as $BaSO_4$ and filtered. However, where dissolved sulphide is present (it can be detected by smell at pH below about 7 to 8), sampling is complicated because oxidation by O_2 from the air will convert the sulphide to SO_4^{2-} and neither component will then be representative of *in situ* conditions. In this case, they must be stabilized or separated in the field.

SAMPLING SYNOPSIS: Sulphate with no sulphide present

VOLUME: Depends on concentration of sulphate in water. About 20 mg $BaSO_4$ is required. The following sample volumes provide a rough guide. Note: 10 g ≅ 1 teaspoon.

Sulphate (mg/L SO_4^{2-})	Volume of sample	$BaCl_2 \cdot 2H_2O$
10	2 L	~100 mg
50	500 mL	~150 mg
100	250 mL	~150 mg
200	100 mL	~100 mg

EQUIPMENT:
- Jug for precipitation of sulphate from low-concentration waters.
- 20-mL bottle for transport, clean of carbonate and residual sulphate.
- Barium chloride ($BaCl_2 \cdot 2H_2O$, reagent grade).
- Dilute HCl and pH paper.
- Filter system (0.45 μ).
- Field kits for alkalinity and sulphate.

TECHNIQUE: Where water volumes of 1 L or less are sufficient (see table above), sulphate can be stripped in the laboratory. Groundwaters with low sulphate contents where precipitation in the field eliminates the need to transport large water samples.
- Measure SO_4^{2-} and HCO_3^- in the sample water, if not known.
- For samples with very low SO_4^{2-} concentrations, ion exchange resins are recommended (e.g. Bio-Rad AG-1-X8 Anion Exchange Resin). These resins must be prepared according to manufacturer's instructions.
- Samples should be filtered (<0.45 µ).
- Lower pH to ~ 4 to 5 to prevent co-precipitation of $BaCO_3$.
- Add barium chloride. Note: barium chloride is toxic so don't inhale it.
- Allow to settle for 3-6 hours, then decant or filter clear water and transfer precipitate into the shipping bottle.

ANALYSIS Analyses of ^{34}S and ^{18}O are done by mass spectrometry (IRMS). $\delta^{34}S$ values are determined on SO_2 generated by high-temperature dissociation of sulphate under vacuum. Some labs prefer conversion to SF_6 to avoid correction for ^{18}O in the SO_2, although the preparation is rather complicated. Measurement of ^{18}O in sulphate is done on CO_2 produced by combustion of sulphate-graphite mixtures at high temperature.

SAMPLE SYNOPSIS: Sulphate and sulphide present

EQUIPMENT As above for sulphate plus:
- Cadmium acetate (CdAc) or ZnAc to precipitate sulphide.
- Tygon tubing to fill bottle out of contact with air.

TECHNIQUE:
- Put a 1 to 2 g of CdAc or ZnAc into a clean 2-L bottle.
- Fill the bottle with the tubing at the bottom of the bottle to prevent air contact.
- As CdAc dissolves, bright yellow CdS forms, indicating even trace amounts of sulphide. Note: cadmium acetate is toxic and must be used with care, especially as far as disposal is concerned. Zinc acetate is a more environmentally safe chemical but forms a less obvious white precipitate.
- Let precipitate settle and decant, or filter. Store as slurry in small bottle.
- Use filtrant for SO_4^{2-} precipitation with $BaCl_2$ as described above.

ANALYSIS
- For analysis of $\delta^{34}S$, the wet CdS is converted to AgS by titration with $AgNO_3$. The silver sulphide is mixed with cupric oxide and combusted to SO_2 under vacuum, for analysis by IRMS.
- The ^{34}S and ^{18}O contents of the $BaSO_4$ are determined by IRMS on SO_2 and CO_2, as described above.

Nitrate and organic nitrogen

Nitrate is now routinely analysed for ^{15}N and ^{18}O in many laboratories (e.g. Bötcher et al., 1990). Before sampling, contact the laboratory for their protocol. ^{15}N can also be analysed in dissolved ammonium (NH_4^+) and organic matter. All analyses are carried out by standard IRMS.

Groundwater samples are generally collected as bulk water samples although in low nitrate waters, ion exchange resins can be used. Water samples must be preserved by killing biological activity. This can be achieved by adding 1 mL/L of chloroform, or by acidification to pH 2 with HCl. Obviously, Na-azide (NaN_2) and nitric acid are not appropriate for preservation.

Less than 1 mmol N is required for ^{15}N. The dissolved nitrate is reduced to NH_3 by addition of Devarda alloy, liberated with MgO, and oxidation to N_2 with lithium hypobromite. The N_2 is then analysed directly on the mass spectrometer against AIR. For measurement of ^{18}O (less than 1 mmol NO_3^- required) the sample is passed through a column to exchange cations for H^+. PO_4^{3-} and SO_4^{2-} are removed with $BaCl_2$ and the filtrate then dried and cooked at 550°C with $Hg(CN)_2$ to produce CO_2 for mass spectrometer analysis.

Chloride

Sampling for chlorine isotope analyses is extremely simple since it involves only the collection and shipment of one to several litres of water to the laboratory. Chloride is a conservative species in water, and so no treatment or preservative is required. Two analyses can be considered, although the number of laboratories undertaking either is limited.

^{36}Cl by AMS: In the early days of ^{36}Cl research, ^{36}Cl was measured by β^- counting, which required extraction from huge quantities of water. It is now measured by AMS, which requires less than 25 mg Cl^-. Chlorine is separated from other dissolved constituents, especially sulphur, since ^{36}S interferes with ^{36}Cl determinations. It is usually precipitated as Ag_2Cl_2 for preparation of the target to be mounted in the source of the accelerator.

^{37}Cl by IRMS: The measurement of stable chlorine isotopes ($^{37}Cl/^{35}Cl$) is now carried out using methyl chloride gas (CH_3Cl), which is synthesized by reaction of methyl iodide (CH_3I) with AgCl. The sample must be filtered to 0.45 μm. An excess of $AgNO_3$ is then added to precipitate AgCl. This precipitate is filtered from solution for conversion to methyl chloride.

Uranium series nuclides

The radionuclide ^{238}U has an exceedingly long half-life (4.49 · 10^9 yrs) and decays in a complex sequence of α- and β^--emissions and through a series of shorter-lived daughter elements to stable lead, ^{206}Pb (Fig. 8-25). Dating groundwaters with isotopes in this decay chain is limited by the complexity of the aqueous geochemistry that controls the movement of these nuclides. Particular attention is given to the activity ratios: $^{234}U/^{238}U$, $^{230}Th/^{234}U$, $^{230}Th/^{232}Th$, $^{226}Ra/^{230}Th$, and $^{222}Rn/^{226}Ra$.. The geochemical conditions must also be studied. The solubility of uranium increases under oxidizing conditions, and in waters with high HCO_3^- and high P_{CO_2}. Radium solubility increases with salinity, and thorium is insoluble.

Sampling for U-series isotopes is nearly as complicated as their analysis and interpretation. Only a few micrograms of uranium are required for analysis by α-spectrometry. Even less is required for analysis by thermal ionization mass spectrometry (TIMS). Samples should be filtered to assure that no detrital material contributes to the analysis, then acidified with HCl to prevent any uranium from plating out on the walls of the bottle. Nitric acid is to be avoided, as it can oxidize organics or other species. Consult the laboratory for their sampling protocol.

Water in the Unsaturated Zone

As seen in Chapters 4 and 5, a great deal can be learned about recharge and groundwater geochemistry from the unsaturated zone. Sampling in this zone involves both water and gas samples. The latter will have to be taken *in situ* by extracting gases from probes placed to the depth of interest. Soil moisture can be obtained by either taking samples of the soil itself (with augers, split spoon sample or coring, depending on the medium) or by installation of suction lysimeters with porous porcelain tips.

Soil samples that cannot be cored and transported within the capped core barrel must be immediately packed into high-density plastic bags or, better, in large-mouth plastic bottles with tight caps. The bottles should be filled in order to minimize evaporation into the open air. Extraction of the fluid can be done by squeezing in a special arrangement with non-wetting liquids, by centrifuging or by heating under vacuum. The method of choice depends on water content and grain size. Note that water samples extracted by centrifuge or squeezing can be used for chemistry and isotopes of solutes, while heating under vacuum extracts only pure water.

Lysimeters are porous porcelain cups to which negative pressure can be applied in order to draw soil moisture into the cup. They can also be used below the water table in low-permeability sediments. However, suction lysimeters can draw water only above a certain soil moisture content, which will be determined by the type of the cup used (see Freeze and Cherry, 1979). Water collected in lysimeters can be used for chemistry and solute isotopes, although the negative pressure can affect the carbonate system. The use of lysimeters is described in most textbooks on groundwater (Freeze and Cherry, 1979; Fetter, 1988).

Precipitation

Precipitation samples are required in most hydrogeological studies that use ^{18}O, ^{2}H and ^{3}H analyses. For some studies, one can "adopt" the appropriate meteoric water line from one of the global network of IAEA monitoring stations. In areas where IAEA monitoring is inadequate (i.e. areas with strong local variations in precipitation such as regions of high relief or complicated weather patterns) or where greater detail is needed, a program of precipitation monitoring must be established. Note that while infrequent "samples of opportunity" may help constrain the meteoric signal, consideration should be given to setting up sampling stations.

Rain samples for ^{18}O, ^{2}H and ^{3}H

Monitoring programs for precipitation generally integrate event-samples on a monthly basis, giving just twelve samples per year. Daily rain gauge samples are poured into a monthly sample bottle and sealed. The cumulative amount is noted for weighting calculations. Event samples can also be measured, but are less meaningful for groundwater studies. The composition of rain evolves during a storm, and so, unless it is this variation that is being studied, the rain sample must integrate the whole storm. A simple sample of roof-runoff or an individual hailstone does not characterize a rain storm and is not representative of the potential recharge. Local meteorological stations at airports or other government sites can often be used for monthly monitoring. Contact the station personnel and supply them with a box of pre-labelled 25-mL bottles (250 mL if you want ^{3}H, too); they may be happy to oblige.

Temporary samplers may be set up in remote areas. However, if they cannot be emptied immediately after a rainfall, then precautions against evaporation are essential. In remote areas, a funnel-and-jug type sampler can be set up and left unattended for several days or even weeks if provisions are made to prevent heating by the sun. However, evaporation may still be a problem and, therefore, a layer of paraffin or silicon oil (at least 2-mm thick) should be poured into the collection bucket. This floats over the sample and will prevent evaporation after the rainfall. When pouring off the sample; use a second bottle to minimize the liquid paraffin getting into the sample storage bottle. For this reason, it is necessary to test in the laboratory not only the physical separation, but also the solubility of the liquid to be used since even partial solubility can affect the measurement of $\delta^{18}O$ and especially δ^2H.

In some climates, precipitation may occur from fog or mist that condenses on vegetation. Samplers should be designed to intercept this "occult" precipitation. Such an intercepter can be made from a triangular piece of galvanized sheet metal or nylon mesh (approximately 1 m in length per side) and wired, point down, between two stakes and facing the dominant wind direction to intercept the mist and collect it in a glass bottle beneath the lower point. As long as the air moisture is close to 100%, no liquid paraffin is required.

Sampling, storage and analyses have been described above under "groundwater."

Snow and ice ^{18}O, ^2H and tritium

Frozen precipitation is collected for many isotope studies in northern climates where either snow melt or ablation of glaciers represents a major input to groundwater systems. Depending on the accessibility, such samples can be quite easily collected in sealable plastic bags or similar containers. Once melted, the water can then be transferred to a more transportable plastic bottle.

To be representative of the storm event, snow should be sampled shortly after it has fallen. Sublimation, recrystallization, partial melting, rainfall on snow and redistribution by wind are all processes that alter the primary isotope content of the snowfall.

Glaciers and thick snow packs can be drilled using hand augers to retrieve cores of ice and snow. Sampling more than a few metres depth may require power-assisted drills. Samples are sealed in plastic in the field, then later can be subdivided, melted and bottled. No preservation is needed for these isotopes of water, although evaporation must be avoided. If bottles are not full, store them at 4°C to minimize vapour losses when opened for analysis. Analytical methods are described above under "groundwater."

Gases

Soil CO_2

Soil carbon dioxide is where geochemical evolution in the carbonate system begins. It is often appropriate to measure its partial pressure and ^{13}C contents, particularly in studies of radiocarbon ages. P_{CO_2} can be measured with commercially available Draeger® tubes (colour indicator of volumetric CO_2 content for fixed volume of air passed through) or by sampling the gas and analyzing it by gas chromatography. For isotope analysis, the gas is extracted at specified depths through a thin stainless steel probe, and collected either in evacuated flasks or as barium carbonate precipitate in a $BaCl_2$-NaOH solution trap. Use the latter with caution, as strong isotope effects occur if not all the CO_2 is trapped. After purification (or acidification of the $BaCO_3$), the gas is analyzed directly by IRMS for both ^{13}C and ^{18}O. Radiocarbon can also be determined on such samples to determine what $a^{14}C_o$ truly is for a given recharge area.

Gas in groundwater

Gases are an integral part of most geochemical systems, and are particularly important in redox reactions. Gas analyses can provide additional insights into the isotopic and geochemical processes taking place in the subsurface, and must be considered in thermodynamic models. Major gases should be considered. Helium and other noble gases may also be important. In a natural spring discharge, as the water approaches surface pressure, degassing will occur and this separate gas phase can be sampled. In most confined aquifers, a gas phase does not naturally exist, although degassing can occur after drilling and pumping. Sampling gas for chemical and isotopic analyses can be done on both dissolved and exsolved phases.

Downhole samplers allow samples to be collected at formation pressures with minimal contamination. Note that gas samples are usually collected as water samples. The gas phase is then separated in the laboratory by flushing the water sample with helium, which carries the sample gas through a chromatograph. *In situ* pressures should be maintained for such samples to prevent losses prior to sample preparation and analysis.

A free gas phase accompanying flow from artesian wells and springs can be sampled, although it may not fully reflect the *in situ* gas composition. Nonetheless, it provides an approximate gas composition and is suitable for most isotope analyses (e.g. CH_4 and H_2). Sampling a separate gas phase in a spring can be done quite simply, providing a small pool exists at the discharge point and gas bubbles are forming. A well-rinsed bottle is filled with sample water and inverted under water in the spring vent. Fit a funnel into the neck of the bottle to collect the rising gas bubbles. After displacing at least 50% of the volume with gas, remove the funnel and cap the bottle under water. The sample is transported upside-down to keep the cap wet and minimize diffusion. Reversing the procedure in the laboratory in a container of boiled, deionized water will allow extraction of a portion of gas for analysis. Alternatively, put on a septum cap in the field or in the laboratory. If flowing boreholes exist (including those drilled from underground levels in a mine), a simple flow-through arrangement based on the displacement technique can be installed. You can purchase high quality gas bottles, but an empty soda pop bottle with a screw-on cap works great.

- *Major gases for volumetric ratios and isotopes (N_2, O_2, CO_2, CH_4)*: Major gases can be sampled in glass or thick-walled polypropylene bottles. They can also be collected in glass soda pop bottles with screw-on caps. Near-formation pressures can be maintained using a polypropylene tube (~12-mm diameter) attached to the discharge pipe of a well, that is kinked over and crimped at both ends after filling under pressure. This allows extraction in the laboratory with a syringe through the wall. Major gases are stripped from solution with He, which serves as the carrier gas in the gas chromatograph. The isotopic composition of the separated gases can also be measured (usually ^{13}C, ^{18}O and ^{2}H in CO_2 and CH_4). This is simplified by continuous flow mass spectrometry, which allows entry of the gas as it exits the GC column (or after combustion in the case of CH_4).

- *Hydrogen (^{2}H)*: This gas is highly diffusive, and must be sampled in glass or stainless steel. The best arrangement is a vessel with stopcocks at both ends that can be flushed under positive pressure and closed. The method below using copper tubes for He is also appropriate for H_2. In the laboratory, the H_2 gas is purified and analysed by IRMS for ^2H.

- *Helium ($^3He/^4He$)*: This is also a highly diffusive gas and is properly sampled only in soft copper tubes (10-mm diameter) that can be cold welded using refrigerator clamps. A crib is used to hold a 50-cm tube with the clamps partially tightened at both ends. The copper tubing is connected to the borehole and the system is pressurized by restricting the flow through the tubing. After several sample volumes have been discharged to remove adsorbed gases on the copper, the clamps at the exit and then the entrance to the tubing are closed. The sampling equipment and protocol is usually provided by the laboratory doing the analyses. In the laboratory, He is separated from other gases cryogenically, by adsorbing on activated charcoal at liquid nitrogen temperatures, and then measured by IRMS using a mass spectrometer with large-radius flight tube capable of measuring the isotopes of noble gases.

- *Argon (^{39}Ar)*: Due to the low activity of ^{39}Ar in the atmosphere and, hence, in groundwaters, argon must be degassed from a large volume of water in the field. Normally, a 2-L (STP) sample is required, which involves extraction by vacuum-degassing or boiling up to 15 m^3 of water. Specialized equipment and trained personnel from the analytical laboratory, are required. In the laboratory the gas is purified of water vapour, CO_2, O_2, N_2 and other trace gases using physical and chemical methods. Measurement is carried out by counting β^- decay events in a high-pressure gas proportional counter over a period of about 1 month.

- *Krypton (^{81}Kr and ^{85}Kr)*: Krypton has two isotopes, ^{81}Kr and ^{85}Kr, that can be used as groundwater age dating tools. Analysis of ^{85}K, a β^- emitter, is carried out on gas extracted from about 100 L water. It is vacuum-extracted in the field under sealed conditions to prevent atmospheric contamination, and measured by proportional counting. By contrast, the extremely low concentrations of ^{81}Kr require measurement by counting atoms with AMS.

Geochemistry

The importance of understanding the geochemical evolution of groundwaters as a basis for isotope hydrogeology cannot be overstated. Interpretations of stable and radioactive isotopes in hydrogeological systems must be complemented by an evaluation of the inorganic and organic geochemistry of groundwater. A variety of special sampling and preservation techniques are required for the analysis of major and certain trace species in natural waters. The following discussion outlines some methods and precautions.

A general practice in sampling water for chemical analysis is to filter the sample through a 0.45-micron pore diameter filter. This will remove most bacteria, almost all suspended clays and a proportion of iron and manganese oxyhydroxides, but will not retain viruses or some organic molecules such as fulvic and humic acids. Filtering is required to assure that the laboratory analysis represents dissolved species (which take part in most geochemical reactions and are used in chemical equilibrium equations) and not suspended constituents, which may be contributed from the wells.

Field measurements

Geochemical studies are based on the measurement of the inorganic constituents or species in a groundwater and a series of parameters that control the interactions of these species. Certain parameters such as temperature and pH are difficult to preserve during storage and should be measured in the field. A synopsis of field measurement protocols follows on page 288.

Temperature: The temperature of groundwater is a fundamental measurement and is required in all thermodynamic calculations and models. It is also needed to correct EC measurements and for calibrating the pH meter. It must be measured as close to *in situ* conditions as possible. Use a standard glass thermometer, or the temperature probe that is a component of most pH meters.

Electrical conductivity (EC): Electrical conductivity is proportional to the quantity of dissolved ions present in solution and can provide a rough idea of the total dissolved solids (TDS). For most groundwaters, the EC value, in µS/cm corrected to 25°C, is about 50% greater than the TDS expressed as mg/L, and can be estimated according to:

$$TDS = A \times EC \text{ (µS/cm)}$$
$$A \cong 0.55 \text{ in bicarbonate waters}$$
$$\cong 0.75 \text{ in high sulphate waters}$$
$$\cong 0.9 \text{ in high chloride waters}$$

pH: A reliable pH measurement is one of the most important field parameters to be measured, and must be made with care and patience. pH is an expression of the negative log of H^+ activity (pH = $-\log [H^+]$). It is fundamental to thermodynamic calculations, and to the interpretation of carbonate and $\delta^{13}C$ data. In natural waters, pH is generally between about 6.5 and 8. Measurement at a sampling site requires the calibration of the meter with reference buffer solutions of known pH according to the recommended protocol in the synopsis below.

REDOX potential: Redox measurements provide a measure of the electromotive force of a water, or the relative dominance of oxidized vs. reduced species in solution. Electromotive potential (E) is expressed in volts, relative to the standard hydrogen electrode (Eh), which by this convention has zero potential (Eh = 0 volts). It is generally measured with a silver-silver chloride electrode, which has its own standard potential, E_R.

As electrodes are simply measuring an electromotive force, there is no calibration. The electrode can, however, be checked against a solution of fixed Eh to determine whether it is functioning. Two solutions with fixed Eh are Zobell's (ferric-ferrous cyanide solution) and quinhydrone [pH 4 and 7 buffers saturated with quinhydrone powder (~0.2 g/100 mL]:

Zobell's Solution (Eh = 0.430 V @ 25°C)	Quinhydrone solution
0.0033 M K^+-Fe^{2+}-CN^- solution	pH 7 solution = 86 ± 20 mV
0.0033 M K^+-Fe^{3+}-CN^- solution	pH 4 solution = 263 ± 20 mV
0.01 M K^+/Cl^- solution	Δ3 pH (difference between the two solutions) = 177 ± 4 mV

Alkalinity: HCO_3^- and CO_3^{2-} concentrations are measured by an alkalinity titration. In most natural waters, total alkalinity can be expressed as bicarbonate concentration. CO_3^{2-} only becomes an important component of DIC above pH 8.5. Other bases can act as proton acceptors, and so alkalinity is defined as:

$$\text{Alkalinity} = m\text{HCO}_3^- + m\text{CO}_3^{2-} + m\text{H}_3\text{SiO}_4^- + m\text{H}_2\text{BO}_3^- + m\text{HS}^- + m\text{OH}^-$$

A variety of organic compounds and colloids can also contribute to alkalinity, although HCO_3^- remains the major contributor in most natural waters. It is determined by titrating with an acid (generally H_2SO_4) to an end point near pH 4.3. Carbonic acid is a weak acid, and so the titration end point increases to about pH 5 in low alkalinity waters. The volume of acid (and normality N) required is proportional to the alkalinity, which is most often expressed as equivalent concentration of HCO_3^-:

$$\text{Alkalinity (mg-}HCO_3^-/L) = \frac{\text{mL of acid} \times N}{\text{mL of sample}} \times 61{,}000$$

Field measurements are greatly aided by using field kits (e.g. Hach, Merck or others) that replace pipettes, burettes and bottles of acid with a microtitrator and syringes of acid. The titration end-point can be identified with a colour indicator (usually Bromcresol), but it is better to use a pH meter and record about 5 to 10 readings below pH 6 (Fig. 10-1).

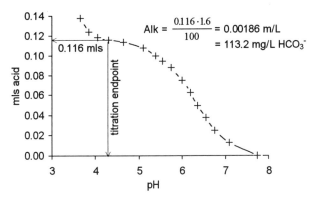

Fig. 10-1 Field titration curve using 1.6 N H_2SO_4 acid and 100- mL sample. The titration end-point is identified by inflection of titration curve.

It is recommended to perform an alkalinity titration in the field, although many will argue that alkalinity is conservative, even if degassing or calcite precipitation takes place. Nonetheless, reactions such as oxidation of Fe^{2+}, degassing of H_2S or equilibration with atmospheric CO_2 can have an effect. Field titrations with commercially available kits are simple to do and rival laboratory bench auto-titrators in accuracy.

Sulphate should be measured if samples are collected for either sulphate or carbonate isotope analyses, to ensure that sufficient sample is collected, and for calculation of the amount of barium chloride needed to fully precipitate the sulphate and carbonate. A field kit can be used. Alternatively, a simple test for sulphate is to add barium chloride to an aliquot of sample water, acidified to pH < 6 (to avoid carbonate reaction). The water will turn cloudy if SO_4^{2-} is present. Pour slowly into a graduated cylinder (100 mL will do) with a black X on the bottom until the X is no longer visible. The sulphate concentration is roughly determined from the height of water in the cylinder (h_w in cm), using the equation:

$$\text{Concentration } SO_4^{2-} \text{ (mg/L)} = -130 \cdot \log(h_w) + 190$$

SYNOPSIS OF FIELD GEOCHEMICAL MEASUREMENTS:

SAMPLE: Ideally, field measurements are made in a flow-through cell with a low, constant flow of groundwater. Measurements can also be made in aliquots taken from a well or spring, although one must consider the effects of atmospheric contact (CO_2 degassing, oxidation with O_2, etc.).

EQUIPMENT:
- Flow-through cell, beaker or sample container.
- pH-mV meter (with or without temperature electrode).
- pH, redox (ORP) electrodes and spare filling solutions.
- pH buffers, thermometer.
- Eh standard solution.
- EC meter and probe.
- Alkalinity test kit, sulphuric acid cartridges
- Filter system (positive pressure) with 0.45-µm filter disks.
- Deionized water.

pH:
- Select two buffers that bracket the anticipated pH of the groundwater, i.e. pH 4 and 7, or 7 and 10.
- Equilibrate electrodes and buffers to the temperature of the water to be measured. Failure to do this will result in slowly drifting pH readings as the electrode thermally equilibrates with the sample.
- Follow manufacturer instructions for specific instrument calibration procedures, with corrections for temperature of buffers:

T°C	0	5	10	15	20	25	30
Buffer pH	4.00	4.00	4.00	4.00	4.00	4.01	4.02
Buffer pH	7.12	7.09	7.06	7.04	7.02	7.00	6.99
Buffer pH	10.2	10.1	10.1	10.1	10.0	10.0	9.96

- Rinse the electrode and shake gently or touch the electrode with an absorbent paper; do not wipe.
- During a pH measurement, move the electrode slowly to allow it to "see" a different portion of the water at all times. Still water or rapid flow causes drift due to voltage potentials.
- pH measurements made in poorly buffered solutions (low TDS, low bicarbonate) may show considerable drift, taking 5 to 10 minutes for a stable reading. Instability indicates that the electrode has aged and should be reconditioned or discarded.
- Store electrode in 2 M KCl solution or in pH 7 buffer solution (short term), or dry (long term).

Eh:
- Test electrode and meter for functioning against standard solution.
- Stir sample gently in measuring if there is no flow in the measurement container, to allow electrode to "see" all of solution.
- Wait for a stable reading (1 to 5 minutes). If readings are dropping and begin to rise, record lowest reading as atmospheric O_2 contamination may be occurring.
- Correct the measured E to Eh by adding the standard potential, E_R, to the measured value, according to the equation: $Eh = E_{measured} + E_R$. The following are the standard potentials for a Ag-AgCl electrode with 3.8 N KCl filling solution (according to the equation $E_R = -1.39T(°C) + 305$):

T°C	0	5	10	15	25	35	50
E_R	220	216	213	209	202	195	184

	• Clean electrode if there are problems with reference readings or time for stabilization. Use liquid hand soap and a soft brush. For scale, dip in 10% HCl for a couple of minutes and rinse with deionized water. • Do not use the Eh aliquot as a sample due to K^+ and Cl^- contamination. • Store electrode in a soaker bottle (bottle with O-ring seal), filled with 2 M KCl.
ALKALINITY:	• Measurement should be made as close to point of sampling as possible due to alteration of the sample alkalinity from CO_2 degassing, Fe^{2+} oxidation or other reactions. If this is not possible, take filtered sample in a HDLP or glass bottle (fill completely), and analyze as soon as reasonably possible (store cool). • As little as 5 to 10 mL can be analysed by most equipment. Micro-analytical methods are available in some labs. Standard sample size is 100 mL. • Filter sample with 0.45-μm paper, using positive pressure to avoid degassing of CO_2. Avoid contact with air with high pH waters. • Titrate to an endpoint of pH 4.3, and record acid normality, amount used, and sample volume.

Major anions (Cl^-, SO_4^-, NO_3^-, F^-, Br^-)

Analysis of major anions is normally done in a laboratory by liquid chromatography, although field kits and ion-specific electrodes exist for most anions of interest. Most major anions can be analysed in less than 5 mL of sample, although collecting 25- to 50-mL samples allows repeat measurements and dilutions if required. HCO_3^- and CO_3^{2-} are determined by alkalinity titration on 50 to 100 mL samples. Use thoroughly rinsed HDPE or PP bottles. Samples should be filtered (0.45 μm), stored cool, and — obviously — not acidified.

Sulphate analyses on water that contains reduced sulphur species (H_2S, HS^- or S^{2-}) can be erroneous since these sulphide species oxidize very fast and sulphate readings are then too high. It is mandatory to remove dissolved sulphide, preferably as CdS since its yellow colour is indicative of the presence of reduced S-species. Filter before analysing for sulphate. A second 25-mL sample is required for total S (H_2S and SO_4^{2-}), filtered out of contact with atmosphere. Preserve with 0.25 mL of 30% H_2O_2 to oxidize all sulphide to sulphate. This measurement for SO_4^{2-}, less that of the first sample, gives total sulphide.

Major, minor and trace metals

All geochemical analyses require measurement of major metals or cations (Ca^{2+}, Mg^{2+}, Na^+, K^+ and SiO_2). Minor (Fe, Mn, Sr and Ba) and trace metals may be required as well. Cations are best analysed in the laboratory by routine methods, including ICP-AES (inductively coupled plasma-atomic emission spectrometry), ICP-MS (ICP-mass spectrometry), AA (atomic absorption) or FE (flame emission), although several field methods are also available. Only a few millilitres of sample are required for ICP instruments that measure multiple wavelengths or masses simultaneously. A normal sample size is 25 to 50 mL.

Carbonates and oxides can precipitate after sampling, and so samples are acidified to keep metals in solution. To ensure that only dissolved species are analysed, the solution must be filtered first. Ultrapure nitric acid is used to acidify to below pH 2. Reagent grade is adequate for major and minor cations only. HCl should be used for brines as it releases hydrogen ions faster. The water must be filtered and should be stored in fresh HDPE or PP bottles. Used bottles are fine too, but should be acid rinsed.

Dissolved organic carbon (DOC)

Sampling for DOC concentration is recommended for any study where redox processes and the carbonate system are important. It is measured in the laboratory by oxidation of an acidified sample (initially degassed of CO_2) and measuring the evolved CO_2. The sample must be filtered to 0.45 μm at the point of sampling, and acidified to stop bacterial activity. Acidify with HCl, because HNO_3 can oxidize organics in the sample prior to analysis. Collect a 10- to 20-mL sample in a glass vial with a non-organic cap seal (aluminum foil, for example). Scintillation vials are ideal and cheap.

SAMPLING SYNOPSIS: Geochemistry — Anions, Cations and DOC:

SAMPLE VOLUME:
- 25 to 50 mL for anions.
- 25 to 50 mL for cations.
- 10 mL for DOC.

EQUIPMENT:
- HDPE or PP bottles. Glass for DOC.
- Filter system (positive pressure) with 0.45 μm filter disks.
- Reagent-grade HNO_3 acid (or ultrapure for trace metals).
- Acid delivery system: micropipette or dropper bottle.
- 10% HCl for DOC samples.

METHOD:
- Take the required field measurements of T, pH etc.
- Filter sample water into sample bottles, rinsing bottle and cap at least twice with filtered sample water.
- Fill anion bottle completely (pre-labelled with sample code and marked as filtered) and cap.
- Rinse and fill cation bottle (pre-labelled with sample code and marked as filtered, acidified) leaving 2-mL head space. Acidify with ~5 drops (120 μL) 5 N HNO_3 acid for a 50-mL sample, and cap.
- If collecting a DOC sample, rinse and fill the 20-mL glass vial with filtered sample water, leaving space for addition of 5 drops HCl.

REFERENCES

Aggarwal, P.K. and Hinchee, R.E., 1991. Monitoring in situ biodegradation of hydrocarbons by using stable carbon isotopes. *Environmental Science Technology*, 25: 1173-1180.

Aiken, G.R., McKnight, D.M., Wershaw, R.L. and MacCarthy, P. (Eds.), 1985. *Humic Substances in Soil, Sediment and Water.* Wiley Interscience, New York.

Al-Aasm, I.S., Taylor, B.E. and South, B., 1990. Stable isotope analysis of multiple carbonate samples using selective acid extraction. *Chemical Geology (Isotope Geoscience Section)*, 80: 119-125.

Alburger, D.E., Harbottle, G. and Norton, E.F., 1986. Half-life of ^{32}Si. *Earth and Planetary Science Letters*, 78: 168-176.

Allison, G.B., 1982. The relationship between ^{18}O and deuterium in water in sand columns undergoing evaporation. *Journal of Hydrology*, 55: 163-169.

Allison, G.B., Barnes, C.J., Hughes, M.W. and Leaney, F.W.J., 1984. Effect of climate and vegetation on oxygen-18 and deuterium profiles in soils. *In: Isotope Hydrology 1983*, IAEA Symposium 270, September 1983, Vienna: 105-123.

Amberger, A. and Schmidt, H.-L., 1987. Natürliche isotopegehalte von nitrat als indicatoren für dessen herkunft. *Geochimica et Cosmochimica Acta*, 51: 2699-2705.

Andersen, L.J. and Sevel, T., 1974. Profiles in the unsaturated and saturated zones, Grønhøj, Denmark. *In: Isotope Techniques in Groundwater Hydrology, Vol. 1*, IAEA Symposium, Vienna: 3-20.

Andrews, J.N., 1985. The isotopic composition of radiogenic helium and its use to study groundwater movement in confined aquifer. *Chemical Geology*, 49: 339-351.

Andrews, J.N., 1987. Noble gases in groundwaters from crystalline rocks. *In: P.Fritz and S.K.Frape (Eds.) Saline water and gases in crystalline rocks*. Geological Association of Canada, Special Paper 33: 234-244.

Andrews, J.N., 1992. Mechanisms for noble gas dissolution by groundwaters. *In: Isotopes of Noble Gases as Tracers in Environmental Studies*, IAEA Symposium, May 1989, Vienna: 87-110.

Andrews, J.A. and Fontes, J.-Ch., 1992. Importance of the in situ production of ^{36}Cl, ^{36}Ar and ^{14}C in hydrology and hydrogeochemistry. *In: Isotope Techniques in Water Resources Development 1991*, IAEA Symposium 319, March 1991, Vienna: 245-269.

Andrews, J.N. and Lee, D.J., 1979. Inert gases in groundwater from the Bunter Sandstone of England as indicators of age and palaeoclimatic trends. *Journal of Hydrology*, 41: 233-252.

Andrews, J.N., Balderer, W., Bath, A., Clausen, H.B. Evans, G.V., Florkowski, T., Goldbrunner, J.E., Ivanovich, M., Loosli, H. and Zojer, H., 1984. Environmental isotope studies in two aquifer systems. *In: Isotope Hydrology 1983*, IAEA Symposium, 270, September 1983, Vienna: 535-576.

Andrews, J.N., Davis, S.N., Fabryka-Martin, J., Fontes, J.-Ch., Lehmann, B.E., Loosli, H.H., Michelot, J-L., Moser, H., Smith, B. and Wolf, M., 1989. The *in situ* production of radioisotopes in rock matrices with particular reference to the Stripa granite. *Geochimica et Cosmochimica Acta*, 53: 1803-1815.

Andrews, J.N., Edmunds, W.M. Smedley, P.L., Fontes, J.-Ch., Fifield, L.K. and Allan, G.L., 1994. Chlorine-36 in groundwater as a paleoclimatic indicator: the East Midlands Triassic Sandstone aquifer (UK). *Earth and Planetary Science Letters*, 122: 159-171.

Andrews, J.N., Fontes, J.-Ch., Michelot, J-L. and Elmore, D., 1986. In-situ neutron flux, ^{36}Cl production and groundwater evolution in crystalline rocks at Stripa, Sweden. *Earth and Planetary Science Letters*, 77: 49-58.

Andrews, J.N., Giles, I.S., Kay, R.L.F., Lee, D.J., Osmond, J.K., Cowart, J.B., Fritz, P., Barker, J.F. and Gale, J., 1982. Radioelements, radiogenic helium and age relationships for groundwaters from the granites at Stripa, Sweden. *Geochimica et Cosmochimica Acta*, 46: 1533-1543.

Apps, J.A., 1985. Methane formation during hydrolysis by mafic rocks. Lawrence Berkley Laboratories, Earth Sciences Division, Annual Report, p. 13-17.

Araña, V. and Panichi, C., 1974. Isotopic composition of steam samples from Lanzarote, Canary Islands. *Geothermics*, 3: 142-145.

Aranyossy, J.-F., Filly, A., Tandia, A.A., Louvat, D., Ousmane, B., Joseph, A. and Fontes, J.-Ch., 1992. Estimation des flux d'évaporation diffuse sous couvert sableux en climat hyper-aride (Erg de Bilma, Niger). *In: Isotope Techniques in Water Resources Development 1991*, IAEA Symposium 319, March 1991, Vienna: 309-324.

Aravena, R. and Wassenaar, L.I., 1993. Dissolved organic carbon and methane in a regional confined aquifer: Evidence for associated subsurface sources. *Applied Geochemistry*, 8: 483-493.

Aravena, R., Evans, M.L. and Cherry, J.A., 1996. Stable isotopes of oxygen and nitrogen in source identification of nitrate from septic systems. *Ground Water*, 31: 180-186.

Aravena, R., Schiff, S.L., Trumbore, S.E., Dillon, P.J. and Elgood, R., 1992. Evaluating dissolved inorganic carbon cycling in a forested lake watershed using carbon isotopes. *Radiocarbon*, 34: 1-10.

Aravena, R., Suzuki, O. and Polastri, A., 1989. Coastal fogs and their relation to groundwater, IV region, Chile. *Chemical Geology (Isotope Geoscience Section)*, 79: 83-91.

Aravena, R., Warner, B.G., Charman, D.J., Belyea, L.R., Mathur, S.P. and Dinel, H., 1993. Carbon isotopic composition of deep carbon gases in an ombrogenous peatland, Northwestern Ontario, Canada. *Radiocarbon*, 35: 271-276.

Aravena, R., Wassenaar, L.I. and Plummer, L.N., 1995. Estimating ^{14}C groundwater ages in a methanogenic aquifer. *Water Resources Research*, 31: 2307-2317.

Arnason, B., 1969. Equilibrium constant of the fractionation of deuterium between ice and water. *Journal of Physical Chemistry*, 79: 3491.

Arnason, B., 1977. The hydrogen and water isotope thermometer applied to geothermal areas in Iceland. *Geothermics*, 5: 75-80.

Artinger, R., Buckau, G., Kim, J.I., Geyer, S. and Wolf, M., 1996. Influence of sedimentary organic matter on dissolved fulvic acids in groundwater. Significance for groundwater dating with ^{14}C in dissolved organic matter. *In: Isotopes in Water Resources Management*, Vol. I, IAEA Symposium 336, March 1995, Vienna: 57-72.

Baertschi, P., 1976. Absolute ^{18}O content of Standard Mean Ocean Water. *Earth and Planetary Science Letters*, 31: 341-344.

Bajjali, W., Clark, I.D. and Fritz, P., 1997. The artesian thermal groundwaters of northern Jordan: insights to their recharge history and age. *Journal of Hydrology*, 187: 355-382.

Bard, E., Arnold, M., Fairbanks, R.G. and Hamelin, B., 1993. ^{230}Th-^{234}U and ^{14}C dates obtained by mass spectrometry on corals. *Radiocarbon*, 35: 191-199.

Bard, E., Hamelin, B., Fairbanks, R.G. and Zindler, A., 1990. Calibration of the ^{14}C time scale over the past 30,000 years using mass spectrometric U-Th ages from Barbados corals. *Nature*, 345: 405.

Barker, J.F. and Fritz, P., 1981a. Carbon isotope fractionation during microbial methane oxidation. *Nature*, 293: 289-291.

Barker, J.F. and Fritz, P., 1981b. The occurrence and origin of methane in some groundwater flow systems. *Canadian Journal of Earth Sciences*, 18: 1802-1816.

Barker, J.F. and Pollock, S.J., 1984. The geochemistry and origin of natural gases in southern Ontario. *Bulletin of Canadian Petroleum Geology*, 32: 313-326.

Barnes, I., Irwin, W.P. and White, D.E., 1978. Global distribution of carbon dioxide discharges and major zones of seismicity. USGS Water Resource Investigations Open File Report 78-39, 12 p.

Barry, R.G. and Chorley, R.J., 1987. *Atmosphere, Weather and Climate*. 5th ed., Methuen, 460 p.

Bartholomew R.M., Brown F. and Lounsbury M., 1954. Chlorine isotope effect in reactions of tert-butyl chloride. *Canadian Journal of Chemistry*, 32: 979-983.

Bassett, R.L., 1990. A critical evaluation of the available measurements for the stable isotopes of boron. *Applied Geochemistry*, 5: 541-554.

Bath, A.H., Darling, W.G., George, I.A. and Milodowski, A.E., 1987. $^{18}O/^{16}O$ and $^2H/^1H$ changes during progressive hydration of a Zechstein anhydrite formation. *Geochimica et Cosmochimica Acta*, 51: 3113-3118.

Bath, A.H., Edmunds, W.M. and Andrews, J.N., 1979. Paleoclimatic trends deduced from the hydrochemistry of a Triassic Sandstone aquifer, U.K. *In: Isotope Hydrology 1978*, Vol. II, IAEA Symposium 228, June 1978, Neuherberg, Germany: 545-568.

Bein, A. and Dutton, A.R., 1993. Origin, distribution, and movement of brine in the Permian Basin (U.S.A.): A model for displacement of connate brine. *Geological Society of America Bulletin*, 105: 695-707.

Benson, B.B. and Krause, D., 1976. Empirical laws for dilute aqueous solutions of non-polar gases. *Journal of Chemical Physics*, 64: 639-709.

Bentley, H.W., Phillips, F.M. and Davis, S.N., 1986. Chlorine-36 in the terrestrial environment. Chapter 10, *In: P. Fritz and J.-Ch. Fontes, (Eds.) Handbook of Environmental Isotope Geochemistry, Vol. 2, The Terrestrial Environment, B.* Elsevier, Amsterdam, The Netherlands: 427-480.

Berner, R.A. and Lasaga, A.C., 1989. Modeling the geochemical carbon cycle. Natural geochemical processes that result in the slow buildup of atmospheric carbon dioxide may have caused past geologic intervals of global warming through the greenhouse effect. *Scientific American*: 74-81.

Bortolami, G.C., Ricci, B., Susella, G.F. and Zuppi, G.M., 1979. Isotope hydrology of the Val Corsaglia, Maritime Alps, Piedmont, Italy. *In: Isotope Hydrology 1978*, Vol. I, IAEA Symposium 228, June 1978, Neuherberg, Germany: 327-350.

Böttcher, J , Strebel, O., Voerkelius, S, and Schmidt, H.-L., 1990. Using isotope fractionation of nitrate nitrogen and nitrate oxygen for evaluation of denitrification in a sandy aquifer. *Journal of Hydrology*, 114: 413-424.

Böttcher, J., Strebel, O. and Kölle, W., 1992. Redox conditions and microbial sulphur reduction in the Fuhrberger Feld sandy aquifer. *In: G. Matthess et al. (Eds.) Progress in Hydrogeochemisty*, Springer Verlag, Berlin: 219-226.

Bottinga Y., 1968. Calculation of fractionation factors for carbon and oxygen in the system calcite - carbon dioxide - water. *Journal Physical Chemistry*, 72: 800-808.

Bottinga, Y., 1969. Calculated fractionation factors for carbon and hydrogen isotope exchange in the system calcite-CO_2-graphite-methane-hydrogen and water vapour. *Geochimica et Cosmochimica Acta*, 33: 49-64.

Bottomley, D.J. and Veizer, J. 1992. The nature of groundwater flow in fractured rock: evidence from the isotopic and chemical evolution of recrystallized fracture calcites from the Canadian Precambrian Shield. *Geochimica et Cosmochimica Acta*, 56: 369-388.

Bottomley, D.J., Douglas, M., Clark, I.D., Katz, A., Starinsky, A., Chan, L., Gregoire, D.C. and Raven, K.G., 1997. The origin of marine brines in archean volcanic rocks of the Yellowknife greenstone belt, Slave Province, Northwest Territories, Canada. Submitted to *Geochimica et Cosmochimica Acta*.

Bottomley, D.J., Gregoire, D. and Raven, K.G., 1994. Saline groundwaters and brines in the Canadian Shield: Geochemical and isotopic evidence for a residual evaporite brine component. *Geochimica et Cosmochimica Acta*, 58: 1483-1498.

Brown, R.M., 1961. Hydrology of tritium in the Ottawa Valley. *Geochimica et Cosmochimica Acta*, 21: 199-216.

Bùason, Th., 1972. Equations of isotope fractionation between ice and water in a melting snow column with continuous rain and percolation. *Journal of Glaciology*, 11: 387-405.

Buffle, J., 1977. Les substances humiques et leurs interactions avec les ions minéraux. *In: Conference Proceedings, La Commission d'Hydrologie Appliquée de l'A.G.H.T.M.*, L'Université de Paris-Sud, Orsay, France: 3-10.

Bullen, T.D., Krabbenhoft, D.P. and Kendall, C., 1996. Kinetic and mineralogic controls on the evolution of groundwater chemistry and $^{87}Sr/^{86}Sr$ in a sandy silicate aquifer, northern Wisconsin, USA. *Geochimica et Cosmochimica Acta*, 60: 1807-1821.

Burgman, J.O., Calles, B. and Westman, F., 1987. Conclusions from a ten year study of oxygen-18 in precipitation and runoff in Sweden. *In: Isotope Techniques in Water*

Resources Development, IAEA Symposium 299, March 1987, Vienna: 579-590.

Buttle, J.M., 1994. Isotope hydrograph separations and rapid delivery of pre-event water from drainage basins. *Progress in Physical Geography*, 18: 16-41.

Cane, G., 1996. Groundwater Quality and Contaminant Pathways in the Raisin River Agricultural Watershed; Susceptibility to Groundwater Contamination, Cornwall, Ontario. Unpublished M.Sc. Thesis, University of Ottawa, Canada, 231 p.

Carothers, W.W., Adami, L.H. and Rosenbauer, R.J., 1988. Experimental oxygen isotope fractionation between siderite-water and phosphoric acid liberated CO_2-siderite. *Geochimica et Cosmochimica Acta*, 52: 2445-2450.

Carrillo-Rivera, J.J., Clark, I.D. and Fritz, P., 1992. Investigating recharge of shallow and paleo-groundwaters in the Villa de Reyes basin, SLP, Mexico with environmental isotopes. *Applied Hydrogeology*, 4: 35-48.

Carslaw, H.S. and Jaeger, J.C., 1959. *The Conduction of Heat in Solids*. 2nd ed. Oxford Science Publications, Oxford University Press, London, 510 p.

Cerling, T.E., Solomon, D.K., Quade, J. and Bowman, J.R., 1991. On the isotopic composition of carbon in soil carbon dioxide. *Geochimica et Cosmochimica Acta*, 55: 3403-3405.

Chan, L.H., Edmond, J.M., Thompson, G. and Gillis, K., 1992. Lithium isotopic composition of submarine basalts: implications for the lithium cycle in the oceans. *Earth and Planetary Science Letters*, 108: 1551-160.

Charman, D.J., Aravena, R., and Warner, B.G., 1994. Carbon dynamics in a forested peatland in north-eastern Ontario, Canada. *Journal of Ecology*, 82: 55-62.

Chiba, H., Kusakabe, M., Hirano, S.-I., Matsuo, S. and Somiya, S., 1981. Oxygen isotope fractionation factors between anhydrite and water from 100 to 550°C. *Earth and Planetary Science Letters*, 53: 55-62.

Chiba, H. and Sakai, H., 1985. Oxygen isotope exchange rate between dissolved sulphate and water at hydrothermal temperatures. *Geochimica et Cosmochimica Acta*, 49: 993-1000.

Clark, I.D. 1987. Groundwater resources in the Sultanate of Oman: origin, circulation times, recharge processes and paleoclimatology. Isotopic and geochemical approaches. Unpublished doctoral thesis, Université de Paris-Sud, Orsay, France, 264 p.

Clark, I.D. and Fontes, J-Ch., 1990. Palaeoclimatic reconstruction in northern Oman based on carbonates from hyperalkaline groundwaters. *Quaternary Research*, 33: 320-336.

Clark, I.D. and Lauriol, B., 1992. Kinetic enrichment of stable isotopes in cryogenic calcites. *Chemical Geology (Isotope Geoscience Section)*, 102: 217-228.

Clark, I.D., Bajjali, W.T. and Phipps, G.Ch., 1996. Constraining ^{14}C ages in sulphate reducing groundwaters: two case studies from arid regions. *In: Isotopes in Water Resources Management*, IAEA Symposium 336, March 1995, Vienna: 43-56.

Clark I.D., Fritz, P., Michel, F.A. and Souther, J.G., 1982. Isotope hydrogeology and geothermometry of the Mount Meager geothermal area. *Canadian Journal of Earth Sciences*, 19: 1454-1473.

Clark, I.D., Fritz, P., Quinn, O.P., Rippon, P., Nash H. and bin Ghalib al Said B., 1987. Modern and fossil groundwater in an arid environment, A look at the hydrogeology of Southern Oman. *Use of Stable Isotopes in Water Resources Development*, IAEA Symposium. 299, March 1987, Vienna: 167-187.

Clarke, W.B. and G. Kugler, 1973. Dissolved helium in groundwater: as possible method for U and Th prospecting. *Economic Geology*, 68: 243-251.

Claypool, G.E., Holser, W.T., Kaplan, I.R., Sakai, H. and Zak, I., 1980. The age curves of sulfur and oxygen isotopes in marine sulphate and their mutual interpretation. *Chemical Geology*, 28: 199-260.

Clayton, R.N., Friedman, I., Graff, D.L., Mayeda, T.K., Meents, W.F. and Shimp, N.F., 1966. The origin of saline formation waters, 1. Isotopic composition. *Journal of Geophysical Research*, 71: 3869-3882.

Cole, D.R. and Ohmoto, H., 1986. Kinetics of isotope exchange at elevated temperatures and pressures. Chapter 2 *In: J.W. Valley, H.P. Taylor and J.R. O'Neil, (Eds.) Stable Isotopes in High Temperature Geological*

Processes, Reviews in Mineralogy Vol. 16, Mineralogical Society of America: 41-87.

Coleman, M.L., 1992. Water composition variation within one formation. In: Y. Kharaka and A.S. Maest (Eds.) Proceedings of the 7th International Symposium on Water-Rock Interaction, July 1992, Park City, Utah, Balkema: 1109-1112.

Coleman, M.L., Shepherd, T.J., Durham, J.J., Rouse, J.E. and Moore, G.R., 1982. Reduction of water with zinc for hydrogen isotope analysis. Analytical Chemistry, 54: 993-995.

Coleman, M., Eggenkamp, H., Matray, J.M and Pallant, M., 1993. Origins of oil-field brines by Cl stable isotopes. Terra Abstacts, 5: 638.

Cook, P.G., Jolly, I.D., Leaney, F.W., Walker, G.R., Allan, G.L., Fifield, L.K. and Allison, G.B., 1994. Unsaturated zone tritium and chlorine-36 profiles from southern Australia: their use as tracers of soil water movement. Water Resources Research, 30: 1709-1719.

Cook, P.G., Solomon, D.K., Plummer, L.N., Busenberg, E. and Schiff, S.L., 1995. Chlorofluorocarbons as tracers of groundwater transport processes in a shallow, silty sand aquifer. Water Resources Research, 31: 425-434.

Coon, J.H., 1949. ^{3}He isotopic abundance. Physical Review, 75: 1355-1357.

Coplen, T.B., 1996. New guidelines for reporting stable hydrogen, carbon and oxygen isotope-ratio data. Geochimica et Cosmochimica Acta, 60: 3359-3360.

Coplen, T.B. and Hanshaw, B.B., 1973. Ultrafiltration by a compacted clay membrane. I. Oxygen and hydrogen isotope fractionation. Geochimica et Cosmochimica Acta, 37: 2295-2310.

Coplen, T.B., Kendall, C. and Hopple, J., 1983. Comparison of isotope reference samples. Nature, 302: 236-236.

Craig, H., 1957. Isotopic standards for carbon and oxygen and correction factors for mass-spectrometric analysis of carbon dioxide. Geochimica et Cosmochimica Acta, 12: 133-149.

Craig, H., 1961a. Standard for reporting concentrations of deuterium and oxygen-18 in natural water. Science, 133: 1833-1834.

Craig, H., 1961b. Isotopic variations in meteoric waters. Science, 133: 1702-1703.

Craig, H., 1963. The isotopic geochemistry of water and carbon in geothermal areas. In: E. Tongiorgi, (Ed.), Nuclear Geology on Geothermal Areas, Spoleto, 1963. Consiglio Nazionale delle Ricerche, Laboratorio di Geologia Nucleare, Pisa: 17-53.

Craig, H., 1966. Isotopic composition and origin of the Red Sea and Salton Sea brines. Science, 154: 1544-1547.

Craig, H. and Gordon, L., 1965. Deuterium and oxygen-18 variation in the ocean and the marine atmosphere. In: E. Tongiorgi, (Ed.), Stable Isotopes in Oceanographic Studies and Paleotemperatures, Spoleto 1965: 9-130.

Criss, R.E., 1991. Temperature dependence of isotopic fractionation factors. In: H.P. Taylor, J.R. O'Neil and I.R. Kaplan, (Eds.) Stable Isotope Geochemistry: A Tribute to Samuel Epstein, The Geochemical Society Special Publication No. 3, San Antonio, Texas: 11-16.

Dakin, R.A., Farvolden, R.N., Cherry, J.A. and Fritz, P., 1983. Origin of dissolved solids in groundwaters of Mayne Island, British Columbia, Canada. Journal of Hydrology, 63: 233-270.

Damon, P.E., Cheng, S. and Linick, T.W., 1989. Fine and hyperfine structure in the spectrum of secular variations of atmospheric ^{14}C. Radiocarbon, 31: 704-718.

Dansgaard, W., 1953. The abundance of O^{18} in atmospheric water and water vapour. Tellus, 5: 461-469.

Dansgaard, W., 1964. Stable isotopes in precipitation. Tellus, 16: 436-468.

Dansgaard, W., Johnsen, S.J., Clausen, H.B., Dahl-Jensen, D., Gundestrup, N.S., Hammer, C.U., Hvidberg, C.S., Steffensen, J.P., Sveinbjörnsdottir, A.E., Jouzel, J. and Bond, G., 1993. Evidence for general instability of past climate from a 250-kyr ice-core record. Nature, 364: 218-220.

Darling, W.G. and Bath, A.H., 1988. A stable isotope study of recharge processes in the English chalk. Journal of Hydrology, 101: 31-46.

Deines, P., Langmuir, D. and Harmon, R.S., 1974. Stable carbon isotope ratios and the existence of a gas phase in the evolution of carbonate groundwaters. Geochimica et Cosmochimica Acta, 38: 1147-1164.

Delmore, J.E. 1982. Isotopic analysis of iodine using negative surface ionization.

International Journal Mass Spectrometry and Ion Physics, 43: 273-281.

Desaulniers, D.E., Cherry, J.A. and Fritz, P., 1981. Origin, age and movement of porewater in argillaceous Quaternary deposits at four sites in southwestern Ontario. *Journal of Hydrology*, 50: 231-257.

Desaulniers, D.E., Kaufmann, R.S., Cherry, J.A. and Bentley, H.W., 1986. ^{35}Cl-^{37}Cl variations in a diffusion-controlled groundwater system. *Geochimica et Cosmochimica Acta*, 50: 1757-1746.

Dimitrakopoulos, R. and Muehlenbachs, K., 1987. Biodegradation of petroleum as a source of ^{13}C-enriched carbon dioxide in the formation of carbonate cement. *Chemical Geology (Isotope Geoscience Section)*, 65: 283-291.

Dinçer, T., 1968. The use of oxygen-18 and deuterium concentrations in the water balance of lakes. *Water Resources Research*, 4: 1289-1305.

Dinçer, T., Al-Mugrin, A. and Zimmermann, U., 1974. Study of the infiltration and recharge through the sand dunes in arid zones with special reference to the stable isotopes and thermonuclear tritium. *Journal of Hydrology*, 23: 79-109.

Dodson, M.H., 1973. Closure temperature in cooling geochronological and petrological systems. *Contributions to Mineral Petrology*, 40: 259-274.

Domenico, P.A. and Schwartz, F.W., 1990. *Physical and Chemical Hydrogeology*. John Wiley & Sons, New York, 824 p.

Dowuona, G.N., Mermut, A.R. and Krouse, H.R., 1992. Isotopic composition of salt crusts in Saskatchewan, Canada. *Chemical Geology (Isotope Geoscience Section)*, 94: 205-213.

Douglas, M., 1997. Mixing and temporal variations of groundwater inflow at the Con Mine, Yellowknife, Canada; An analogue for a radioactive waste repository. Unpublished M.Sc. thesis, Department of Geology, University of Ottawa, Canada, 101 p.

Dray, M., Gonfiantini, R. and Zuppi, G.M., 1983. Isotopic composition of groundwater in the southern Sahara. *In: Paleoclimates and Paleowaters: A Collection of Environmental Isotope Studies*, Proceedings of an Advisory Group Meeting, November 1980, Vienna: 187-199.

Drever, J.I., 1997. *The Geochemistry of Natural Waters: Surface and Groundwater Environments*, 3rd ed., Prentice Hall, New Jersey, 436 p.

Drimmie, R.J., Aravena, R., Wassenaar, L.I., Fritz, P., Hendry, M.J. and Hut, G., 1991. Radiocarbon and stable isotopes in water and dissolved constituents of the Milk River Aquifer. *Applied Geochemistry*, 6: 381-392.

Dubois, J.D. and Flück, J., 1984. Geochemistry: utilisation of geothermal resources of the Baden area. Basel, Swiss National Energy Research Foundation, NEFF 165-1B-032, 165 p. (cited in Pearson et al., 1991).

Dyke, A.S. and Prest, V.K., 1987. Late Wisconsinan and Holocene retreat of the Laurentide ice sheet. *Geological Survey of Canada*, Map 1702A, scale 1:5,000,000.

Eastoe C.J. and Guilbert J.M., 1992. Stable chlorine isotopes in hydrothermal processes. *Geochimica et Cosmochimica Acta*, 56: 4247-4255.

Edmunds, W.M. and Wright, E.P., 1979. Groundwater recharge and paleoclimate in the Sirte and Kufra Basins, Libya. *Journal of Hydrology*, 40: 215.

Edwards, L.R., Chen, J.H. and Wasserburg, G.J. 1987. ^{238}U-^{234}U-^{230}Th-^{232}Th systematics and the precise measurement of time over the past 500,000 years. *Earth and Planetary Science Letters*, 81: 175-192.

Eggenkamp, H.G.M., 1994. $\delta^{37}Cl$, *The Geochemistry of Chlorine Isotopes*. Doctoral Thesis, Faculteit Aardwetenschappen, Universiteit Utrecht, The Netherlands, 150 p.

Eggenkamp, H.G.M., Kreulen, R. and Koster van Groos, A.F., 1994a. Chlorine stable isotope fractionation in evaporites. Chapter 10, *In: H.G.M. Eggenkamp, $\delta^{37}Cl$, The Geochemistry of Chlorine Isotopes*. Doctoral Thesis, Faculteit Aardwetenschappen, Universiteit Utrecht, The Netherlands: 111-122.

Eggenkamp, H.G.M., Coleman, M.L., Matray, J.M. and Scholten, S.O., 1994b. Variations of chlorine stable isotopes in formation waters. Chapter 7, *In: H.G.M. Eggenkamp, $\delta^{37}Cl$, The Geochemistry of Chlorine Isotopes*. Doctoral Thesis, Faculteit Aardwetenschappen, Universiteit Utrecht, The Netherlands: 77-91.

Ehleringer, J.R., Sage, R.F., Flanagan, L.B. and Pearcy, R.W., 1991. Climate change and the evolution of C_4 photosynthesis. *Trends in Ecology and Evolution*, 6: 95-99.

Eichinger, L., Merkel, B., Nemeth, G., Salvamoser, J. and Stichler, W., 1984. Seepage velocity determinations in unsaturated Quarternary gravel. *Recent Investigations in the Zone of Aeration*, Symposium Proceedings, Munich, Oct. 1984: 303-313.

Ekwurzel, B., Schlosser, P., Smethie, Jr., W.M., Plummer, N., Busenberg, E., Michel, R.L., Weppernig, R. and Stute, M., 1994. Dating of shallow groundwater: comparison of the transient tracers $^3H/^3He$, chlorofluorocarbons and ^{85}Kr. *Water Resources Research*, 30: 1693-1708.

El Bakri, A. Tantawi, A., Blavoux, B. and Dray, M., 1992. Sources of water recharge identified by isotopes in El Minya Governate (Nile Valley, Middle Egypt). *In: Isotope Techniques in Water Resources Development 1991*, IAEA Symposium 319, March 1991, Vienna: 643-645.

Ellis, A.J. and Mahon, W.A.J., 1967. Natural hydrothermal systems and experimental hot water/rock interaction, (part II). *Geochimica et Cosmochimica Acta*, 37: 1255-1275.

Elmore, D., Gove, H.E., Ferraro, R., Kilius, L.R., Lee, H.W., Chang, K.H., Beukens, R.P., Litherland, A.E., Russo, C.J., Purser, K.H., Murrell, M.T. and Finkel, R.C. 1980 Determination of ^{129}I using tandem accelerator mass spectrometry. *Nature*, 286: 138-140.

Environment Canada, 1995. *Canadian Water Quality Guidelines*. Canadian Council of Resource and Environment Ministers, Water Quality Branch, Environment Canada, Ottawa.

Epstein, S. and Mayeda T.K., 1953. Variations of the $^{18}O/^{16}O$ ratio in natural waters. *Geochimica et Cosmochimica Acta*, 4: 213.

Epstein, S.E., Buchsbaum, R., Lowenstam, H.A. and Urey, H.C., 1953. Revised carbonate-water isotopic temperature scale. *Geological Society of America Bulletin*, 64: 1315-1326.

Eriksson, E., 1966. Major pulses of tritium in the atmosphere. *Tellus*, 17: 118-130.

Fabryka-Martin, J., Curtis, .D.B., Dixon, P., Rokop, D., Roensch, F., Aguilar, R. and Attrep, M., 1994. Natural nuclear products in the Cigar Lake deposit. *In: J.J. Cramer and J.A.T. Smellie (Eds.), Final Report of the AECL/SKB Cigar Lake Analog Study*, Report AECL-10851, Atomic Energy of Canada Limited, Whiteshell Laboratories, Pinawa, Manitoba.

Faure, G., 1986. *Principles of Isotope Geology*, 2nd Edition, Wiley, New York, 589 p.

Ferronsky, V.I. and Brezgunov, V.S., 1989. Stable isotopes and ocean dynamics. *In: P. Fritz, P. and J.-Ch Fontes (Eds.), Handbook of Environmental Isotope Geochemistry, Vol. 3, The Marine Environment*, A., Elsevier, Amsterdam, The Netherlands: 1-26.

Fetter, C.W., 1988. *Applied Hydrogeology*. 2nd Edition, Merrill, Columbus, OH, 592 p.

Fisher, D.A., 1991. Remarks on the deuterium excess in precipitation in cold regions. *Tellus*, 43: 401-407.

Fleischer, E., Goldberg, M., Gat, J.R. and Magaritz, M., 1977. Isotopic composition of formation waters from deep drillings in southern Israel. *Geochimica et Cosmochimica Acta*, 41: 511-525.

Flipse, W. J., Jr. and Bonner, F. T., 1985. Nitrogen isotope ratios of nitrate in groundwater under fertilized fields, Long Island, New York. *Ground Water*, 32: 59-67.

Florkowski, T., 1985. Sample preparation for hydrogen isotope analysis by mass spectrometry. *Journal of Applied Radiation and Isotopes*, 36: 991-992.

Fontes, J.-Ch., 1965. Fractionnement isotopique dans l'eau de cristallisation du sulfate de calcium. *Geol. Rundsch.*, 55: 172-178

Fontes, J.Ch., 1985. Méthode au chlore-36 datation des eaux. *In: E. Roth and B. Poty (Eds.) Méthodes de Datation par les Phénomènes Nucléaires Naturels, Applications.*, Masson, Paris: 399-420.

Fontes, J.-Ch. and Garnier, J-M., 1979. Determination of the initial ^{14}C activity of total dissolved carbon: A review of existing models and a new approach. *Water Resources Research*, 15: 399-413.

Fontes, J.-Ch., Gonfiantini, R., 1967a. Comportement isotopique au cours de l'évaporation de deux bassins sahariens. *Earth and Planetary Science Letters*, 7: 325-329.

Fontes, J.-Ch. and Gonfiantini, R., 1967b. Fractionnement isotopique de l'hydrogène dans l'eau de cristallisation du gypse. *Comptes Rendus, Academie des .Sciences de Paris, Séries. D.* 265: 4-6.

Fontes, J.-Ch., Gonfiantini, R. and Roche, M.A., 1970. Deutérium et oxygène-18 dans les eaux du lac Tchad. *In: Isotope Hydrology 1970*, IAEA Symposium 129, March 1970, Vienna: 387-404.

Fontes, J.-Ch. and Olivry, J.C., 1977. Gradient isotopique entre 0 et 4000 m dans les précipitations du Mont Cameroun. *Comptes Rendus Réunion Annuelle Sciences de la Terre*, Société géologie française, Paris: 171.

Förstel, H., 1982. $^{18}O/^{16}O$ ratio of water in plants and their environment. *In: H.L. Schmidt, H. Förstel and K. Heinzinger (Eds.), Stable Isotopes*. Elsevier, Amsterdam, The Netherlands: 503-516.

Forster, M., Moser, H. and Loosli, H.H., 1984. Isotope hydrological study with carbon-14 and argon-39 in the Bunter Sandstones of the Saar Region. *In: Isotope Hydrology 1983*, IAEA symposium 270, September 1983, Vienna: 515-533.

Fouillac, C. and Michard, G., 1981. Sodium/lithium ratio in water applied to geothermometry of geothermal reservoirs. *Geothermics*, 10: 55-70.

Fournier, R.O., 1981. Application of water chemistry to geothermal exploration and reservoir engineering. *In: L. Rybach and L.J.P. Muffler (Eds.) Geothermal Systems: Principles and Case Histories*. John Wiley & Sons: 109-143.

Fournier, R.O. and Potter, R.W., 1979. Magnesium correction to the Na-K-Ca chemical geothermometer. *Geochimica et Cosmochimica Acta*, 43: 1543-1550.

Fournier, R.O. and Potter, R.W., 1982. A revised and expanded silica (quartz) geothermometer. *Geothermal Resources Council Bulletin*, 11: 3-9.

Fournier, R.O. and Truesdell, A.H., 1973. An empirical Na-K-Ca geothermometer for natural waters. *Geochimica et Cosmochimica Acta*, 37: 1255-1275.

Frape, S.K. and Fritz, P., 1982. The chemistry and isotopic composition of saline groundwaters from the Sudbury basin, Ontario. *Canadian Journal of Earth Sciences*, 19: 645-661.

Frape, S.K. and Fritz, P., 1987. Geochemical trends for groundwaters from the Canadian Shield. *In: P. Fritz and S. Frape (Eds.) Saline Water and Gases in Crystalline Rocks*. Geological Association of Canada, Special Paper 33: 19-38.

Freeze, R.A. and Cherry, J.A., 1979. *Groundwater*. Prentice-Hall, 604 p.

Friedli, H., Lotscher, H., Oeschger, H., Siegenthaler, U. and Stauffer, B., 1986. Ice core record of the $^{13}C/^{12}C$ ratio of atmospheric CO_2 in the past two centuries. *Nature*, 324: 237-238.

Friedman, I., 1953. Deuterium content of natural waters and other substances. *Geochimica et Cosmochimica Acta*, 4: 89-103.

Friedman, I. and O'Neil, J.R., 1977. Compilation of stable isotope fractionation factors of geochemical interest. *In: M. Fleischer (Ed.) Data of Geochemistry*, U.S. Geological Survey Professional Paper 440-KK, 6th ed., U.S.G.S., Reston VA.

Friedman, I., Benson, C. and Gleason, J., 1991. Isotopic changes during snow metamorphism. *In: H.P. Taylor, J.R. O'Neil and I.R. Kaplan (Eds.) Stable Isotope Geochemistry: A Tribute to Samuel Epstein*, The Geochemical Society Special Publication No. 3, San Antonio, Texas: 211-221.

Friedman, I., Machta, L. and Soller, R., 1962. Water vapour exchange between a water droplet and its environment. *Journal of Geophysical Research.*, 67: 2761-2766.

Friedman, I., O'Neil, J.R. and Cebula, G., 1982. Two new carbonate stable isotope standards. *Geostandard Newsletter*, 6: 11-12.

Fritz, P. and Fontes, J.-Ch. (Eds.), 1980. *Handbook of Environmental Isotope Geochemistry, Vol. 1, The Terrestrial Environment., A.* Elsevier, Amsterdam, The Netherlands, 545 p.

Fritz, P. and Fontes, J.-Ch. (Eds.), 1986. *Handbook of Environmental Isotope Geochemistry, Vol. 2, The Terrestrial Environment., B.* Elsevier, Amsterdam, The Netherlands, 557 p.

Fritz, P. and Frape, S.K., 1982. Saline groundwaters in the Canadian Shield - a first overview. *Chemical Geology*, 36: 179-190.

Fritz, P. and Lodemann, M., 1990. Die salinaren tiefenwässer der KTB-vorbohrung. *Die Geowissenschaften*, 9: 281-285.

Fritz, P. and Reardon, E.J., 1979. Isotopic and chemical characteristics of mine water in the Sudbury area. *AECL Technical Report 35*, Atomic Energy of Canada Limited, Chalk River, Ontario, Canada, 37 p.

Fritz, P., Basharmal, G.M., Drimmie, R.J., Ibsen, J. and Qureshi, R.M., 1989. Oxygen isotope exchange between sulphate and water during bacterial reduction of sulphate. *Chemical Geology (Isotope Geoscience Section)*, 79: 99-105.

Fritz, P., Cherry, J., Weyer, K.U. and Sklash, M., 1976. Storm runoff analyses using environmental isotopes and major ions. *In: Interpretation of Environmental Isotope and Hydrochemical Data in Groundwater Hydrology 1975*, Workshop Proceedings, IAEA, Vienna: 111-130.

Fritz, P., Clark, I.D., Fontes, J.-Ch., Whiticar, M.J. and Faber, E. 1992. Deuterium and ^{13}C evidence for low temperature production of hydrogen and methane in a highly alkaline groundwater environment in Oman. *In: Y. Kharaka and A.S. Maest (Eds.) Proceedings of the 7th International Symposium on Water-Rock Interaction*, July 1992, Park City, Utah, Balkema: 793-796.

Fritz, P., Drimmie, R.J., Frape, S.K. and O'Shea, O., 1987a. The isotopic composition of precipitation and groundwater in Canada. *In: Isotope Techniques in Water Resources Development*, IAEA Symposium 299, March 1987, Vienna: 539-550.

Fritz, P., Frape, S.K., Drimmie, R.J., Appleyard, E.C. and Hattori, K., 1994. Sulfate in brines in the crystalline rocks of the Canadian Shield. *Geochimica et Cosmochimica Acta*, 58: 57-66.

Fritz, P., Frape, S.K. and Miles, M., 1987b. Methane in the crystalline rocks of the Canadian Shield, *In: P. Fritz and S.K. Frape (Eds.) Saline Water and Gases in Crystalline Rocks*, Geological Association of Canada, Special Paper 33: 211-223.

Fritz, P., Lapcevic, A., Miles, M., Frape, S.K., Lawson, D.E. and O'Shea, K.J., 1988. Stable isotopes in sulphate minerals from the Salina formation in southwestern Ontario. *Canadian Journal of Earth Sciences*, 25: 195-205.

Fritz, S.J., Drimmie, R.J. and Fritz, P., 1991. Characterizing shallow aquifers using tritium and ^{14}C: periodic sampling based on tritium half-life. *Applied Geochemistry*, 6: 17-33.

Fritz, S., Hinz, D.W. and Grossman, E.L., 1987. Hyperfiltration induced fractionation of carbon isotopes. *Geochimica et Cosmochimica Acta*, 51: 1121-1134.

Fry, B., Cox, J., Gest, H. and Hayes, J.M., 1986. Discrimination between ^{34}S and ^{32}S during bacterial metabolism of inorganic sulfur compounds. *Journal of Bacteriology*, 165: 328-330.

Fry, B., Ruf, W., Gest, H. and Hayes, J.M., 1988. Sulphur isotope effects associated with oxidation of sulphide by O_2 in aqueous solution. *Chemical Geology (Isotope Geoscience Section)*, 73: 205-210.

Galley, M.R., Miller, A.I., Atherley, J.F. and Mohn, M., 1972. GS process-physical properties: Chalk River, Ontario, Canada, Atomic Energy of Canada Limited, AECL-4225.

Games, L.M. and Hayes, J.M., 1974. Carbon in ground water at the Columbus, Indiana landfill. *In: Waldrip and Ruhe (Eds.) Solid Waste Disposal by Land Burial in Southern Indiana*. Indiana University Water Resources Research Centre, Bloomington, IN: 81-110.

Garrels, R.M. and Christ, C.L., 1965. *Solutions, Minerals, and Equilibria*. Freeman, Cooper & Company, San Francisco, 450 p.

Gascoyne, M, and Kotzer, T., 1995. Isotopic methods in hydrogeology and their application to the Underground Research Laboratory, Manitoba. Report AECL-11370, Atomic Energy of Canada Limited, Pinawa, Manitoba, 100 p.

Gat, J.R., 1971. Comments on the stable isotope method in regional groundwater investigations. *Water Resourses Research*, 7: 980-993.

Gat, J.R., 1980. The isotopes of hydrogen and oxygen in precipitation. *In: P. Fritz and J.-Ch. Fontes (Eds.) Handbook of Environmental Isotope Geochemistry, Vol. 1, The Terrestrial Environment.*, A. Elsevier, Amsterdam, The Netherlands: 21-48.

Gat, J.R. and Carmi, I., 1970. Evolution of the isotopic composition of atmospheric waters in the Mediterranean Sea area. *Journal of Geophysical Research*, 75: 3039-3048.

Gat, J.R. and Dansgaard, W., 1972. Stable isotope survey of fresh water occurrences in Israel and the Northern Jordan Rift Valley. *Journal of Hydrology*, 16: 177.

Gelwicks, J.T., 1989. The stable carbon isotope biogeochemistry of acetate and methane in freshwater environments. Unpublished Ph.D. thesis, Indiana University. Data cited in Lansdown et al., 1992.

General Electric, 1989. *Nuclides and Isotopes and Chart of the Nuclides.* 14th Edition, General Electric Nuclear Company, Nuclear Energy Operations, San Jose, CA, 57 p. plus chart.

Geyer, S., Wolf, M., Wassenaar, L.I., Fritz, P., Buckau, G. and Kim, J.I., 1993. Isotope investigations on fractions of dissolved organic carbon for ^{14}C dating. *In: Isotope Techniques in the Study of Past and Current Environmental Changes in the Hydrosphere and the Atmosphere,* IAEA Symposium 329, April 1993, Vienna: 359-380.

Ghomshei, M.M. and Clark, I.D. 1992. Oxygen and hydrogen isotopes in deep thermal waters from the South Meager Creek geothermal area, British Columbia. *Geothermics,* 22: 79-89.

Giggenbach, W.F., 1971. Isotopic composition of waters of the Broadlands geothermal field (New Zealand). *New Zealand Journal of Science,* 14: 959-970.

Giggenbach, W.F., 1992. Isotopic shifts in waters from geothermal and volcanic systems along convergent plate boundaries and their origin. *Earth and Planetary Science Letters,* 113: 495-510.

Glueckauf, E., 1946. A micro-analysis of the He and Ne contents of air. *Proceedings of the Royal Society,* A185: 98-119.

Gonfiantini, R., 1965. Effetti isotopici nell'evaporazione di acque salate. *Atti Soc. Toscana Sci. Nat. Pisa, Ser. A,* 72: 550-569.

Gonfiantini, R., 1978. Standards for stable isotope measurements in natural compounds. *Nature,* 271: 534-536.

Gonfiantini, R., 1981. The δ-notation and the mass spectrometric measurement techniques. *In: J.R. Gat and R. Gonfiantini (Eds.) Stable Isotope Hydrology, Deuterium and Oxygen-18 in the Water Cycle,* IAEA Technical Report Series No. 210, IAEA, Vienna: 35-84.

Gonfiantini, R., 1986. Environmental isotopes in lake studies. *In: P. Fritz and J.-Ch. Fontes (Eds.) Handbook of Environmental Isotope Geochemistry, Vol. 2, The Terrestrial Environment., B.* Elsevier, Amsterdam, The Netherlands: 113-168.

Gonfiantini, R. 1996. On the isotopic composition of precipitation. In: Jean Charles Fontes (1936-1994). *Un souvenir, Proceedings, International Symposium,* December 1995. European Geologist, 2: 5-8.

Gonfiantini, R., Conrad, G., Fontes, J.-Ch., Sauzay, G. and Payne, B.R., 1974. Étude isotopique de la nappe du Continental Intercalaire et de ses relations avec les autres nappes du Sahara Septentrional. *In: Isotope Techniques in Groundwater Hydrology 1974,* Vol. 1, IAEA Symposium, Vienna: 227-241.

Graves, B., 1988. *Radon in groundwater.* Lewis Publishers, Chelsea, MI, 546 p.

Green, M. and Taube, H., 1963. Isotopic fractionation in the OH^--H_2O exchange reaction. *Journal of Physical Chemistry,* 67: 1565.

Gregoire, D.C., 1987. Determination of boron isotope ratios in geological materials by inductively coupled plasma mass spectrometry. *Analytical Chemistry,* 59: 2479-2484.

Gregory, A. and Clay, R.W., 1988. Cosmic Radiation. *In: R.C. Weast (Ed.) CRC Handbook of Chemistry and Physics,* 69th Ed., CRC Lewis, Boca Raton, FA: 160-163.

Grossman, E.L., Coffman, B.K., Fritz, S.J. and Wada, H., 1989. Bacterial production of methane and its influence on ground-water chemistry in east-central Texas aquifers. *Geology,* 17: 495-499.

Guha, J. and Kanwar, R. 1987. Vug brines-fluid inclusions: a key to the understanding of secondary gold enrichment processes and the evolution of deep brines in the Canadian Shield, *In: P. Fritz and S.K. Frape (Eds.) Saline Water and Gases in Crystalline Rocks.* Geological Association of Canada, Special Paper 33: 95-101.

Gunter B.D. and Musgrave B.C., 1971. New evidence on the origin of methane in hydrothermal gases. *Geochimica Cosmochimica Acta,* 35: 113-118.

Hageman, R., Nief, G. and Roth, E., 1970. Absolute isotopic scale for deuterium analysis of natural waters. Absolute D/H ratio for SMOW. *Tellus,* 22: 712-715.

Hamid, S., Dray, M., Fehri, A., Dorioz, J.M., Normand, M. and Fontes, J.Ch., 1989. Étude des transfers d'eau à l'intérieur d'une formation moraïnique dans le Bassin du Leman-transfer d'eau dans la zone non-saturée. *Journal of Hydrology,* 109: 369-385.

He, S. and Morse, J.W., 1993. The carbonic acid system and calcite solubility in aqueous Na-K-Ca-Mg-Cl-SO_4 solutions from 0 to 90°C.

Geochimica et Cosmochimica Acta, 57: 3533-3554.

Heinzinger, K. and Weston, R.E., 1964. Isotopic fractionation of hydrogen between water and the aqueous hydroxide ion. *Journal of Physical Chemistry*, 68: 2179-2183.

Hendry, M.J. and Schwartz, F.W., 1988. An alternative view on the origin of chemical and isotopic patterns in groundwater from the Milk River Aquifer, Canada. *Water Resources Research*, 24: 1747-1763.

Hendry, M.J., Schwartz, F.W., 1990. The chemical evolution of ground water in the Milk river aquifer, Canada. *Groundwater*, 28: 253-261.

Herczeg, A.L., Simpson, H.J., Anderson, R.F., Trier, R.M., Mathieu, G.G. and Deck, B.L., 1988. Uranium and radium mobility in groundwaters and brines within the Delaware Basin, southeastern New Mexico, U.S.A. *Chemical Geology (Isotope Geoscience Section)*, 72: 181-196.

Hillaire-Marcel, C., 1986. Isotopes and food. Chapter 12 In: P. Fritz and J.-Ch. Fontes, (Eds.) *Handbook of Environmental Isotope Geochemistry, Vol. 2, The Terrestrial Environment, B.* Elsevier, Amsterdam, The Netherlands: 507-548.

Hillaire-Marcel, C. and Ghaleb, B., 1997. Thermal ionization mass spectrometry measurements of ^{226}Ra and U-isotopes in surface and groundwaters - porewater/matrix interactions revisited and potential dating implications. *Isotope Techniques in the Study of Past and Current Environmental Changes in the Hydrosphere and Atmosphere*, IAEA Symposium 349, April, 1997, Vienna.

Hinton, M.J., Schiff, S.L. and English, M.C., 1994. Examining the contribution of glacial till water to storm runoff using two and three-component hydrograph separations. *Water Resources Research*, 30: 983-993.

Hitcheon, B. and Friedman, I., 1969. Geochemistry and origin of formation waters in the western Canada sedimentary basin. I. Stable isotopes of hydrogen and oxygen, *Geochimica et Cosmochimica Acta*, 33: 1321-1349.

Hoefs, J., 1987. *Stable Isotope Geochemistry*, 3rd ed. Springer-Verlag, 236 p.

Hoehn, E. and Gunten, H.R., 1989. Radon in groundwater: a tool to assess infiltration from surface waters to aquifers. *Water Resources Research*, 28: 1795-1803.

Hoehn, E. and Santschi, P.H., 1987. Interpretation of tracer displacement during infiltration of river water to groundwater. *Water Resources Research*, 23: 633-640.

Hoehn, E., Willme, U., Hollerung, R., Schulte-Ebbert, U. and Gunten, H.R., 1992. Application of the ^{222}Rn technique for estimating the residence times of artificially recharged groundwater. *In: Isotope in Water Resources Development 1991.* IAEA Symposium 319, March 1991, Vienna: 712-714.

Hollocher, T.C., 1984. Source of oxygen atoms in nitrate in the oxidation of nitrite by *Nitrobacter agilis* and evidence against a P-O-N anhydrite mechanism in oxidative phosphorylation. *Archives of Biochemistry and Biophysics*, 233: 721-27.

Holt, B.D., Engelkemeir, A.G. and Venters, A., 1972. Variations of sulfur isotope ratios in samples of water and air near Chicago. *Environmental Science and Technology*, 6: 338-341.

Horibe, D.Y. and Craig, H., 1975. written communication, cited in Friedman, I. and O'Neil, J.R., 1977. Compilation of stable isotope fractionation factors of geochemical interest. *Data of Geochemistry, U.S.G.S. Prof. Paper 440-KK (6th ed.).*

Horita, J. and Gat, J.R., 1989. Deuterium in the Dead Sea: remeasurement and implications for the isotopic activity correction in brines. *Geochimica et Cosmochimica Acta*, 53: 131-133.

Horita, J. and Wesolowski, D.J., 1994. Liquid-vapor fractionation of oxygen and hydrogen isotopes of water from the freezing to the critical temperature. *Geochimica et Cosmochimica Acta*, 58: 3425-3437.

Hötzl, H., Job, C., Moser, H., Rauert, W., Stichler, W. and Zötl, J.G., 1980. Isotope methods as a tool for Quaternary studies in Saudi Arabia. *In: Arid Zone Hydrology: Investigations with Isotope Techniques*, Proceedings of an IAEA Advisory Group Meeting, November 1978, Vienna: 215-235.

Hoyt, D.V., Kyle, H.L., Hickey, J.R. and Maschhoff, R.M., 1992. The Nimbus 7 solar total irradiance: a new algorithm for its derivation. *Journal of Geophysical Research*, 97:51.

Hutcheon, I., Abercrombie, H.J., Putnam, P., Gardner, R. and Krouse, H.R., 1989. Diagenesis and sedimentology of the Clearwater Formation at Tucker Lake. *Bulletin of Canadian Petroleum Geology*, 37: 83-97.

IAEA, 1983. *Isotope Techniques in Hydrogeological Assessment of Potential Sites for the Disposal of High-Level Radioactive Wastes*. IAEA Technical Report Series No. 228, IAEA, Vienna, 151p.

IAEA, 1995. Reference and intercomparison materials for stable isotopes of light elements. *Proceedings of a Consultants Meeting held in Vienna, 1-3 December, 1993*. IAEA-TECDOC-825, IAEA, Vienna. 165 p.

Imbrie, J.J., Hays, J.D., Martinson, D.G., McIntyre, A., Mix, A.C., Morley, J.J., Pisias, N.G., Prell, W.L. and Shackelton, N.J., 1984. The orbital theory of Pleistocene climate: support from a revised chronology of the marine $d^{18}O$ record. *In: Berger et al. (Eds.) Milankovitch and Climate*, D. Reidel, Dordrecht Publishing Co., The Netherlands: 269-305.

Ingraham, N.L. and Matthews, R.A., 1988. Fog drip as a source of groundwater recharge in northern Kenya. *Water Resources Research*, 24: 1406-1410.

Ingraham, N.L. and Matthews, R.A., 1990. A stable isotopic study of fog: the Point Reyes Peninsula, California, U.S.A. *Chemical Geology (Isotope Geoscience Section)*, 80: 281-290.

Irwin, H., Curtis, Ch. and Coleman, M., 1977. Isotopic evidence for source of diagenetic carbonates formed during burial of organic-rich sediments. *Nature*, 269: 209-213.

Ivanovich, M., Frohlich, K., Hendry, M.J. andrews, J.N., Davis, S.N. ands Drimmie, R.J., 1992. Evaluation of isotopic methods for the dating of very old groundwaters. A case study of the Milk River Aquifer. *In: Isotope Techniques in Water Resources Development 1991*, IAEA Symposium 319, March 1991, Vienna: 229-244.

Ivanovich, M. and Harmon, R.S., 1982. *Uranium Series Disequilibrium*, Clarendon Press, Oxford, 571p.

James, A.T. and Baker, D.R., 1976. Oxygen isotope exchange between illite and water at 22°C. *Geochimica et Cosmochimica Acta*, 40: 235-240.

Kakiuchi, M. and Matsuo, S., 1979. Direct measurements of D/H and $^{18}O/^{16}O$ fractionation factors between vapor and liquid water in the temperature range from 10° to 40°. *Geochemical Journal*, 13: 307-311.

Kamineni, D.C., 1987. Halogen-bearing minerals in plutonic rocks: a possible source of chlorine in saline groundwater in the Canadian Shield. *In: P. Fritz and S.K. Frape (Eds.) Saline Water and Gases in Crystalline Rocks*. Geological Association of Canada, Special Paper 33: 69-80.

Kaufmann R.S., 1989. Equilibrium exchange models for chlorine stable isotope fractionation in high temperature environments. *In: D.M. Miles (Ed.) Proceedings, Water-Rock Interaction 6*, Balkema: 365-368..

Kaufmann R.S., Frape S., Fritz P. and Bentley H., 1987. Chlorine stable isotope composition of Canadian Shield brines. *In: P. Fritz and S.K. Frape (Eds.) Saline Water and Gases in Crystalline Rocks*. Geological Association of Canada Special Paper, 33: 89-93.

Kaufmann, R.S., Frape, S.K., McNutt, R. and Eastoe, C., 1993. Chlorine stable isotope distribution of Michigan Basin formation waters. *Applied Geochemistry*, 8: 403-407.

Kaufmann R.S., Long A., Bentley H. and Davis S., 1984. Natural chlorine isotope variations. *Nature*, 309: 338-340.

Kaufman, S. and Libby, W.F., 1954. The natural distribution of tritium. *Physical Review*, 93: 1337-1344.

Kawabe, I., 1978. Calculation of oxygen isotope fractionation in quartz-water system with special reference to the low temperature fractionation. *Geochimica et Cosmochimica Acta*, 42: 613-621.

Kendall, C., Sklash, M. and Bullen, Th. D., 1995. Isotope tracers of water and solute sources in catchments. Chapter 10. *In: S.T. Trudgill (Ed.), Solute Modelling in Catchment Systems*, Wiley, New-York.

Khoury H.N., Salameh E. and Abdul-Jaber Q., 1985. Characteristics of an unusual highly alkaline water from the Maqarin Area, northern Jordan. *Journal of Hydrology*, 81: 79-91.

Kim, J.I., Artinger, R., Buckau, G., Kardinal, Ch., Geyer, S., Wolf, M., Halder, H. and Fritz, P., 1995. Grundwasserdatierung mittels ^{14}C-

Bestimmung and gelösten Humin- und Fulvinsäuren. *Final Report BMFT project 02 E 8331 6 and Rep. Techn. Univ. Munich, RCM 00895*, Munich, Germany, 1995: 221.

Kita, I., Taguchi, S. and Matsubaya, O., 1985. Oxygen isotope fractionation between amorphous silica and water at 34-93°C. *Nature*, 314: 83-84.

Klass, D.L., 1984. Methane from anaerobic fermentation. *Science*, 223: 1021-1028.

Koerner, R.M., 1989. Ice core evidence for extensive melting of the Greenland Ice Sheet in the last interglacial. *Science*, 244: 964-968.

Kreitler, C.W., 1975. Determining the source of nitrate in ground water by nitrogen isotope studies. *Report of Investigations No. 83*, Bureau of Economic Geology, University of Texas, Austin.

Krouse, H.R., Cook, F.D., Sasaki, A. and Šmejkal, V., 1970. Microbial isotope fractionation in springs in western Canada. In: K. Ogata and T. Hayakawa (Eds.) *Recent Developments in Mass Spectroscopy*. Proceedings, International Conference on Mass Spectroscopy, Kyoto, Japan: 629-639.

Krouse, H.R., 1980. Sulphur isotopes in our environment. In: P. Fritz and J.-Ch. Fontes (Eds.) *Handbook of Environmental Isotope Geochemistry I, The Terrestrial Environment, A*. Elsevier, Amsterdam, The Netherlands: 435-472.

Kurtz, M.D. and Jenkins, W.J., 1981. The distribution of helium in oceanic basalt glasses. *Earth and Planetary Science Letters*, 53: 41-54.

Kutschera, M., Henning, W., Paul., M., Smither, R.K., Stephenson, E.J., Yentna, Y. IL., Alburger, D.E. and Harbottle, G., 1980. Measurement of the half-life of ^{32}Si via accelerator mass spectrometry. *Physics Review Letters*, 45: 592.

Lakey, B. and Krothes, N.C., 1996. Stable isotopic variation of storm discharge from a perennial karst spring, Indiana. *Water Resources Research*, 32: 721-731.

Land, L.S., 1991. Dolomitization of the Hope Gate Formation (north Jamaica) by seawater: Reassessment of mixing-zone dolomite. In: H.P. Taylor, J.R. O'Neil and I.R. Kaplan, (Eds.) *Stable Isotope Geochemistry: A Tribute to Samuel Epstein*, The Geochemical Society Special Publication No. 3, San Antonio, Texas: 121-133.

Lansdown, J.M., Quay, P.D. and King, S.L., 1992. CH_4 production via CO_2 reduction in a temperate bog: A source of ^{13}C-depleted CH_4. *Geochimica et Cosmochimica Acta*, 56: 3493-3503.

Langmuir, D., 1997. *Aqueous Environmental Geochemistry*. Prentice Hall, New Jersey. 600 p.

Lauriol, B. and Clark, I.D., 1993. An approach to determine the origin and age of massive ice blockages in two Arctic caves. *Permafrost and Periglacial Processes*, 4: 77-85.

Lauriol, B., Duchesne, C. and Clark, I.D., 1995. Systématique du remplissage en eau des fentes de gel: les résultats d'une étude oxygène-18 et deutérium. *Permafrost and Periglacial Processes*, 16: 47-55.

Lauriol, B., Clark, I.D., Hillaire-Marcel, C. and Ghalab, B., 1996. Biogenic fissure carbonates from the Yukon: A new material to date climatic optimums and the last interglacial in the north. Proceedings, Workshop on Climate History and Dynamics, Canadian Geophysical Union Annual Meeting, May 5-11, 1996, Banff.

Lawrence, J.R., 1989. The stable isotope geochemistry of deep-sea pore water. In: P. Fritz and J.-Ch. Fontes (Eds) *Handbook of Environmental Isotope Geochemistry, Vol. 3, The Marine Environment, A.*, Elsevier, Amsterdam, The Netherlands: 317-356.

Lawrence, J.R. and Taylor, H.P., Jr., 1972. Hydrogen and oxygen isotope systematics in weathering profiles. *Geochimica et Cosmochimica Acta*, 33: 1377-1393.

Létolle, R., 1980. Nitrogen-15 in the natural environment. Chapter 10, In: P. Fritz and J.-Ch. Fontes (Eds.) *Handbook of Environmental Isotope Geochemistry, Vol. 1, The Terrestrial Environment, A*. Elsevier, Amsterdam, The Netherlands: 407-433.

Lloyd, J.W., 1980. Aspects of environmental isotope geochemistry in groundwaters in Eastern Jordan. *Arid Zone Hydrogeology: Investigations with Isotope Techniques*. Proceedings, IAEA Advisory Group Meeting, November, 1978, Vienna:193-204.

Lloyd, R.M., 1967. Oxygen-18 composition of oceanic sulfate. *Science*, 156: 1228-1231.

Lloyd, R.M., 1968. Oxygen-18 behavior in the sulfate-water system. *Journal of Geophysical Research*, 73: 6099-6110.

Lodemann, M., Fritz, P., Wolf, M., Hansen, B.T, Ivanovich, M. and Nolte, E., 1997. On the origin of saline fluids in the KTB (Continental Deep Drilling Project of Germany). *Applied Geochemistry*, submitted for publication March 1997.

Longinelli, A., 1989. Oxygen-18 and sulphur-34 in dissolved oceanic sulphate and phosphate. Chapter 7 *In: P. Fritz and J.-Ch. Fontes (Eds.) Handbook of Environmental Isotope Geochemistry, Vol. 3, The Marine Environment, A.* Elsevier, Amsterdam, The Netherlands. 219-256.

Loosli, H.H., 1983. A dating method with ^{39}Ar. *Earth and Planetary Science Letters*, 63: 51-62.

Loosli, H.H., Lehmann, B.E. and Däppen, G., 1991. Dating by radionuclides. Chapter 4 *In: J. Pearson et al. (Eds.) Applied Isotope Hydrology. A Case Study In Northern Switzerland.* Studies in Environmental Science, 43, Elsevier, Amsterdam: 153-174.

Loosli, H.H. and Oeschger, H., 1979. Argon-39, carbon-14 and krypton-85 measurements in groundwater samples. *In: Isotope Hydrology 1978*, Vol. II, IAEA Symposium 228, June 1978, Neuherberg, Germany: 931-997.

Magenheim, A.J., Spivack, A.J., Volpe, C. and Ransom, B., 1994. Precise determination of stable chlorine isotopic ratios in low-concentration natural samples. *Geochimica et Cosmochimica Acta*, 58: 3117-3121.

Majoube, M., 1971. Fractionnement en oxygène-18 et en deutérium entre l'eau et sa vapeur. *Journal of Chemical Physics*, 197: 1423-1436.

Maloszewski, P. and Zuber, A., 1984. Interpetation of artificial and environmental tracers in fissured rocks with a porous matrix. *In: Isotope Hydrology 1983.* IAEA Symposium 270, September 1983, Vienna: 635-651.

Maloszewski, P. and Zuber, A., 1991. Influence of matrix diffusion and exchange reactions on radiocarbon ages in fissured carbonate aquifers. *Water Resources Research*, 27: 1937-1945.

Maloszewski, P., Moser, H., Stichler, W., Bertleff, B. and Hedin, K., 1987. Modelling of groundwater pollution by riverbank filtration using oxygen-18 data. *In: Groundwater Monitoring and Management*, Proceedings, Dresden Symposium, March, 1987, IAHS Publ. No. 173: 153-161.

Mariotti, A. and Letolle, R., 1977. Application de l'étude isotopique de l'azote en hydrologie et en hydrogéologie — Analyse des résultats obtenus sur un example précis: le bassin de Mélarchez (Seine et Marnes, France). *Journal of Hydrology*, 33: 157-172.

Marty, B. and A. Jambon, 1987. C/^{3}He in volatile fluxes from the solid earth: implications for carbon geodynamics. *Earth and Planetary Science Letters*, 83: 16-26.

Mathieu, R. and Bariac, T., 1996. An isotopic study (^{2}H and ^{18}O) on water movements in clayey soils under a semiarid climate. *Water Resources Research*, 32: 779-789.

McClure, H.A., 1976. Radiocarbon chronology of Late Quaternary lakes in the Arabian Peninsula. *Nature*, 263: 755-756.

McCrea, J.M., 1950. On the isotope chemistry of carbonates and a paleotemperature scale. *Journal of Chemical Physics*, 18: 849-857.

McDermott, F., Ivanovich, M., Frape, S.K. and Hawkesworth, C.J., 1996. Paleoclimatic controls on hydrological systems: evidence from U-Th dated calcite veins in the Fennoscandian and Canadian shields. *In: Isotopes in Water Resources Management*, IAEA Symposium 336, March 1995, Vienna: 401-416.

McKenzie, W.F. and Truesdell, A.H., 1977. Geothermal reservoir temperatures estimated from the oxygen isotope compositions of dissolved sulfate and water from hot springs and shallow drillholes. *Geothermics*, 5: 51-61.

Merlivat, L., 1970. L'Étude quantitative de bilans de lacs à l'aide des concentrations en deutérium et oxygene-18 dans l'eau. *In: Isotope Hydrology 1970*, IAEA Symposium 129, March 1970, Vienna: 89-107.

Merlivat, L. and Jouzel, J., 1979. Global climatic interpretation of the deuterium - oxygen-18 relationship for precipitation. *Journal of Geophysical Research*, 84: 5029-5033.

Michel, F.A., 1977. Hydrogeologic studies of springs in the Central Mackenzie Valley, N.W.T., Canada. M.Sc. Thesis, Department of Earth Sciences, University of Waterloo, Waterloo, Ontario, Canada, 185 p.

Michel, F.A., 1982. Isotope investigations of permafrost waters in northern Canada. Ph.D. thesis, University of Waterloo, Waterloo, Ontario, Canada, 424 p.

Michel, F.A., Kubasiewicz, M., Patterson, R.J. and Brown, R.M., 1984. Ground water flow velocity derived from tritium measurements at the Gloucester landfill site, Gloucester, Ontario. *Water Pollution Research Journal of Canada*, 19: 13-22.

Michel, F.A., 1986. Isotope geochemistry of frost-blister ice, North Fork Pass, Yukon, Canada. *Canadian Journal of Earth Sciences*, 23: 543-549.

Minagawa, M. and Wada, E., 1984. Stepwise enrichment of ^{15}N along food chains: further evidence and the relation between $\delta^{15}N$ and animal age. *Geochimica et Cosmochimica Acta*, 48: 1135-1140.

Mook, W.G., Bommerson, J.C. and Staverman, W.H., 1974. Carbon isotope fractionation between dissolved bicarbonate and gaseous carbon dioxide. *Earth and Planetary Science Letters*, 22:169-176.

Moorman, B.J., Michel, F.A. and Drimmie, R.J., 1996. Isotopic variability in Arctic precipitation as a climatic indicator. *Geoscience Canada*, 23: 189-194.

Morgenstern, U., Gellerman, R., Hebert, D., Börner, I., Stolz, W., Vaikmäe R. and Putnik, H., 1995. ^{32}Si in limestone aquifers. *Chemical Geology*, 120: 127-134.

Moser, H. and Stichler, W., 1970. Deuterium measurements on snow samples from the Alps. *Isotope Hydrology 1970*, IAEA Symposium 129, March, 1970, Vienna: 43-57.

Moser, H. and Stichler, W., 1975. Deuterium and oxygen-18 contents as index of the properties of snow blankets. *In: Snow Mechanics*, Proceedings, Grindelwald Symposium, April 1974, IAHS Publication 114: 122-135.

Moser, H., Wolf, M., Fritz, P., Fontes, J-Ch., Florkowski, T. and Payne, B.R., 1989. Deuterium, oxygen-18, and tritium in Stripa groundwater. *Geochimica et Cosmochimica Acta*, 53: 1757-1764.

Moser, H., Pak E., Rauert W., Stichler W. and Zötl J.G., 1978. Isotopic composition of waters of Al Qatif and Al Hasa areas. *In: S.S. Al-Sayari and J.G. Zotl (Eds.) Quaternary Period in Saudi Arabia 1: Sedimentological, Hydrogeological, Hydrochemical, Geomorphological, and Climatological Investigations in Central and Eastern Saudi Arabia*, Springer Verlag, Vienna: 153-163.

Musgrove, M. and Banner, J.L., 1993. Regional ground-water mixing and the origin of saline fluids: Midcontinent, United States. *Science*, 259: 1877-1882.

Murphy, E.M., Davis, S.N., Long, A., Donahue, D. and Jull, A.J.T., 1989. Characterization and isotopic composition of organic and inorganic carbon in the Milk River Aquifer. *Water Resources Research*, 25: 1893-1905.

Nicolet, M., 1957. The aeronomic problem of helium. *Annals of Geophysics*, 13: 1-21.

Nordstrom, D.K., Burchard, J.M. and Alpers, N., 1990. The production and seasonal variability of acid mine drainage from Iron Mountain, California: A superfund site undergoing rehabilitation. *In: J.W. Gadsby J.A. Malick and S.J. Day (Eds.) Acid Mine Drainage: Designing for Closure*, Proceedings, GAC/MAC Joint Annual Meeting, May, 1990, Vancouver, B.C., BiTech Publishers, Vancouver, Canada: 13-22.

Northrop, D.A. and Clayton, R.N., 1966. Oxygen isotope fractionation in systems containing dolomite. *Journal of Geology*, 74: 174-196.

Nriagu, J.O., 1975. Sulfur isotopic variations in relation to sulfur pollution of Lake Erie. *In: Isotope Ratios as Pollutant Source and Behaviour Indicators*. IAEA Symposium, Vienna: 77-93.

Nriagu, J.O. and Coker, R.D., 1978. Isotopic composition of sulphur in precipitation within the Great Lakes Basin. *Tellus*, 30: 365-375.

Oeschger, H. and Loosli, H.H., 1977. New developments in sampling and low level counting of natural radioactivity. *In: P. Povinec and S. Usacev (Eds.) Low Radioactivity Measurements and Applications*. International Conference, October 1975, High Tatras: 13-22.

O'Leary, M.H., 1988. Carbon isotopes in photosynthesis. *BioScience*, 38: 328.

O'Neil, J.R., 1968. Hydrogen and oxygen isotope fractionation between ice and water. *Journal of Physical Chemistry*, 72: 3683-3684.

O'Neil, J.R. and Taylor, H.P., 1967. The oxygen isotope and cation exchange chemistry of feldspars. *American Mineralogist*, 52: 1414-1437.

O'Neil, J.R. and Truesdell, A.H., 1991. Oxygen isotope fractionation studies of solute-water interactions. *In: H.P. Taylor , J.R. O'Neil and I.R. Kaplan (Eds.) Stable Isotope Geochemistry: A Tribute to Samuel Epstein*, The Geochemical Society Special Publication No. 3, San Antonio, Texas: 17-25.

O'Neil, J.R., Clayton, R.N. and Mayeda, T.K., 1969. Oxygen isotope fractionation in divalent metal carbonates. *Journal of Chemical Physics*, 51: 5547-5558.

Orlov, D.S., 1995. *Humic Substances of Soils and General Theory of Humification*. Russian Translation Series 111, A.A. Balkema: Rotterdam, 323 p.

Panichi, C., Ferrara, G.C. and Gonfiantini, R., 1977. Isotope thermometry in the Larderello (Italy) geothermal field. *Geothermics*, 5: 81-88.

Parkhurst, D.L., Plummer, L.N. and Thorstenson, D.C., 1980. PHREEQE — a computer program for geochemical calculations. *U.S. Geological Survey Water Resources Investigation 80-96*. (Revised, 1985).

Pearson, F.J., 1965. Use of C-13/C-12 ratios to correct radiocarbon ages of material initially diluted by limestone. *In: Proceedings of the 6th International Conference on Radiocarbon and Tritium Dating*, Pulman, Washington, 357.

Pearson, F.J., Balderer, W., Loosli, H.H., Lehmann, B.E., Matter, A., Peters, Tj., Schmassmann, H. and Gautschi, A., 1991. *Applied Isotope Hydrology. A Case Study In Northern Switzerland*. Studies in Environmental Science, 43, Elsevier, Amsterdam 439 p.

Pearson, F.J. and Hanshaw, B.B., 1970. Sources of dissolved carbonate species in groundwater and their effects on carbon-14 dating. *In: Isotope Hydrology 1970*, IAEA Symposium 129, March 1970, Vienna: 271-286.

Pearson, G.W., Pilcher, J.R., Baillie, M.G.L. Corbett, D.M. and Qua, F., 1986. High-precision ^{14}C measurement of Irish oaks to show the natural ^{14}C variations from AD 1840 to 5210 BC. *Radiocarbon*, 28: 911-934.

Pierre, C., 1989. Sedimentation and diagenesis in restricted marine basins. *In P. Fritz and J.-Ch. Fontes (Eds.) Handbook of Environmental Isotope Geochemistry, Vol. 3, The Marine Environment, A*, Elsevier, Amsterdam, Netherlands: 257-316.

Phillips, F.M., Bentley, H.W., Davis, S.N., Elmore, D. and Swanick, G., 1986. Chlorine-36 dating of very old groundwater. 2. Milk River aquifer, Alberta, Canada. *Water Resources Research*, 22: 2003-2016.

Plummer, N.L., Prestemon, E. C. and Parkhurst, D. L., 1994. *An interactive code (NETPATH) for modeling net geochemical reactions along a flow path, version 2.0*. U.S. Geological Survey Water Resources Investigations Report 94-4169, U.S.G.S., Reston, VA.

Qureshi, R.M., Aravena, R., Fritz, P. and Drimmie, R., 1989. The CO_2 absorption method as an alternative to benzene synthesis method for ^{14}C dating. *Applied Geochemistry*, 4: 625-633.

Rank, D., Völkl, G., Maloszewski, P. and Stichler, W., 1992. Flow dynamics in an alpine karst massif studied by means of environmental isotopes. *Isotope Techniques in Water Resources Development 1991*, IAEA Symposium 319, March 1991, Vienna: 327-343.

Rath, H.K., 1988. Simulation of the Global ^{85}Kr and $^{14}CO_2$ Distribution by Means of a Time Dependent Two-dimensional Model of the Atmosphere (in German). Ph.D. thesis, University of Heidelberg, Germany.

Reardon, E.J. and Fritz, P., 1978. Computer modelling of groundwater ^{13}C and ^{14}C isotope compositions. *Journal of Hydrology*, 36: 210-224.

Remenda, V.H., Cherry, J.A. and Edwards, T.W.D., 1994. Isotopic composition of old groundwater from Lake Agassiz: Implications for late Pleistocene climate. *Science*, 266: 1975-1978.

Richet, P., Bottinga, Y. and Javoy, M., 1977. A review of hydrogen, carbon, nitrogen, oxygen, sulphur, and chlorine stable isotope fractionation among gaseous molecules. *Annual Review of Earth Planetary Science*, 5: 65-110.

Robinson, B.W., 1973. Suphur isotope equilibrium during sulphur hydrolysis at high temperatures. *Earth and Planetary Science Letters*, 18: 47-50.

Rodhe, A., 1984. Groundwater contribution to stream flow in Swedish forested till soil as estimated by oxygen-18. *Isotope Hydrology*

1983, IAEA Symposium 270, September 1983, Vienna: 55-66.

Rosenbaum, J. and Sheppard, F.M.F. 1986. An isotopic study of siderites, dolomites and ankerites at high temperature. *Geochimica et Cosmochimica Acta*, 50: 1147-1150.

Rozani, C. and Tamers, M.A., 1966. Low-level chlorine-36 detection with liquid scintillation techniques. *Radiochimica Acta*, 6: 206-210.

Rozanski, K., 1985. Deuterium and Oxygen-18 in European groundwaters-links to atmospheric circulation in the past. *Chemical Geology (Isotope Geoscience Section)*, 52: 349-363.

Rozanski, K., Araguás-Araguás, L. and Gonfiantini, R., 1993. Isotopic patterns in modern global precipitation. *In: Continental Isotope Indicators of Climate*, American Geophysical Union Monograph.

Salati, E., Dall'olio, A., Matsui, E. and Gat, J.R., 1979. Recycling of water in the Amazon Basin: an isotopic study. *Water Resources Research*, 15: 1250.

Saliège, J.-F. and Fontes, J.-Ch., 1984. Essai de détermination expérimentale du fractionnement des isotopes ^{13}C et ^{14}C du carbone au cours de processus naturels. *International Journal of Applied Radiation and Isotopes*, 35: 55-62.

Sakai, H., 1968. Isotopic properties of sulphur compounds in hydrothermal processes. *Geochemical Journal*, 2: 29-49.

Savin, S.M. and Epstein, S., 1970. The oxygen and hydrogen isotope geochemistry of clay minerals. *Geochimica et Cosmochimica Acta*, 34: 25-42.

Schlosser, P., Stute, M., Sonntag, C. and Munnich, K.O., 1989. Tritogenic ^{3}He in shallow groundwater. *Earth and Planetary Science Letters*, 94: 245-254.

Schoell, M., 1980. The hydrogen and carbon isotopic composition of methane from natural gas of various origins. *Geochimica et Cosmochimica Acta*, 44: 649-662.

Scholtis, A., Pearson, F.J. Jr., Loosli, H.H., Eichinger, L., Waber, H.N. and Lehmann, B.E., 1996. Integration of environmental isotopes, hydrochemical and mineralogical data to characterize groundwaters from a potential repository site in central Switzerland. *In: Isotopes in Water Resource Management*, IAEA Symposium 336, March, 1995, Vienna: 263-280.

Schwarcz, H.P., 1986. Geochronology and isotopic geochemistry of speleothem. Chapter 7 *In: P. Fritz and J.-Ch. Fontes, (Eds.) Handbook of Environmental Isotope Geochemistry, Vol. 2, The Terrestrial Environment, B.* Elsevier, Amsterdam, The Netherlands: 271-303.

Seuss, H.E., 1980. The radiocarbon record in tree rings of the last 8000 years. *In: M. Stuiver and R.S. Kra (Eds.) Proceedings, 10th International ^{14}C Conference, Radiocarbon*, 22: 200-209.

Shackleton, J.N. and Opdyke, N.D., 1973. Oxygen isotope and palaeomagnetic stratigraphy of equatorial Pacific core V28-238: oxygen isotope temperatures and ice volumes on a 10^5 and 10^6 year scale. *Quaternary Research*, 3: 39-55.

Sharma, T. and Clayton, R.N., 1965. Measurement of ^{18}O/^{16}O ratios of total oxygen of carbonates. *Geochimica et Cosmochimica Acta*, 29: 1347-1353.

Shemesh, A., Ron, H., Erel, Y., Kolodny, Y. and Nur, A., 1992. Isotopic composition of vein calcite and its fluid inclusions: implications to paleohydrological systems, tectonic events and vein formation processes. *Chemical Geology (Isotope Geoscience Section)*, 94: 307-314.

Sherwood, B., Fritz, P., Frape, S.K., Macko, S.A., Weise, S.M. and Welhan, J.A., 1988. Methane occurrences in the Canadian Shield. *Chemical Geology*, 71: 223-236.

Sherwood Lollar, B., Frape, S.K., Weise, S.M., Fritz, P., Macko, S.A. and Welhan, J.A., 1993. Abiogenic methanogenesis in crystalline rocks. *Geochimica et Cosmochimica Acta*, 57: 5087-5097.

Shieh Y.N. and Taylor H.P., 1969. Oxygen and carbon isotope studies of contact metamorphism of carbonate rocks. *Journal of Petrology*, 10: 307-331.

Shiro, Y. and Sakai, H., 1972. Calculation of the reduced partition function ratios of alpha-beta quartz and calcite. *Japan Chemical Society Bulletin*, 45: 2355-2359.

Siegel, D.T., Stoner, D., Byrnes, T. and Bennett, P., 1990. A geochemical process approach to identify inorganic and organic ground-water contamination. *Ground Water Management, Number 2*, Fourth National Outdoor Action Conference on Aquifer Restoration, Ground Water Monitoring and Geophysical Methods,

Las Vegas, Nevada, National Ground Water Association, Columbus, OH: 1291-1301.

Siegenthaler, U., Schotterer, U. and Oeschger, H., 1983. Sauerstoff-18 und Tritium als natürliche Tracer für Grundwasser: *Gas-Wasser-Abwasser*, 63: 477-483.

Simon, K. and Hoefs, J., 1993. O, H and C isotope study of rocks from the KTB pilot hole: crustal profile and constraints on fluid evolution. *Contributions to Mineralogy and Petrology*, 114: 42-52.

Simpson, H.J., Hamza, M.S., White, J.W.C., Nada, A. and Awad, M.A., 1987. Evaporative enrichment of deuterium and ^{18}O in arid zone irrigation. *In: Isotope Techniques in Water Resources Development*, IAEA Symposium 299, March 1987, Vienna: 241-256.

Sklash, M.G., Farvolden, R.N. and Fritz, P., 1976. A conceptual model of watershed response to rainfall, developed through the use of oxygen-18 as a natural tracer. *Canadian Journal of Earth Sciences*, 13: 271-283.

Sofer, Z. and Gat, J.R., 1972. Activities and concentrations of oxygen-18 in concentrated aqueous salt solutions: analytical and geophysical implications. *Earth and Planetary Science Letters*, 15: 232-238.

Sofer, Z. and Gat, J.R., 1975. The isotopic composition of evaporating brines: effect on the isotopic activity ratio in saline solutions. *Earth and Planetary Science Letters*, 26: 179-186.

Sofer, Z., 1978. Isotopic composition of hydration water in gypsum. *Geochimica et Cosmochimica Acta*, 42: 1141-1149.

Sonntag, C., Klitzsch, E., Lohnert, E.P., El-shazli, E.M., Münnich, K.O., Junghans, Ch., Thorweihe, U., Weistoffer, K. and Swailem, F.M., 1979. Palaeoclimatic information from deuterium and oxygen-18 in carbon-14 dated North Saharan groundwaters. *In: Isotope Hydrology 1978*, Vol. II, IAEA Symposium 228, June 1978, Neuherberg, Germany: 569-581.

Stahl, W.J., 1980. Compositional changes and $^{13}C/^{12}C$ fractionations during the degradation of hydrocarbons by bacteria. *Geochimica et Cosmochimica Acta*, 44: 1903-1907.

Stanger, G., 1986. The Hydrogeology of the Oman Mountains. Unpublished. Ph.D. thesis, Open University, U.K., 572 p.

Stevenson, F.J., 1985. Geochemistry of soil humic substances. Chapter 2, *In: G.R.Aiken, D.M. McKnight, R.L.Wershaw and P. MacCarthy (Eds.), Humic Substances in Soil, Sediment, and Water*, John Wiley & Sons, New York: 13-52.

Stichler, W., 1980. Modell zur Berechnung der Verweilzeit des infiltrierten Niederschlags. *GSF Bericht R 240*, Munich, Germany.

Strebel, O., Böttcher, J. and Fritz, P., 1990. Use of isotope fractionation of sulphate sulphur and sulphate oxygen to assess bacterial desulphication in a sandy aquifer. *Journal of Hydrology*, 121: 155-172.

Street, F.A. and Grove, A.T., 1979. Global maps of lake-level fluctuations since 30,000 B.P. *Quaternary Research*, 12: 83-118.

Stuiver, M. and Quay, P.D., 1980. Changes in atmopsheric ^{14}C attributed to a variable sun. *Science*, 9: 1-20.

Stumm, W. and Morgan, J.J., 1996. *Aquatic Chemistry, An Introduction Emphasizing Chemical Equilibria in Natural Waters*. 2nd Edition, Wiley Interscience, 780 p.

Suchomel, K.H., Kreamer, D.K. and Long, A., 1990. Production and transport of carbon dioxide in a contaminated vadose zone: a stable and radioactive isotope study. *Environmental Science Technology*, 24: 1824-1831.

Suzuoki, T. and Kumura, T., 1973. D/H and $^{18}O/^{16}O$ fractionation in ice-water systems. *Mass Spectroscopy*, 21: 229-233.

Suzuoki, T. and Epstein, S., 1976. Hydrogen isotope fractionation between OH-bearing minerals and water. *Geochimica et Cosmochimica Acta*, 40: 1229-1240.

Szabo, Z., Rice, D.E., Plummer, L.N., Busenberg, E. Drenkard, S. and Schlosser, P., 1996. Age dating of shallow groundwater with chlorofluorocarbons, tritium/helium-3, and flow path analyses, southern New Jersey coastal plain. *Water Resources Research*, 32: 1023-1038.

Szapiro, S. and Steckel, F., 1967. Physical properties of heavy oxygen water. 2. Vapour pressure. *Transactions Faraday Society*, 63: 883.

Szaran, J., 1996. Experimental investigation of sulphur isotopic fractionation between dissolved and gaseous H_2S. *Chemical Geology (Isotope Geoscience Section)*, 127: 223-228.

Tamers, M.A., 1975. The validity of radiocarbon dates on groundwater. *Geophysical Survey*, 2: 217-239.

Tan, F.C., 1989. Stable carbon isotopes in dissolved inorganic carbon in marine and estuarine environments. Chapter 5, *In: P. Fritz and J.-Ch. Fontes (Eds.), Handbook of Environmental Isotope Geochemistry, Vol. 3, The Marine Environment, A.*: 177-190.

Tanaka N. and Rye D., 1991. Chlorine in the stratosphere. *Nature*, 353: 707.

Taylor, B.E., Wheeler, M.C. and Nordstrom, D.K., 1984. Stable isotope geochemistry of acid mine drainage: experimental oxidation of pyrite. *Geochimica et Cosmochimica Acta*, 48: 2669-2678.

Taylor, C.B., Wilson, D.D., Brown, L.J., Stewart, M.K., Burden, R.J. and Brailsford, G.W., 1989. Sources and flow of North Canterbury Plains groundwater, New Zealand. *Journal of Hydrology*, 106: 311-340.

Thode, H.G., Shima, M., Rees, C.E. and Krishnamurty, K.V., 1965. Carbon-13 isotope effects in systems containing carbon dioxide, bicarbonate, carbonate, and metal ions. *Canadian Journal of Chemistry*, 43: 582-595.

Thompson, G.M. and Hayes, J.M., 1979. Trichlorofluoromethane in groundwater — a possible tracer and indicator of groundwater age. *Water Resources Research*, 15: 546-554.

Thurman, E.M., 1985. Humic substances in groundwater. Chapter 4, *In: G.R.Aiken, D.M. McKnight, R.L.Wershaw and P. MacCarthy (Eds.), Humic Substances in Soil, Sediment, and Water*. John Wiley & Sons:, New York: 87-103.

Toran, L. and Harris, R.F., 1989. Interpretation of sulphur and oxygen isotopes in biological and abiological sulfide oxidation. *Geochimica et Cosmochimica Acta*, 53: 2341-2348.

Torgersen, T., Clarke, W.B. and Jenkins, W.J., 1979. The tritium/helium-3 method in hydrology. *Isotope Hydrology 1978*, Vol. II, IAEA Symposium 228, June 1978, Neuherberg, Germany: 917-30.

Truesdell, A.H. and Hulston, J.R., 1980. Isotopic evidence of environments of geothermal systems, Chapter 5. *In: P. Fritz and J.-Ch. Fontes (Eds.), Handbook of Environmental Isotope Geochemistry, Vol. 1, The Terrestrial Environment., A.* Elsevier, Amsterdam, The Netherlands: 179-226.

Truesdell, A.H., Nathenson, M. and Rye, R.O. 1977. The effects of subsurface boiling and dilution on the isotopic composition of Yellowstone thermal waters. *Journal of Geophysical Research*, 82: 3694-3703.

Turi, B., 1986. Stable isotope geochemistry of travertines. *In: P. Fritz and J.-Ch. Fontes, (Eds.) Handbook of Environmental Isotope Geochemistry, Vol. 2, The Terrestrial Environment, B.* Elsevier, Amsterdam, The Netherlands: 207-238.

Turner, J.V., 1982. Kinetic fractionation of carbon-13 during calcium carbonate precipitation. *Geochimica et Cosmochimica Acta*, 46: 1183-1191.

Turner, J.V., Bradd, J.M. and Waite, T.D., 1992. Conjunctive use of isotopic techniques to elucidate solute concentration and flow processes in dryland salinized catchments. *In: Isotope Techniques in Water Resources Development 1991*, IAEA Symposium 319, March 1991, Vienna: 33-60.

Unterweger, M.P., Coursey B.M., Shima F.J. and Mann W.B., 1980. Preparation and calibration of the 1978 National Bureau of Standards tritiated water standards. *International Journal of Applied Radiation and Isotopes*, 31: 611-614.

Urey, H.C., 1947. The thermodynamic properties of isotopic substances. *Journal of Chemical Society*, 1947: 562-581.

Urey, H.C., Lowenstam, H.A., Epstein, S. and McKinney, C.R., 1951. Measurement of paleotemperatures and temperatures of the Upper Cretaceous of England, Denmark and the Southeastern United States. *Geological Society of America Bulletin*, 62: 399-416.

Valley J.W., 1986. Stable isotope geochemistry of metamorphic rocks. *In: P. H. Ribbe (Ed.) Stable Isotopes in High Temperature Geologic Processes*, Mineralogical Society America, Reviews in Mineralogy, 16: 445-487.

Van Everdingen, R.O. and Krouse, H.R., 1985. Isotope composition of sulphates generated by bacterial and abiological oxidation. *Nature*, 315: 395-396.

Van Warmerdam, E.M., Frape, S.K., Aravena, R., Drimmie, R.J., Flatt, H. and Cherry, J.A., 1995. Stable chlorine and carbon isotope measurements of selected chlorinated organic solvents. *Applied Geochemistry*, 10: 547-552.

Veizer, J., 1989. Strontium isotopes in seawater through time. *Annual Review of Earth and Planetary Science,* 17: 141-167.

Veizer, J., Bruckschen, P., Pawellek, F., Podlaha, O., Jasper, T., Korte, C., Strauss, H., Azmy, K. and Ala, D., 1997. Oxygen isotope evolution of Phanerozoic seawater. *Paleogeography, Paleoclimatology and Paleoecology*, in press.

Veizer, J., Hoefs, J., Lowe, D.R. and Thurston, P.C., 1989. Geochemistry of Precambrian carbonates. 2. Archean greenstone belts and Archean seawater. *Geochimica et Cosmochimica Acta*, 53: 859-871.

Veizer, J., Plumb, K.A., Clayton, R.N., Hinton, R.W. and Grotzinger, J.P., 1992. Geochemistry of Precambrian carbonates.V. Late Paleoproterozoic (1.8 ± 0.2 Ga) seawater. *Geochimica et Cosmochimica Acta*, 56: 2487-2501.

Velderman, B.-J., 1993. Groundwater Recharge and Contamination: Sensitivity Analysis for Carbonate Aquifers in South-Eastern Ontario — The Jock River Basin Study. Unpublished M.Sc. Thesis, University of Ottawa, Ottawa, Canada, 126 p.

Vogel, J.C., 1970. Carbon-14 dating of groundwater. *In: Isotope Hydrology 1970*, IAEA Symposium 129, March 1970, Vienna: 225-239.

Vogel, J.C., 1993. Variability of carbon isotope fractionation during photosynthesis. *In: J.R. Ehleringer, A.E. Hall and G.D. Farquhar (Eds.) Stable Isotopes and Plant Carbon - Water Relations*, Academic Press, San Diego, CA: 29-38.

Vogel, J.C., Grootes, P.M. and Mook, W.G., 1970. Isotope fractionation between gaseous and dissolved carbon dioxide. *Z. Phys.*, 230: 255-258.

Wada, E., Kadonaga, T. and Matsuo, S., 1975. ^{15}N abundance in nitrogen of naturally ocurring substances and global assessment of denitrification from isotopic view point. *Geochemical Journal*, 9: 139-148.

Wadleigh, M.A. and Veizer, J., 1992. $^{18}O/^{16}O$ and $^{13}C/^{12}C$ in Lower Paleozoic articulate brachiopods: implications for the isotopic composition of seawater. *Geochimica et Cosmochimica Acta*, 56: 431-443.

Walker, R.G., Jolly, I.D., Stadter, M.H., Leaney, F.W., Davie, R.F. Fifield, L.K., Ophel, T.R. and Bird, J.R., 1992. Evaluation of the use of ^{36}Cl in recharge studies. *In: Isotope Techniques in Water Resources Development 1991*, IAEA Symposium 319, March 1991, Vienna: 19-32.

Warner, M.J. and Weiss, R.F., 1985. Solubilities of chlorofluorocarbons 11 and 12 in water and seawater. *Deep Sea Research*, 32: 485-497.

Wassenaar, L., Aravena, R., Fritz, P. and Barker, J., 1990. Isotopic composition (^{13}C, ^{14}C, ^{2}H) and geochemistry of aquatic humic substances from groundwater. *Organic Geochemistry*, 15: 383-396.

Wassenaar, L., Aravena, R., Hendry, J. and Fritz, P., 1991. Radiocarbon in dissolved organic carbon, a possible groundwater dating method: Case studies from western Canada. *Water Resources Research*, 27: 1975-1986.

Welhan, J.A., 1987. Characteristics of abiotic methane in rocks. *In: P. Fritz and S.K. Frape (Eds.) Saline water and gases in crystalline rocks*, Geological Association of Canada Special Paper 33: 225-233.

Welhan, J.A. and Fritz, P., 1977. Evaporation pan isotopic behaviour as an index of isotopic evaporation condtions. *Geochimica et Cosmochimica Acta*, 41: 682-686.

White, D.E., 1974. Diverse origins of hydrothermal ore fluids. *Economic Geology*, 69: 954-973.

Whiticar, M.J., Faber, E. and Schoell, M., 1986. Biogenic methane formation in marine and freshwater environments: CO_2 reduction vs. acetate fermentation — Isotope evidence. *Geochimica et Cosmochimica Acta*, 50: 693-709.

Winograd, I.J., Coplen, T.B., Landwehr, J.M., Riggs, A.C., Ludwig, K.R., Szabo, B.J., Kolesar, P.T. and Revesz, K.M., 1992. Continuous 500,000-year climate record from vein calcite in Devils Hole, Nevada. *Nature*, 258: 255-260.

Yamamoto, S., Alcauskas, J.B. and Crozier, T.E., 1976. Solubility of methane in distilled water and seawater. *Journal of Chemical and Engineering Data*; 21: 78-80.

Yang, C., Telmer, K. and Veizer, J., 1996. Chemical dynamics of the St. Lawrence riverine system. *Geochimica et Cosmochimica Acta*, 60: 851-866.

Yonge, C.J., Ford, D.C., Gray, J. and Schwarcz, H.P., 1985. Stable isotope studies of cave seepage water. *Chemical Geology (Isotope Geoscience Section)*, 58: 97-105.

Yonge, J.C., Goldenberg, L. and Krouse, H.R., 1989. An isotope study of water bodies along a traverse of southwestern Canada. *Journal of Hydrology*, 106: 245-255.

You, C.-F. and Chan, L.-H., 1996. Precise determination of lithium isotopic composition in low concentration natural samples. *Geochimica et Cosmochimica Acta*, 60: 909-915.

Zimmermann, U., Münnich, K.O. and Roether, W., 1967. Downward movement of soil moisture traced by means of hydrogen isotopes. *In: Isotope Techniques in the Hydrologic Cycle*, Geophysical Monograph Series 11, American Geophysical Union.

INDEX

(General Index Begins on Page 316)

Isotopes (by mass)

^2H
- analysis in water 9
- deuterium excess 43
- evolution during water-rock exchange 249
- fractionation during methanogenesis 157
- hydrological cycle 21
- in H-bearing minerals 247
- natural abundance **6**
- range in rocks and water types **248**
- shift in shield brines 257
- shift in water, exchange with H_2S 262–63
- shift in water, methanogenesis 158
- water, sampling and analysis 273

^3H
- ^3H–^3He dating 187–88
- concentration units, TU, defined 175
- decay equation 181
- decay mode, to ^3He 174, 187
- decay model for recharge input **183**
- decay, on graph **182**
- defined, measurement 16–17
- electrolytic enrichment 17
- fallout 16
- gas proportional counting 17
- half-life, production **17**
- in groundwater .... See ^3H in groundwater
- in pre-1954 vintage wine 175
- in precipitation 174–79
- latitudinal controls, precipitation 177
- measurement 16
- nuclear reactor tritium 178
- Ottawa record in precipitation **178**
- pre-bomb, natural levels in precipitation .. 175
- production, cosmogenic 174–75
- production, reactors 178
- production, subsurface 186
- production, subsurface 179
- production, thermonuclear 175
- propane synthesis 17
- radiological hazards 177
- sampling and analysis 273
- spring leak from stratosphere 177
- thermonuclear .... See thermonuclear ^3H

^3H in groundwater
- dating with ^3H 179–86
- input function 183–84
- qualitative estimates of age 184
- subsurface production 179
- time series dating 184

^3He
- ^3H–^3He dating modern groundwaters . 187–88
- natural abundance 6
- tritogenic 188

$^3He/^4He$
- atmospheric ratio 187
- expression 242
- in air **242**
- in global reservoirs **242**
- primordial, mantle 242
- sampling, analysis 285

^4He
- atmospheric concentration 187
- concentration in global reservoirs 242
- groundwater dating, case studies 242
- production rate 242
- subsurface production 241

^6Li **6**, 12
- fission, ^3H production 179
- fission, and ^4He production 241
- in brines 156
- origin of shield brines 257

^{10}Be 19

^{11}B **6**, 12
- and shield brines 258

^{13}C
- $10^3\ln\alpha^{13}C_{CO2-DIC}$, temperature equations .. 121
- and biodegradation **160**
- atmospheric CO_2 115
- C_3 plants 119
- C_4 plants 119
- coexisting CO_2 and methane **130**
- DIC, sampling and analysis 274
- DOC, sampling and analysis 279
- during biodegradation 160
- fractionation with DIC 121
- in ^{14}C dating .... See ^{14}C correction models
- in CO_2–DIC exchange 120
- in DIC, during recharge 205
- in organic contaminants, DNAPL 159
- natural abundance 6
- ranges in natural compounds **113**
- values for $10^3\ln\alpha^{13}C_{CO2-DIC}$ **121**
- variations in DIC **162**
- vegetation, photosynthesis 119–20

^{14}C
- ^{14}C-free carbon 202, 204, 207
- AMS 19

Index

analysis 18
anthropogenic sources......................204–5
as tracer for carbon 132, 158
atmospheric activity...............................18
atmospheric CO_2, 1960 to present............204
atmospheric CO_2, past 30 ka...................**203**
atmospheric, natural production...........202–4
Carbasorb...19
CO_2..18
dating...*See* ^{14}C dating
decay..**201**
decay constant 201
decay, measure of time 201
dilution and the decay equation................206
dilution factor, q, defined.........................206
fractionation during uptake 205
global reservoirs.....................................**202**
half-life, Godwin 200
half-life, production................................**17**
in water balance studies...........................98
Libby half-life 200
measurement 18
modern carbon...18
natural variations, Pleistocene, Holocene 203–4
normalized 18
nuclear power stations 204
pathway into groundwaters **205**
sampling, analysis 275
sorption in aquifers.................................213
standard, pmC 18
standard, pmC 202
subsurface production..............................202
thermonuclear..204

^{14}C dating of groundwaters
^{14}C pathway into groundwaters 205–6
^{18}O in Pleistocene recharge **216, 217, 223**
Cracow-Silesian carbonates, Poland.........218
Gorleben site.....................................229–31
Milk River sandstone aquifer.............227–29
Triassic Bunter sandstone aquifer, U.K.
....215–17, 215
Umm er Radhuma carbonate aquifer, Oman...
..................................214–15, 222–24, **214**
complications 217–24
DOC and case studies.........................225–31
DOC and DIC compared......................227–31
effect of carbonate dissolution............206–17
effect of geogenic (volcanic, metamorphic)
CO_2..220
effect of methanogenesis....................220–22
fulvic acid, ^{14}C dating.............................**230**
fulvic acid, initial ^{14}C activity.............225–26
groundwater dating range 201
matrix diffusion effect 217–18
sulphate reduction 218–20
sulphate reduction, case study, Oman. 222–24
^{14}C dating, correction models

^{13}C mixing model 210–11
alkalinity model................................. 208–10
chemical mass balance model 209–10, 231
effect of dolomite dissolution..................212
Fontes-Garnier matrix exchange 212–13
geogenic (volcanic, metamorphic) CO_2 220
matrix diffusion................................. 217–18
methanogenesis 220–22
NETPATH reaction path modelling ... 224–25
statistical models............................. 207–8
sulphate reduction 218–20

^{14}N
and ^{14}C production 202

^{15}N **6**
enrichment during denitrification............. 151
fractionation in food webs....................... 150
fractionation, denitrification 154
in AIR... 12
measurement..................................... 11
nitrate 150–51
range in natural compounds.................. **149**
sampling and analysis............................. 280

^{18}O
analysis in water...................................... 9
atmospheric CO_2 25
atmospheric O_2..................................... 142
brine correction 60
dating modern groundwaters.............. 173–74
evolution during water-rock exchange......249
fractionation, denitrification 154
geothermal shift in water 250–52
global precipitation map **66**
hydrograph separations........................... 100
hydrological cycle..................................21
in glacial meltwater................................. 106
in Pleistocene groundwaters 198–200
in Pleistocene groundwaters, compared with
^{14}C ages 217
in precipitation 64–73
in seawater..38
marine sediments, SPECMAP 38
natural abundance **6**
nitrate 11, 150–51
nitrate, sampling and analysis................ 280
range in rocks and water types.................**248**
seasonal effects in precipitation 71
seasonal variations, dating 173
shift in water, exchange with CO_2...........263
silicate minerals 251
sulphate... 11
water, sampling and analysis 273

^{26}Al 19
^{32}P 194
^{32}Si 172
ages, in ice cores 193
half-life 194
submodern groundwaters........................ 192

^{34}S
  measurement 11
  natural abundance **6**
  ranges in nature **139**
  SO_4^{2-}, H_2S, sampling and analysis 279
  sulphate reduction 144–48
  sulphide 11
  sulphides, terrestrial sulphate 142–43
  sulphur cycle 138–48
^{36}Cl 19
  $A^{36}Cl$, defined 232
  activation of marine Cl$^-$ 190
  case study, Great Artesian Basin 235
  concentration in rain, calculation 190
  cosmogenic production 232
  dating old groundwaters 231–37
  decay modes 233
  diffusion from aquitards 237
  epigenic production 232
  fallout, Oman 92
  fallout, with latitude **233**
  half-life 12, 231
  half-life, production **17**
  in groundwater, ingrowth vs. decay **235**
  in precipitation, United States **234**
  measurement 19
  Milk River sandstone 236
  natural atmospheric production 189
  production 92
  $R^{36}Cl$, defined 232
  recharge studies 92
  sampling and analysis 281
  secular equilibrium, by rock type **235**
  subsurface (hypogenic) production 234
  thermonuclear 92
  thermonuclear and modern groundwater 189–90
  thermonuclear, in ice cores 190
  Triassic Bunter sandstone 236
  units and data expression 231
^{36}S 20
^{37}Cl **6**
  diffusion through clay 163
  fractionation 155
  in solvents 159
  measurement 12
  ranges in brines and minerals **156**
  sampling and analysis 281
  source of chloride salinity 155–56
^{37}Ar 172
^{39}Ar
  $^{40}Ar/^{36}Ar$ and groundwater age 193
  ages, in ice cores 193
  atmospheric production 192
  dating submodern groundwaters 173, 192
  half-life, production **17**, 192
  measurement 20
  sampling, analysis 285

subsurface production 192
^{41}Ca 19
^{81}Br **6**, **12**, 257
^{81}Kr
  and groundwater dating 191
^{81}Kr and ^{85}Kr 20
  half-lifes, production **17**
  sampling, analysis 20, 285
^{85}Kr 172
  See also groundwater dating — modern
  atmospheric concentrations, ... **191**
  natural production 191
  sampling and analysis 191
^{87}Rb 260
$^{87}Sr/^{86}Sr$
  abundance **6**
  in Phanerozoic seawater **261**
  measurement 13
  rock-water interaction 260–61
  TIMS analysis 15
^{129}I 19
  half-life, production **17**
^{206}Pb 241
^{222}Rn
  and modern groundwaters 241
  dating groundwater 172
  geochemistry 241
  half-life **17**
  hazard 241
  sampling, analysis 281
^{226}Ra
  geochemistry 240
  half-life **17**
  in Stripa granite 241
  sampling, analysis 281
^{230}Th
  and ^{234}U, dating corals **203**
  groundwater dating, *See* groundwater dating
  half-life **17**
^{232}Th 4
^{234}Th, recoil 239
^{234}U
  and ^{234}Th recoil 239
  and ^{238}U, groundwater dating 238–40
  half-life **17**
$^{234}U/^{238}U$
  sampling, analysis 281
  Triassic Bunter sandstone groundwaters .. 239
^{235}U 4
  fission 176
  uranium bomb 176
^{238}U 3
  and ^{234}U, groundwater dating 238–40
  decay series **239**
  decay series, sampling and analysis 281
  half-life **17**
  TIMS 15

A

α and ^{4}He production 241
α, alpha particle .... *See* radioactive decay production in ^{238}U and ^{232}Th decay series . 241
α, fractionation factor 21
Accelerator mass spectrometry .... *See* AMS
Acetate 127
Acetate fermentation 128
Acetic acid 124
Activities of solutes 112–15
Activity coefficient, defined 113
Activity ratio
  ^{234}U/^{238}U 238, **240**
  uranium series 281
Addis Ababa 72
Adiabatic expansion 47, 70
AECL ..... *See* Atomic Energy of Canada Limited
Africa 86
Agriculture *See also* groundwater contamination
  and nitrate 149
  C_3 and C_4 crops 119
AIR, ^{15}N reference 12
Alaska 45
Alberta 234
Alberta Basin **258**
Albite 119
Algae, elemental composition 149
ALK model ..... *See* ^{14}C dating, correction models
  carbonate 116
  carbonate, in landfill leachates 157
  defined 116
  in ^{14}C dating .... *See* ^{14}C correction models
  measurement 286
Alpine effect 70
Alps, altitude effect 70
Altitude effect 70
  case studies **71**
  in ice cores 76
  Mont Blanc 71
  Mount Cameroun 71
Amazon 67
Amino acid 124
Amorphous silica 254
Amount effect 51
  Bahrain 51
  New Delhi 73
  Oman **51**
AMS, accelerator mass spectrometry 15
  ^{129}I analysis 15
  ^{14}C 19, 225
  ^{14}C analysis 15
  ^{36}Cl 233
  ^{36}Cl analysis 5
  dendrochronology 203
  possible isotopes 19

Anatolia 58
Andesitic volcanism 252
Anhydrite 139
Anorthite 119, 255
Apatite 264
Appalachian Mountains 85
Aqueous geochemistry, fundamentals 112–15
Aquifers
  Alliston, glacio-fluvial, Ontario 224
  alluvial **81**, 132
  alluvial, ^{3}H 185
  alluvial, ^{3}H–^{3}He dating 188
  alluvial, arid 86, 87, 88
  alluvial, Nile River 98
  bedrock, ^{3}H 186
  Blumau, alluvial, Austria 242, **243**
  Bunter Sandstone 194
  Calcaires Carbonifères, France 218
  carbonate 87, 161
  carbonate, arid **88**
  chalk **81**
  confined 162
  contamination potential 161
  Cracow-Silesian carbonates, Poland 218
  Disi sandstone **200**
  fractured 82
  Gorleben, Germany 229
  Great Artesian Basin, Australia 235
  in agricultural watershed 151
  Kufra and Sirte Basins, Libya **200**
  limestone, Estonia 194
  Milk River 198, 199, 227, **228**, 260
  Nubian sandstone 198
  river-connected 96
  Saar sandstone, Germany 194
  sand, tritium peak **180**
  Triassic Bunter sandstone 215–17, **243**
  Umm er Radhuma carbonate, Oman
    .... 147, 222–24, **223**
Aquitards 161, 198, 237
  clay **163**
  groundwater protection 162–65
Arabian Gulf 51
Arctic 52, 75
Argentine Island 45
Arizona 234
Aston, Francis 4
Atmosphere, structure 39
Atmospheric CO_2 .... *See* CO_2
Atmospheric oxygen, $\delta^{18}O$ 144
Atmospheric water vapour, ... 46
Atomic Energy of Canada Limited 16
Atomic mass units (amu) 2
Atomic weight 2
Auto-oxidation, of ^{234}U 238
Avogadro's number 231

B

β⁻, beta particle .... *See* radioactive decay
Bacteria
   and CO_2 in soils 206
   Desulfovibrio desulfuricans 145
   in redox reactions 112, 126
   sulphide oxidation 142
   Thiobacillus concretivorous 142
   Thiobacillus denitrificans 150, 153
   Thiobacillus ferrooxidans 142
   Thiobacillus thiooxidans 142
Bahrain, amount effect 51
Barbados, ^3H 178
Baseflow 99
Basin brines .... *See* sedimentary basin brines
Bicarbonate .... *See* DIC, speciation
Big Otter Creek, hydrograph separation 102, **103**
Biodegradation, ^{13}C and CO_2 .... **160**
Biotite
   ^{18}O exchange 247
   composition 248
Bisulphate 148
Bitumen 125
Bohemian Massif, Germany 249
Boltzmann constant 24
Bomb tritium .... *See* thermonuclear tritium
Bond strength 22
Boundary layer 41, 59, **60**
   in unsaturated zone 89
Bratislava 96
Brines
   ^{18}O correction 59, 60
   activity of water 60
   sedimentary basins 258–59
   shield terrains 256–58
British Columbia 70, 71

C

C_3 plants .... *See* vegetation
C_4 plants .... *See* vegetation
Calcite
   ^{18}O and paleotemperatures 132–34
   dissolution 117
   fracture minerals 122, 223, 257
   secondary .... *See* carbonates, secondary
Calcrete 132
Canada **84**, 199
   Arctic, ice cores **76**
   local T-$\delta^{18}O$ effects 69
   meteoric regimes **53**
   western Cordillera 68
Canadian Shield 81, 256
CANDU reactor 262
Cane sugar 119
Cañon Diablo Troilite .... *See* CDT
Canton Island, central Pacific 45
Cape Hatteras, North Carolina, ^3H 178
Carbasorb 19
Carbohydrate 124
Carbon, global reservoirs, by mass **202**
Carbon cycle 132
   ^{14}C as tracer 132
Carbon fixation 119
Carbonate dissolution
   ^{13}C and open/closed system 122–23
   ^{14}C and open/closed system conditions ... 207
   closed system 118
   $\delta^{13}C$ and pH evolution **123**
   during ^{14}C uptake 206
   evolution of pH and DIC 118
   open system 118
Carbonate geochemistry 115–19
Carbonate reaction constants **117**
Carbonates
   ^{14}C content 207
   fracture 132
   freshwater, ^{13}C and ^{18}O **133**
   marine, ^{13}C and ^{18}O **133**
   marine, $\delta^{13}C$ 122
   secondary 132
Carbonic acid
   carbonate dissolution 117
   carbonate dissolution, ^{14}C dilution 207
   dissociation 115
   production in soils 115
   silicate weathering 119
Carboxylation 119
Carboxylic acid 124
Catabolic reaction 150
Catchment studies 99
CDT
   abundance ratio 6
   defined 11
Cellulose 124
CFCs 39
   analysis 189
   atmospheric concentrations **189**
   dating modern groundwaters 188–89
Chalbi desert 75
Chalcedony 254
Chile, Cordillera de la Costa 74
Chloride
   $\delta^{37}Cl$ 155–56
   sampling, analysis 281
   sources in groundwater 155–57
Chlorinated hydrocarbons ....
   .... *See* groundwater contamination
Chlorine isotopes 12
Chlorofluorocarbons .... *see* CFCs
Cigar Lake 179
Clay minerals 247, **256**
Climate change 76

Closed system conditions..................................
........................*See* carbonate dissolution
CMB model ...*See* ^{14}C dating, correction models
CO_2
    ^{13}C fractionation during recharge **205**
    ^{18}O exchange with water.........................263
    and ^{14}C, atmospheric 200
    atmospheric.............................. 115, 200
    atmospheric, ^{18}O...25
    concentration in soils.................................115
    degassing...132
    diffusion... 115, 120
    dissolution, geochemistry................... 115–19
    geogenic 220, 263
    high P_{CO2} groundwaters 263
    hydration..115
    metamorphic...263
    soil.. 115–17, **113**
    soil, ^{13}C...119
    soil, ^{14}C...207
    soil, sampling and analysis 283
    volcanic, metamorphic, ^{13}C range...........**113**
CO_2 reduction...128
Coast Mountains, B.C........................... 84, 252
Collectors .... *See* mass spectrometer
Common ion effect...219
Concentration units..113
Condensation...26, 75
    and rainout...46
    and temperature..47
    coastal fog 73
    isotopic equilibrium...................................44
Contaminant hydrology.....................................12
Contaminants....*See* groundwater contamination
Continental effect
    Canadian Cordillera................................**69**
    Europe..**69**
    North America..**68**
Continentality
    Conrad's index..67
Continentality...47, **68**
Continuous flow .... *See* mass spectrometry
Corals
    U/Th and ^{14}C dating 203
Coshocton, Ohio 72
Cosmic radiation
    and ^{14}C production..................................202
    and ^{36}Cl production.................................232
    composition..202
Craig, Harmon..8
    meteoric relationship 21, 36
Crassulacean acid metabolism.............*See* CAM
Critical depth...82, 162
Crystalline rocks,.......................*See* weathering
    and ^{14}C dilution 208

D

δ–value, defined ....*See* delta value
Δ *See* separation factor
Dansgaard, T–δ^{18}O 64
Danube River 96
Dead carbon ....*See* ^{14}C, ^{14}C-free carbon
Debye-Hückel equation................................114
Decay............................*See* radioactive decay
Deglaciation 217
Dehydration..25
Delta value, defined..6
Dendrochronology...203
Denitrification 150
    and Eh...**126**
    by sulphate reduction...............................**152**
    Rayleigh effects..55
    Rayleigh enrichment of ^{15}N and ^{18}O........**153**
    via iron oxidation 153
Denmark 180
Desert dams..99
Deserts
    A'Sharqiyah, Oman...............................88, 92
    Chalbi, Kenya 75
    Empty Quarter, Oman and Saudi Arabia .. 199
    groundwater recharge 88
    Najd,, southern Oman...............................223
    Sahara 198
Deuterium 9
Deuterium excess 43–45, 51, 87
    Eastern Mediterranean water line 51
    methanogenesis, ^{2}H shift **158**
Devils Hole, Nevada.......................................134
Dew point..26, 39
Dewatering of clays 258
Dhofar Mountains, Oman **223**
DIC
    ^{13}C and DOC oxidation............................126
    ^{13}C evolution 112, 119–23, 207
    ^{13}C fractionation with CO_2 120–22
    ^{13}C time series monitoring **162**
    and ^{14}C...................................... *See* ^{14}C
    δ^{13}C and pH...**121**
    during denitrification................................151
    during hyperfiltration................................260
    from sulphate reduction 145
    methanogenesis, ^{13}C 222
    reaction constants 117, 116
    sampling, analysis 274
    speciation 115–17
    speciation and pH...................................**116**
    temperature and reaction constants 116
    diffusion................. *See* isotope fractionation
    matrix diffusion of ^{14}C....*See* matrix diffusion
Diffusion profiles
    ^{18}O and ^{2}H in clay..................................... 165

^{37}Cl in clay 164
Dilution factor, q .... *See* ^{14}C, dilution factor, q
  multiple processes 222
Dimethylsulphide 144
Dissociation energy 27
Dissociation energy, of isotopes **23**
Dissolved carbonate .... *See* DIC
Dissolved inorganic carbon .... *See* DIC
Dissolved organic carbon .... *See* DOC
DMS .... *See* dimethylsulphide
DNAPL, dense, non-aqueous phase liquids .. 159
DOC
  Alliston aquifer 224
  and ^{14}C dating .... *See* ^{14}C dating, DOC
  and denitrification 150
  and methanogenesis 157
  and redox reactions **126**
  concentration in groundwater **125**, 161
  fractions 229
  fulvic acid 125
  fulvic acid, ^{14}C 225
  geochemistry of 124–27
  Gorleben aquifer study 229
  humic acid 124, 125
  redox evolution 126
  sampling, analysis 225, 279, 290
Dolomite 123
  incongruent dissolution 123
Dolomite dissolution
  effect on $^{14}C_{DIC}$ 212

E

ε *See* enrichment factor
Eastern Mediterranean water line 51
Edmonton, Alberta 45, 83
Eh
  and redox reactions **126**
Electrical conductivity (E.C.)
  in hydrograph separations 100
Electromotive potential .... *See* Eh
Elemental analyzer 14
Elevation effect .... *See* altitude effect
Enrichment factor ε, defined 31
Environmental isotopes 5
Epstein, Samuel 5
Equal-mass ions 15
Europe 68, 198
Europe, continental effect **69**
Evaporation 26, 80
  deuterium excess 43
  during irrigation 98
  from snowpack 85
  from unsaturated zone 89
  from water table 91
  humidity and isotopes 58
  isotope effects 41–43

loss from soils, calculations 90
meteoric water line, deviation from 43
of brines 59–60
pan evaporation .... *See* pan evaporation
secondary 52, 74, **75**
soil columns 89, **90**
surface waters 57–58
trend in groundwaters **88**
unsaturated zone 87
vs. transpiration 94
Evapotranspiration 67, 80, 92

F

FA, fulvic acid .... *See* DOC, fulvic acid
Falkland Islands, Stanley 72
Faraday cup 13
Fe reduction
  and Eh **126**
Feldspar, ^{18}O exchange 247
Fennoscandian Shield 256
Fermentation
  and methanogenesis 127
Fertilizer 166
  manure, ^{15}N 150
  nitrate 151
  urea 149, 150
Fick's law of diffusion 163
Field measurements
  alkalinity 286
  Eh, redox 286
  electrical conductivity, EC 286
  pH 286
  temperature 286
Field sampling
  ^{14}C in DIC 275
  ^{14}C in DOC 279
  ^{18}O and ^{2}H in water 273
  ^{3}H 273
  chloride 281
  containers 269
  DIC 274
  DOC 279
  nitrogen compounds 280
  snow and ice 283
  sources of water 271
  sulphur compounds 279
  table of methods, requirements **270**
  unsaturated zone 282
Filtering .... *See* field sampling, geochemistry
Fission .... *See* ^{235}U
Flagstaff, Arizona 45
Flight tube .... *See* mass spectrometer
Fog precipitation 73
  See also occult precipitation
  Camanchaca 74
Fontes-Garnier matrix exchange model
  for ^{14}C dilution 212–13

Foraminifera, ^{18}O 37
Formation brines... *See* sedimentary basin brines
Fossil fuels
  ^{13}C range **113**
  and atmospheric CO_2 115
Fossil groundwaters, defined 198
Fraction factors
  carbonate-acid **10**
Fractionation .... *See* isotope fractionation
Fractionation factor
  $10^3 \ln\alpha$, expression defined 28, 29
  α, defined 31
  α, defined 21
  temperature equation 29
Fractionation factors
  See Table 1 inside front cover
  ^{15}N during denitrification 150
  ^{18}O during denitrification 151
  ^{37}Cl diffusion 164
  ^{37}Cl–salts 155
  DIC reactions 120
  during sulphate reduction 147, **148**
  methane–CO_2 130
  sulphate-sulphide **146**
  water-ice **28**
  water-vapour **28**
Fruit juices 120
Ft. Smith 52
Fuhrberger Feld
  nitrate, sulphate in groundwater 152–54
Fuhrberger Feld study 152–54
Fulvic acid, FA 124
  See also DOC or ^{14}C dating, DOC
  in ^{14}C dating .... *See* ^{14}C dating, DOC
  structure **125**

G

Gas concentrations, units 114
Geochemical software .... *See* websites
Geochemistry
  equilibrium reaction 112
  principles, background 112–15
Geogenic CO_2 .... *See* CO_2, geogenic
Geomagnetic field
  and ^{14}C production 203
Geomagnetic latitude 175, 233
Geothermal 70
Geothermal waters
  ^{18}O shift 250–52
  ^{18}O–^2H composition **250**
  ^{34}S in SO_4^{2-}–H_2S exchange 146
  and helium 243
  and methane 131
  geothermometry 253–55
  magmatic 247–49
  mantle fluids 252
  steam separation 253
Geothermometers
  carbon dioxide-methane, ^{13}C 255
  cationic 253
  hydrogen-water, ^2H 255
  silica 254
  sulphate, ^{18}O 254
  sulphate-sulphide, ^{34}S 254
Geothermometry .... *See* geothermal waters
Germany 96
^3H–^3He dating, alluvial groundwaters 188
Gibbsite **256**
Gimli, Manitoba 72, 83
GISP 76
Glacial meltwater **106**, 199
Global meteoric water line
  See also meteoric water lines *and* GMWL
Global Network for Isotopes in Precipitation ......
  .... *See* GNIP
Gloucester, Ontario 181
GMWL 36, 40, 45, 49
GNIP 36, 65
  ^3H data 177
  website 47
Gorleben aquifer, Germany 229
Graham's Law 24, 163
Great Artesian Basin Australia 235
Great Lakes **144**
Greenland, ice core **76**
GRIP 76
Groundwater
  age dating .... *See* groundwater dating
  alluvial aquifer, arid **89**
  Canada 83, **84**
  evaporative enrichment 87, **88**, 98
  evaporative loss, calculation 88
  high pH 264–65
  isotopic variations, seasonal **81**
  loss by evaporation vs. transpiration 94
  mean residence time, defined 172
  mixing 104–8
  salinization 94, 98
  time series monitoring 96
Groundwater age, defined 172
Groundwater contamination
  agricultural watersheds 166–67
  See also Fuhrberger Feld
  chloride salinity 155
  chlorinated hydrocarbons 159–60
  chloro-contaminants 12, 159
  from ^3H 177
  from rivers 96
  H_2S 144
  landfill leachate 157–59
  leaking underground storage tanks 159
  methane 127
  monitoring, time series 161–62

nitrate 148
   septic tank nitrate 151
   summary 165–68
   watersheds, sensitivity 160–65
Groundwater dating — modern
   stable seasonal ^{18}O variations 173–74
   with ^{3}H 179–86
   with ^{3}He–^{3}H 187–88
   with ^{85}K 191
   with bomb ^{14}C 172
   with bomb ^{36}Cl 189–90
   with CFCs 188–89
   with Rn 240–41
Groundwater dating — old
   ^{18}O and ^{2}H, Pleistocene recharge 198–200
   with ^{14}C .... 200–231. See ^{14}C dating
   with ^{36}Cl 231–37
   with ^{4}He 241–43
   with U and Th dissequilbrium 238–40
Groundwater dating — submodern 172, 185, 192
   with ^{32}Si 194
   with ^{39}Ar 192–94
   with Ra 240–41
Groundwater mixing
   and ^{3}H 182
   binary 105
   karst systems 107
   regional flow systems 105
   ternary **106**
Groundwater recharge
   and precipitation 83
   arid regions 86–95
   attenuation of seasonal variations 80, 161
   by snowmelt 83, 85
   carbonate aquifers, arid 87
   critical depth 82
   desert dams 99
   direct infiltration 91
   direct infiltration, agricultural **162**
   direct infiltration, arid 88
   from rivers 96–99
   from rivers, with ^{14}C 98
   mechanisms 80
   rates, arid regions 86
   rates, temperate regions 80
   seasonal effects 80
   soil profiles 89
   spring 83
   temperate regions 80–86
   tritium input function 183–84
   with ^{36}Cl and Cl$^-$ 92
Groundwaters, fossil, defined 198
Groundwaters, paleo, defined 198
Guam 73
Gulf Coast basin **258**
Gulf of Mexico 52
Gypsum 139

   dehydration 259
   fractionation with water **259**
   in sulphate reduction 219
   solubility, calculated 140

H

H_2
   in landfill leachate 157
   in methanogenesis 128
H_2S 138
   See sulphate reduction
   ^{2}H exchange with water 262–63
   dissociation, speciation 145
   in groundwater 144
   sampling and analysis, ^{34}S 279
   speciation, pH **145**
HA, humic acid .... See DOC, humic acid
Half-life
   ^{238}U decay series radionuclides 239
   defined 3
   illustrated **201**
   of environmental radioisopes **17**
Halley Bay, Antarctica 45
Hannover, Germany 152
Hazardous waste facility, siting 168
Helium .... See groundwater dating
   solubility 187
Holocene hypsithermal 217
Hooray sandstone .... See Great Artesian Basin
Hornblende
   ^{18}O exchange 247
   composition 248
Humic acid
   structure **125**
Humic substances 124
   ^{14}C in humic acid **230**
   elemental composition **125**
   from lignite 230
   in sulphate reduction 219
Humidity 26
   effect on evaporation 88
   isotope effects and evaporation **43**
Humification 124
Humin 124
Hydration 25
Hydrocarbons
   biodegradation 160
Hydrogen
   sampling, analysis 284
Hydrogen bomb 176
Hydrogen bond 26
Hydrogen fusion 175
Hydrogen sulphide gas .... See H_2S
Hydrogen-bearing minerals 247
Hydrograph separation 99–104
   Australia 104
   in karst 108

three component **102**
two component **101**
Hydrograph separations
 radon 101
Hydrological cycle 36, 80
 partitioning of isotopes 37
Hydroxide, OH⁻, effect on water 264
Hyperalkaline waters 264–65
Hyperfiltration, of isotopes 260

I

IAEA 7, 47
 ^{3}H data 177
 website for GNIP data 65
 www.iaea.or.at 7
Ice core data website 76
Ice cores 75
Illite **256**
Indus river 94, 96
Interflow, as runoff 100
International Atomic Energy Agency ... *See* IAEA
Inter-tropical convergence zone .... *See* ITCZ
Ion beam .... *See* mass spectrometer
Ion hydration, isotope effects 59
Ionic strength 140
Ionic strength, defined 114
IRMS .... *See* mass spectrometry
Iron Mountain, California 142
Isobaric interferences 15
Isobars, defined **3**
Isotones, defined **3**
Isotope analysis
 radioisotopes .... *See* radioisotopes
 stable isotopes .... *See* stable isotopes
Isotope enrichment factor .. *See* enrichment factor
Isotope equilibrium .... *See* isotope fractionation
Isotope exchange 22
 carbonate-water, ^{18}O 258
 CO_2–water 263
 during snowmelt 85
 effect of temperature 258
 geothermal systems 250
 H_2S–water 262
 high temperature, water-rock 247–55
 hydrated minerals-water 258
 low temperature 258–59
 low temperature, water-rock 255–60
 mechanisms 246–47
Isotope fractionation 21–33
 ^{14}C and ^{13}C during recharge 205
 ^{15}N during denitrification 154
 ^{18}O during denitrification 154
 ^{37}Cl diffusion in clay 163
 bacterial mediation 30
 by diffusion 24, 42
 carbonate-CO_2 263
 clay minerals-water **256**

clay-water 256
CO_2 in soil 120
crystallization water 59
diffusion profiles **164**
diffusion, through air 24
during rainout 48, 49
equations .... *See* Table 1, front cover
equilibrium 25
geothermal fluids .... *See* geothermometers
geothermal systems 250
gypsum-water **259**
H_2S–water 262
hydroxide-water 264
hyperfiltration 260
ice-water 56
in carbonate reactions **121**
ion hydration in brines 59
isotope exchange 22
kinetic 21, 42
methanogenesis 128
mineral-water, high T 249
molecular diffusion 162
physicochemical 21
reaction rate 22, 25
steam separation 253
sulphate–sulphide 146
temperature effect 27
thermodynamic fractionation 21
water–calcite 133
water-silica, ^{18}O **251**
water-vapour 26, 39
Isotope laboratories 8
Isotope partitioning 6, 21
 .... *See* isotope fractionation
Isotope ratio mass spectrometry, IRMS ....
 .... *See* mass spectrometry
Isotope separation factor, Δ *See* separation factor
Isotope, defined 2, 3
Isotopes
 environmental isotopes, defined 5–7
 in landfill leachates 157
 stable valley **3**
Isotopic equilibrium 39
Israel 45, **258**
ITCZ 73

J

Jet streams 39
Jock river basin, Ontario **163**
Jordan 45
Juvenile fluids 247

K

Kabul, afganistan 45
Kaolinite 119, 255, **256**
Karst

and ^{14}C dilution 208
and spelethem 133
carbonate dissolution 118
caves 82
discharge, ^{18}O variations **107**
mixing 107
Kau Bay, Indonesia 164
Kenya 75
Kerogen 229
Kinetic isotope effects
bacterial redox reactions 126
CO_2 diffusion, soils 120
evaporation 42
methanogenesis 128
secondary evaporation 74
sublimation of snow *85*
Krypton isotopes .... 191. *See* ^{81}Kr and ^{85}Kr
KTB deep drillhole, Germany 186, 249, 256

L

Laboratories .... *See* isotope laboratories
Lake Chad 58
Landfills
contamination potential 167
leachate contamination 157–59
Lanzarote geothermal system 251
Lapse rate 39
Latitude effect 66
Lead 15
Leguminosae 149
Libby, Willard Frank 18
Limestone
contact metamorphism, CO_2 263
isotopic composition 132
Limestones .... *See* carbonates, marine
Liquid scintillation counter 17
Loess
and ^{14}C dilution 208
L-SVEC, ^6Li standard **6**
Lysimeters 81, 282

M

MAAT 65, 133
Mackenzie Mountains, Yukon 85
Madrid Basin 234
Magmatic water 247–49, 250
Magnetosphere 232
Maple syrup 120
Marine sulphate 139
isotopes, through geologic time **141**
modern, isotopic composition 141
Mass effect
for ^{14}C fractionation 205
Mass peak 15
Mass spectrograph 5
Mass spectrometer 13

design **14**
dual inlet 13
reference gas, calibration 15
Mass spectrometry
apparent ratio 6
continuous flow 14
history 5
isotope ratio mass spectrometry, IRMS 13–16
machine error 6
measurements 15
references 6
solid source 15
TIMS 15
Matrix diffusion of ^{14}C 217–18
Matrix exchange
effect on ^{14}C 212–13
Meager Creek geothermal area 252
Mean annual air temperature .... *See* MAAT
Mediterranean 70
Metamorphic CO_2 .... *See* CO_2, metamorphic
Meteoric relationship
^{18}O and ^2H in meteoric waters 21
for ^{18}O and ^2H in meteoric waters 36
modification by water-rock interaction ....
.... 250–52, 255–56, 255–60, **246**
Meteoric relationship for ^{18}O and ^2H **37**
Meteoric signature .... *See* meteoric relationship
Meteoric water line 36, 50
^2H shift in shield brines 257
and pluvial climates, Middle East 199
Canada 52, **53**
confidence limits 54
data above line 74
deuterium intercept 36
Eastern Mediterranean 51
evaporation effect 87
global variations **54**
humidity effects 45
local 51
methanogenesis, ^2H shift **158**
Pleistocene, paleoclimate shift 198–200
slope 36, 49
slope and ^2H excess **52**
Methane
^{13}C and coexisting CO_2 130
^{13}C range **113, 129**
abiogenic 131
bacterial oxidation 130, 157
biogenic 127–31
in groundwaters 127–32
mantle 131
solubility 127
sources 127
thermocatalytic 131
Methanogenesis 30, 228
^{13}C and ^2H fractionation 128
and ^{14}C dating .... *See* ^{14}C dating

and ^{14}C dilution ....See ^{14}C dating
and Eh...**126**
H_2 production 128
in landfills..157
kinetic isotope effects 128
Rayleigh effects..................................55
reactions, biological........................127
via CO_2 reduction **158**
Methemoglobinemia 148
Mexico 106, 198
Michgan basin **258**
Microbial activity .... See bacteria
Microfractures, matrix diffusion..................217
Middle east..87
pluvial recharge...................... 199, **200**
Mineral hydration 255
Mineral solubility product........................114
Mixing ....See groundwater mixing
Modern carbon ....See ^{14}C standard
Molality..60
Molality, defined 113
Molarity, defined 113
Monitoring
^{18}O time series................................83, 96
^{18}O time series in karst 107
time series, isotopes 96, 161–62
Monsoon...73
Monsoon, oman 71
Mont blanc, altitude effect **71**
Montmorillonite.....................................**256**
Montreal...164
Mount cameroun, altitude effect...................**71**
MRT .... see groundwater mean residence time
Munich..**81**
Muscovite, ^{18}O exchange 247

N

Najd, southern Oman..............................**223**
National Bureau of Standards .... See NIST
National Institute of Standards and Technology..
 ....See NIST
NBS ....See NIST
NBS 951..**6**
NBS-1 ....**8**
NBS-19 ....**9**
Neodymium 15
NETPATH................See ^{14}C correction models
website for software 224
Neutrons, neutron flux
by rock type.......................................**235**
cosmic..174
cosmogenic......................................202
subsurface 202
subsurface 234
thermonuclear fusion 176
New Delhi 73

New Mexico..234
New Zealand 173
Nier, Alfred...5, 13
Nier mass spectrometer 13
Niger..91
Nile River..94
groundwater recharge..........................98
NIST, National Institute of Standards and
Technology...7
Nitrate
drinking water standards 148
geochemistry 149–50
isotopes.................................... 150–51
isotopic composition................................**151**
sampling, analysis 280
solubility...150
Nitrification..150
and ^{18}O...150
Nitrogen
fixation..149
speciation, redox states.............................149
Nitrogen cycle 148–57
Nitrosamine...148
NOAA................................... 37, 65, 76
Noble gases 16
paleotemperatures **216**
solubility..**187**
North America 68, 198
seasonal effects in $\delta^{18}O$...........................**72**
North Atlantic Drift.....................................66
North Pacific marine vapour 40
Northwest Territories..................................85
Nuclear reactors.......................................178
Nucleosynthesis......................................3, 4
Nucleus..3
Nuclide..2
chart of light nuclides................................**3**
defined..2

O

O_2, atmospheric, $\delta^{18}O$ value 142
Occult precipitation .... See precipitation, occult
sampling 283
Oklo reactors, Gabon 179
Oman
^{36}Cl study..**92**
^3H in alluvial groundwaters 185
amount effect.....................................**51**
case study for ^{14}C dating......................**214**
occult precipitation 74
precipitation............................. 49, **50**
runoff, isotopes.................................**87**
Ontario..82
Open system conditions ....
 ....See carbonate dissolution
Organics
in landfill leachate...................................157

Osmium 15
Ottawa, tritium record 178
Overland flow as runoff 100

P

Paleogroundwaters
    See also Pleistocene recharge and ^{14}C dating
    Middle East, North Africa **200**
    mining, exploitation 199
Paleotemperature 134
    ^{18}O in calcite 132–34
    carbonate scale 10
    ice cores 75
    noble gases **216**
Pampa del Tamuragal 198
Pan evaporation 58
    experiments 87, 94
    Nile Delta **95**
Partial pressure
    CO_2 .... *See* CO_2
    gases, defined 114
Partition functions 22, 27
PDB, defined 9
Peat 125, 132, 218
Pee Dee Belemnite .... *See* PDB and VPDB
PEP carboxylase enzyme 119
Perchloroethylene 159
Perm, Russia 45, 68
Permil, defined 6
Perth, Australia 45
Pesticides 166
pH
    and alkalinity 116
    buffering by weathering 117
    effect on ^{13}C and ^{14}C models **211**
    high pH groundwaters 264–65
Photosynthesis 119
    and ^{14}C uptake 200, 202
    C_3 (Calvin) plants 119
    C_4 (Hatch-Slack) plants 119
    CAM cycle 119
    redox reaction 124
Piston flow, defined 182
Pleistocene, ice cores 76
Pleistocene recharge .... 198–200, 217.
    See also .... ^{18}O, Pleistocene groundwaters
Pluvial climates
    Middle East, North Africa 199
    Oman, early Holocene 224
pmC, defined 18
    .... *See also* ^{14}C
Porosity
    dual 107, 217
    fracture 252
    secondary 217
Portland cement, effect on water 264
Portlandite 264

Precipitation
    ^{18}O, global scale 64
    ^{3}H, natural and thermonuclear 174–79
    altitude effect 70
    amount effect 51
    and groundwater ^{18}O 83
    Camanchaca, Chile 74
    continental effect **68**
    $\delta^{18}O$, global map **66**
    global mean isotopic composition 40, 46
    isotopic evolution 46
    latitude effect 66
    occult 73, 283
    paleoclimate effect 75, 198
    seasonal effects 71, 173
    temperature–$\delta^{18}O$ effect 47, 49, 64–73
    website for isotope data, GNIP monitoring
        stations 47
Precipitation data .... *see* websites, IAEA
Propane 131
Protein 124
Protons 2
    in cosmic radiation 202
Pyrite oxidation 142, 153
    sources of oxygen 143
    via denitrification 154
q, ^{14}C dilution factor .... *See* ^{14}C, dilution factor
Quadropole mass spectrometer 12, 14
Quartz, exchange with water 254

R

Radioactive decay
    ^{3}H decay constant 181
    decay equation 181
    decay modes **4**, 17
    discovery 4
Radioactive waste repository 240, 256
Radiocarbon .... *See* ^{14}C
Radioisotopes 4, 16–20, **17**
    measurement 16–20
Radionuclide, defined 3
Radium 4
Radon .... *See also* ^{222}Rn
    groundwater dating 240
    in hydrograph separations 101
Rainout 36, 37, 66, 84
    ^{18}O and ^{2}H evolution 46–49
    continental effect **69**
    defined 47
    Oman 49
    Rayleigh distillation of ^{18}O **48**
Rayleigh distillation 37
    ^{18}O during rainout **48**
    during denitrification 151, **153**
    during sulphate reduction 146, 224
    equation 55

freezing of water ... 57
general ... 55–57
hydration of silicates ... 256
Recoil effect ... 239
Redox
and DOC oxidation ... 124, **126**
and U solubility ... 238
in landfill leachate ... 157
Reductive dehalogenation ... 160
Resolute, NWT ... 72
Respiration ... 124
and Eh ... **126**
root respiration and ^{14}C ... 205
root respiration of CO_2 ... 115, 202
Rocky Mountains ... 68
Rubisco enzyme ... 119
Runoff ... *See* hydrograph separation
arid regions, evaporation ... 87

S

Sabkha, sulphates ... 139
Sahara desert, pond evaporation ... 59
Salinity
chloride ... 155–57
$\delta^{37}Cl$ ... 155
fluid inclusions ... 257
ion ratio indicators ... 155
shield brines ... 257
sulphate ... 141
Salinization ... *See* groundwater, salinization
Salton Sea geothermal system ... 251
San Juan, Puerto Rico ... 72
Sarnia, Ontario ... 165
Saudi Arabia ... 89
Scandinavia ... 66
Sea surface temperature ... 39
Seasonal effect, precipitation ... 71, **72**
Seawater
^{18}O ... 25
evaporation and isotopic evolution ... **60**
evaporation, salt precipitation ... 155
isotopic composition ... 37
major ions and isotopes ... **140**
Phanerozoic evolution ... 37
Secondary carbonate .. *See* carbonates, secondary
Secondary evaporation ... 52
Secular equilibrium ... 193, 238
^{14}C ... 202
^{3}H ... 175
Sedimentary basin brines ... 258–59, 260
^{37}Cl ... 156
isotopic composition ... **258**
Sedimentary organic carbon, SOC ... 226
Separation factor Δ, defined ... 31
Septic tank effluents ... 151, 166
SF_6 ... 11
Shield brines ... 256–58

^{37}Cl content ... 156
isotopic composition ... **257**
origin ... 256
Silicalite®, DOC sampling ... 229
Silicate minerals
hydration ... 255–56
weathering ... 119
SLAP ... 8
Smectite ... **256**, 260
SMOB ... **6**
SMOC ... **6**, 12
SMOW ... 8
Snow and snowmelt ... 85, **86**
SO_2 ... 11
SOC ... *See* sedimentary organic carbon
Soil CO_2 ... *See* CO_2, soil
sampling and analysis ... 283
Soil moisture ... 90
Solar radiation ... *See also* cosmic radiation
natural variations ... **203**
Solid source mass spectrometry
... *See* mass spectrometry
Solubility product, defined ... 114
Solvents
^{37}Cl and ^{13}C ... **159**
DNAPL contaminants ... 159
Source ... *See* mass spectrometer
South America ... 67
Spallation ... 202
SPECMAP ... 38
Speleothem 133. *See also* carbonates, freshwater
isotopic composition ... 132
SST ... *See* sea surface temperature
Stable isotopes ... **4**
^{18}O and ^{2}H in water ... *See also* ^{18}O and/or ^{2}H
calibration ... 7
defined ... 2
discovery ... 4–5
in groundwater ... *See* groundwater
in precipitation ... *See* precipitation
standards, measurement ... 7–13
Standard Light Antarctic Precipitation *See* SLAP
Standard Mean Ocean Water ... *See* SMOW
STAT model ... *See* ^{14}C dating, correction models
Steam separation, isotope effects ... 253
Stormwater ... 100
Stratosphere ... 39
Stripa granite ... 179, 186, 235, 238
Strontium, geochemistry ... 260
Submodern groundwaters ... 179
See groundwater dating — submodern
Suess wiggles ... 203
Sulphate
^{18}O exhange with H_2O ... 147
^{18}O measurement ... 11
^{18}O, controls on composition ... 142
marine ... 139–41

sampling, analysis 279
terrestrial 138
terrestrial 142–43
terrestrial, range of ^{34}S and ^{18}O **143**
Sulphate reduction 30
 ^{18}O enrichment 146
 ^{18}O exchange with H$_2$O 147
 ^{2}H shift in water 259, 262–63
 ^{34}S enrichment 146
 abiogenic 146
 and ^{14}C dating .... See ^{14}C dating
 and denitrification **152**
 and Eh **126**
 and secondary calcite 224
 bacteria 145
 calcite precipitation 132
 during denitrification 154
 isotope systematics 144–48
 Oman, Umm er Radhuma aquifer **148**
 Rayleigh effects 55
 via methane oxidation 130
Sulphide
 anaerobic oxidation 142
 biological oxidation 142
 dissolved 138
 oxidation, isotope effects 142
 sampling, analysis 279
 weathering of pyrite 142–43
Sulphite 145, 146
Sulphur
 atmospheric 144, 138, **144**
 fallout, central Canada 144
 organic 138
 species, oxidation states 138
Sulphur cycle 138–48
Sulphuric acid
 dissociation constants 148
Sunspot cycles, and ^{14}C 203
Sweden 173
Sweetgrass Hills, Montana 227
Switzerland
 altitude effect 71
 T–δ^{18}O effect 70

T

Tamers model, ^{14}C correction 208
Technetium 3
Temperature – δ^{18}O effects
 altitude 70
 Canada 69
 global 64, 65, 66
 in ice cores 75
 latitude 66
 seasonal 71
Texas 234
The Pas, Manitoba 72, 83
Thermonuclear ^{3}H 175–78

atmospheric bomb testing 177
attenuation by oceans 177
early atmospheric tests **175**
George – 1st fission test 175
how the bomb works 176
in groundwaters 181
in precipitation, 1952 to 1992 177
Ivy-Mike 175
profile in unsaturated zone 180
releases from atmospheric tests **176**
Thermonuclear ^{3}H in groundwater
 bomb peak 180
Thermonuclear fusion reaction 176
Thompson, J.J. 5
Thorium 4
Throughfall 100
Thule, Greenland 45
TIMS .... 20, 203. See mass spectrometry
TIMS, thermal ionization mass spectrometry .. 15
Transpiration 74, 80
 and salinity 94
 vs. evaporation 94
Travertine
 ^{13}C and ^{18}O 132
 paleoclimate reconstruction 223
Tree rings
 and atmospheric ^{14}C **203**
Triassic Bunter sandstone .... See aquifers
 geology,^{14}C dating 215
Trichloroethane 159
Trichloroethylene 159
Tritium .... See ^{3}H, or groundwater dating
Troilite 11
Troposphere 39, 74
TU, defined 17, 175

U

Underground storage tanks ....
 .... See groundwater contamination
Unsaturated zone 81, 88
 arid regions 87
 carbonate dissolution 118
 water sampling, analysis 282
Uranium
 decay series .... See ^{238}U
 geochemistry 238
 sampling and analysis 281
Urey, Harold 5, 6, 21

V

Valentia 68
Vapour diffusion 90
Vapour pressure 27
 isotopic differences 39
Vegetation
 ^{13}C 119–20

^{13}C range **113**
C$_3$ plants 119
C$_4$ plants 119
CAM plants 119
Victoria 52
Vienna Standard Mean Ocean Water ....
.... *See* VSMOW
Volcanism
   andesitic, mantle fluids 252
Vostok ice core **76**, 204
VPDB **6**
   defined 9
   VSMOW conversion 10
VSMOW **6**
   ratio 8
   VPDB conversion 10
VSMOW - VPDB conversion chart **11**

W

Water freezing, isotope effects 55
Water vapour
   concentration in saturated air **39**
   formation **39**
Water-gas isotope exchange reactions 262–63
Water-rock interaction
   ^{18}O and ^{2}H evolution 255–60, 247–55, **249**
   ^{87}Sr/^{86}Sr 260–61
   high temperature 247–55
Water-rock ratios 251, 256, 259
Watershed
   agricultural 162
   agricultural, groundwaters 161
   Fuhrberger Feld 152–54
   nitrate contamination 148–57

Weathering 117–19
   and pH 117
   carbonates 117
   continental 260
   feldspars 247, 255
   silicates 119
Weathership station 68
Wetlands 162
Wisconsin Glaciation 76
Working standard 15
World Wide Websites
   book <**www.science.uottawa.ca/~eih**>
   climate data 65
   geochemical models 115
   IAEA 7
   ice core data 76
   isotope laboratories 8, 268
   mass spectrometer manufacturers 14
   National Ground Water Association 80
   NETPATH 224
   NIST 7
   ocean data 37
   precipitation data 177
   standards for isotope analysis 7
   tritium in precipitation 177
   USEPA software 115
   USGS software 115

X, Y, Z

XAD-8® resin, DOC sampling 229
Yellowknife, Canada 258
Yukon **75**, 85